AF828820

Microplastics in Environment

In recent years, the pervasive presence of microplastics in the environment has emerged as a pressing global concern. As we grapple with the complexities of this issue, it becomes increasingly evident that a comprehensive understanding of microplastics' sources, risks, analytical challenges, and remediation strategies is essential for effective mitigation and safeguarding of our ecosystems and human health.

Microplastics in Environment comprehensively explores the growing concern surrounding microplastic pollution. From their emergence as a significant environmental pollutant to their potential risks and various remediation strategies, this book delves into every aspect of microplastics' impact on ecosystems and human health. With chapters authored by experts in the field, readers gain insights into the sources, classification, properties, and dynamics of microplastics in different environmental compartments, including marine environments, soils, and indoor settings. Analytical challenges related to the identification, quantification, and characterization of microplastics are thoroughly discussed, along with emerging computational approaches to unravel the microplastic mystery. The book highlights recent trends in using bioindicators to monitor microplastic pollution and its adverse impacts on animal health. Additionally, the book explores trends, challenges, and opportunities in plastic recycling and future directions for addressing microplastics.

Microplastics in Environment compiles current knowledge and identifies gaps, serving as a valuable resource for researchers, policymakers, and practitioners working towards a cleaner, healthier planet. It aims to support those dedicated to safeguarding our planet for future generations.

FEATURES

- Focuses on microplastic pollution in the environment and highlights the emergent need to reduce and eliminate microplastic pollution.
- Includes special chapters on microplastic classifications, estimation tools, and the role of computational modeling in microplastic.
- Discusses microplastic occurrence in marine, soil, air, and prevention strategies in different environmental settings.
- Furthermore, it explores different remediation strategies, current trends, challenges, and opportunities in plastic recycling.

Microplastics in Environment

Source, Potential Risks, Analytical Challenges, and Remediation Strategies

Edited by
Ganga Ram Chaudhary, Moondeep Chauhan,
and Bunty Sharma

CRC Press
Taylor & Francis Group
Boca Raton London New York

CRC Press is an imprint of the
Taylor & Francis Group, an **informa** business

Designed cover image: Shutterstock

First edition published 2025
by CRC Press
2385 NW Executive Center Drive, Suite 320, Boca Raton FL 33431

and by CRC Press
4 Park Square, Milton Park, Abingdon, Oxon, OX14 4RN

CRC Press is an imprint of Taylor & Francis Group, LLC

© 2025 selection and editorial matter, Ganga Ram Chaudhary, Moondeep Chauhan, and Bunty Sharma; individual chapters, the contributors

Reasonable efforts have been made to publish reliable data and information, but the author and publisher cannot assume responsibility for the validity of all materials or the consequences of their use. The authors and publishers have attempted to trace the copyright holders of all material reproduced in this publication and apologize to copyright holders if permission to publish in this form has not been obtained. If any copyright material has not been acknowledged please write and let us know so we may rectify in any future reprint.

Except as permitted under U.S. Copyright Law, no part of this book may be reprinted, reproduced, transmitted, or utilized in any form by any electronic, mechanical, or other means, now known or hereafter invented, including photocopying, microfilming, and recording, or in any information storage or retrieval system, without written permission from the publishers.

For permission to photocopy or use material electronically from this work, access www.copyright.com or contact the Copyright Clearance Center, Inc. (CCC), 222 Rosewood Drive, Danvers, MA 01923, 978–750–8400. For works that are not available on CCC please contact mpkbookspermissions@tandf.co.uk

Trademark notice: Product or corporate names may be trademarks or registered trademarks and are used only for identification and explanation without intent to infringe.

ISBN: 978-1-032-74204-5 (hbk)
ISBN: 978-1-032-74206-9 (pbk)
ISBN: 978-1-003-46814-1 (ebk)

DOI: 10.1201/9781003468141

Typeset in Times LT Std
by Apex CoVantage, LLC

Contents

Editor Biographies

Dr. Ganga Ram Chaudhary is Professor at the Department of Chemistry and Centre for advance studies in Chemistry, Panjab University, Chandigarh and has more than 20 years of experience in the fabrication of functionalized nanomaterials/membranes/biopolymers via green methods and exploring their potential applications for the elimination of contaminants from wastewater, as catalysts, electrochemical sensors, and alternatives to plastic materials. He has published over 2004 publications and five patents for reputable journals and publication houses. Presently, he heads the Sophisticated Analytical Instrumentation Facility (SAIF) and Central Instrumentation Laboratory (CIL), which has equipped him with the deeper understanding, knowledge, and skills required to address the analytical challenges associated with detecting microplastics. He is also engaged in consultancy projects in Himachal Pradesh and Panjab industrial regions, evaluating processing parameters to upgrade the common effluent treatment plant and treatment of wastewater from pharmaceutical, electroplating, textile and personal care products.

Dr. Moondeep Chauhan (environmental scientist) received his doctoral degree in environmental studies from Panjab University in 2020. From 2020 to 2022, he worked as a postdoctoral BIRAC innovation fellow at E-YUVA Center, Panjab University. Thereafter, he worked as a research scientist at SAIF/CIL at Panjab University. Currently, he is working as an assistant professor cum scientist in the central instrumentation facility and the Department of Environment Science at Amity University Punjab. His current research interests include developing nano-catalysts and electrochemical sensors for applications in energy and the environment. He is actively engaged in the synthesis of functionalized nanomaterials/membranes via green methods and exploring their potential applications for eliminating contaminants from wastewater. He has published over 30 research articles, including six book chapters and two patents (one granted). He also works on consultancy projects for the treatment of wastewater from a variety of industries, such as the pharmaceutical, electroplating, and textile sectors.

Dr. Bunty Sharma received his doctoral degree in chemistry from Panjab University, India, and during his PhD studies worked at the School of Pharmacy, University of Nottingham, UK (commonwealth scholarship). He has published 20 research articles in reputed peer-reviewed journals including seven book chapters and has filed three Indian patents, two of which were granted. His current research interests focuses on developing metallosurfactants, hybrid nano-colloids, supramolecular structures, and nano-photocatalysts for application in biomedicine and environment pollution remediation. Currently, he is working as a research specialist at Molekule Group Inc., Florida, United States. Before this, he worked as a research officer in the School of Biochemistry and Cell Biology at University College Cork, Ireland. He is also actively engaged in various consultancy for industries.

Contributors

Amit Kumar Sharma
Department of Animal Sciences
Central University of Himachal Pradesh
Himachal Pradesh, India

Anjali Vijeata
Department of Chemistry and Centre of Advanced Studies in Chemistry
Panjab University
Chandigarh, India

Ao Shi,
School of Chemical and Environmental Engineering
China University of Mining and Technology
Beijing, China

Bunty Sharma
Research and Development
Molekule Inc.
Lakeland, Florida, USA

Charuvila T. Aravindakumar
School of Environmental Sciences, Inter University Instrumentation Centre, and International Centre for Polar Studies
Mahatma Gandhi University
Kerala, India

Chunhui Wang
School of Chemical and Environmental Engineering
China University of Mining and Technology
Beijing, China

Chunrong Wang
School of Chemical and Environmental Engineering
China University of Mining and Technology
Beijing, China

Danish Mahajan
Department of Animal Sciences
Central University of Himachal Pradesh
Himachal Pradesh, India

Devlina Das Pramanik
Centre for Biotechnology and Biochemical Engineering
Amity Institute of Biotechnology
Amity University, Noida Campus
Uttar Pradesh- 201313, India

Dipika Narula
Department of Chemistry, School of Basic & Applied Sciences
Maharaja Agrasen University, Baddi
Solan, India

Divya Newar
Department of Biochemistry
North-Eastern Hill University
Meghalaya, India

Dixit Sharma
Department of Animal Sciences
Central University of Himachal Pradesh
Himachal Pradesh, India

Dongbei Yue
School of Environment
Tsinghua University
Beijing, China

Emma Magdalene Nonglang
Department of Zoology
Assam Don Bosco University
Assam, India

Flavius Phrangsngi Nonglang
Department of Biochemistry
North-Eastern Hill University
Meghalaya, India

Ganga Ram Chaudhary
Department of Chemistry, Centre of Advanced Studies in Chemistry, and Sophisticated Analytical Instrumentation Facility (SAIF)/ Central Instrumentation Laboratory (CIL)
Panjab University
Chandigarh, India

Haorangbam Arundeep Singha
Department of Environmental Science
North-Eastern Hill University
Meghalaya, India

Hishani Kumari
Department of Animal Sciences
Central University of Himachal Pradesh
Himachal Pradesh, India

Jaskirat Suri
Towson University
Towson, Maryland

Jaspreet Dhau
Research and Development
Molekule Inc.
Lakeland, Florida

Jianbing Wang
School of Chemical and Environmental Engineering
China University of Mining and Technology
Beijing, China

Jianchao Wang
School of Chemical and Environmental Engineering
China University of Mining and Technology
Beijing, China

Kanika Choudhary
Department of Animal Sciences
Central University of Himachal Pradesh
Himachal Pradesh, India

Komal Saini
Chandigarh College of Pharmacy
Chandigarh Group of Colleges
Punjab, India

Kushal Thakur
Department of Animal Sciences
Central University of Himachal Pradesh
Himachal Pradesh, India

Moondeep Chauhan
Central Instrumentation Facility and Department of Environmental Sciences
Amity University Punjab
Panjab, India

Navya Rana
Department of Animal Sciences
Central University of Himachal Pradesh
Himachal Pradesh, India

Neeraj Chauhan
Department of Chemistry & Centre of Advanced Studies in Chemistry
Panjab University
Chandigarh, India
School of Geography, Earth and Environmental Sciences
University of Birmingham
Birmingham, United Kingdom

Pin Hou
School of Chemical and Environmental Engineering
China University of Mining and Technology
Beijing, China

Pradeep Kaushik
Department of Chemistry and Centre of Advanced Studies in Chemistry
Panjab University
Chandigarh, India

Preeti Garg
Department of Chemistry and Centre of Advanced Studies in Chemistry
Panjab University
Chandigarh, India

Prerna Attri
Department of Chemistry and Centre of Advanced Studies in Chemistry
Panjab University
Chandigarh, India

Rakesh Kumar
Department of Animal Sciences
Central University of Himachal Pradesh
Himachal Pradesh, India

Ravi Kumar
Department of Chemistry and Centre of Advanced Studies in Chemistry
Panjab University
Chandigarh, India

Roudrashwa Nag
Amity Institute of Biotechnology
Amity University, Noida Campus
Uttar Pradesh-201313, India

Savita Chaudhary
Department of Chemistry and Centre of Advanced Studies in Chemistry
Panjab University
Chandigarh, India

Sunil Kumar
Department of Animal Sciences
Central University of Himachal Pradesh
Himachal Pradesh, India

Sushma Negi
Department of Zoology & Environmental Sciences, School of Basic & Applied Sciences
Maharaja Agrasen University, Baddi
Solan, India

Thies Thiemann
Department of Chemistry College of Science
United Arab Emirates University
Al Ain, UAE

Usha K. Aravind
School of Environmental Studies
Cochin University of Science & Technology
Kerala, India

Vishnu S. Moorchilot
School of Environmental Sciences
Mahatma Gandhi University
Kerala, India

Yanli Shi
School of Chemical and Environmental Engineering
China University of Mining and Technology
Beijing, China

Yuan Liu
School of Chemical and Environmental Engineering
China University of Mining and Technology
Beijing, China

Abbreviations

AOPs	Advanced oxidation processes
AFM	Atomic force microscopy
ALFs	Antilipopolysaccharide factors
ATR	Attenuated total reflectance
ATR-FTIR	Attenuated total reflection Fourier transform infrared spectroscopy
AutoML	Automated machine learning
BT	Billion ton
BNF	Biological nitrogen fixation
BiOCl	Bismuth oxychloride
BPA	Bisphenol A
$CaCl_2$	Calcium chloride
CO_2	Carbon dioxide
CI	Carbon index
CNTs	Carbon nanotubes
CAT	Catalase
CNS	Central nervous system
CFU	Colony forming units
CFD	Computational fluid dynamics
C&D	Construction and demolition
PCPs	Cosmetics and personal care products
DCA	Dichloroaniline
DDT	Dichloro-diphenyl-trichloroethane
DSC	Differential scanning calorimetry
DART-MS	Direct analysis in real time—mass spectrometry
DFM	Disposable face masks
DLS	Dynamic light scattering
EDX	Energy dispersive X-ray spectroscopy
ESEM	Environmental scanning electron microscopy
EVA	Ethyl vinyl acetate
FE-SEM	Field emission–scanning electron microscopy
FT-IR	Fourier transform-infrared spectroscopy
GC-MS	Gas chromatography–mass spectrometry
GC-ECD	Gas chromatography with electron capture detector
HDPE	High-density polyethylene
H_2S	Hydrogen sulfide
HI	Hydroxyl index
HZ	Hyporheic zone
IoT	Internet of Things
LDIR	Laser direct infrared
LC	Liquid chromatography
LC-MS	Liquid chromatography–mass spectrometry
LDPE	Low-density polyethylene

ML	Machine learning
MCNTs	Magnetic carbon nanotubes
MBR	Membrane bioreactors
CH4	Methane
MFC	Microbial fuel cells
μg	Microgram
MPs	Microplastic
mL	Milliliter
mm	Millimeter
MT	Million tons
MD	Molecular dynamics
MRU-Net	MultiResUNet
MSW	Municipal solid waste
nm	Nanometer
NPs	Nanoplastics
NOM	Natural organic matter
NIR	Near infrared spectroscopy
N_2	Nitrogen
NMR	Nuclear magnetic resonance
PCPs	Personal care products
PPE	Personal protective equipment
PAEs	Phthalic acid esters
POM	Polarized optical microscopy
PMMA	Poly (methyl methacrylate)
PHBV	Poly(3-hydroxybutyrate-co-3-hydroxyvalerate))
PA	Polyamide
PBDEs	Polybrominated diphenyl ethers
PBD	Polybutadiene
PC	Polycarbonate
PCBs	Polychlorinated biphenyls
PAHs	Polycyclic aromatic hydrocarbons
PES	Polyester
PE	Polyethylene
PET	Polyethylene terephthalate
PHB	Polyhydroxybutyrate
PP	Polypropylene
PS	Polystyrene
PUR	Polyurethane
PVC	Polyvinyl chloride
PFE	Pressurized fluid extraction
Py-GC-MS	Pyrolysis gas chromatography–mass spectrometry
ROS	Reactive oxygen species
rDNA	Recombinant DNA
RO	Reverse osmosis
SEM	Scanning electron microscopy
SWIR	Short-wave infrared

AgNP	Silver nanoparticles
NaCl	Sodium chloride
NaI	Sodium iodide
TPA	Terephthalic acid
TGA	Thermogravimetric analysis
TEM	Transmission Electron Microscopy
TCA	Tricarboxylic acid
UF	Ultrafiltration
UV	Ultraviolet
WCA	Water contact angle
$ZnBr_2$	Zinc bromide
$ZnCl_2$	Zinc chloride

1 Microplastics Pollution
An Overview of the Global Crisis

Moondeep Chauhan, Bunty Sharma, and Ganga Ram Chaudhary

1.1 MICROPLASTICS POLLUTION: AN OVERVIEW

Global annual production of plastic has surged exponentially since the 1950s, reaching about 460 million tons in 2019. Because of their low cost, lightweight, durability, and versatility, plastics have become a necessary component of modern life. The global plastic market has experienced significant growth over the years: In 2020, it was valued at approximately USD 580 billion. If the current trends continue, the world's annual primary plastic production is projected to reach 11,000 million tons (MT) by 2050, and the total cumulative production between 1950 and 2050 is forecast to reach 34,000 MT, all nonbiodegradable [1].

Three quarters of all plastic manufactured is placed or discarded in landfills, becomes part of mismanaged waste streams, or is disposed of or abandoned in the environment, posing serious threats to ecosystems, wildlife, and human health. The expected worldwide cost of municipal solid waste management is projected to rise from USD 38,000 million in 2019 to USD 61,000 billion in 2040 under the business-as-usual scenario. Plastics account for more than 80% of all marine waste, and about 10% of annual production ends up in the oceans [2]. If no intervention is adopted and current trends continue, the quantity of plastic waste entering aquatic environments is expected to increase nearly threefold, from an estimated 9–14 MT per year in 2016 to about 23–37 MT per year in 2040 [3]. Global economic costs of marine plastic pollution were estimated to be USD 6,000–19,000 in 2018. These costs include the impacts of plastic waste on tourism, fisheries, and aquaculture, as well as other costs related to clean-up efforts.

According to a technical report published by the United Nations Environment Programme, over 13,000 chemicals are associated with plastics and their manufacturing processes. Among these, 7,000 chemicals have been screened for hazardous properties, and more than 3,200 have been identified as chemicals of potential concern: plastic monomers/oligomers, additives (e.g., lubricants, antioxidants, colorants, plasticizers, flame retardants, UV, and heat stabilizers), processing aids (e.g., polymerization catalysts, lubricants, or solvents), and non-intentionally added substances (e.g., breakdown products, degraded products, by-products, contaminants). Many of

DOI: 10.1201/9781003468141-1

these chemicals of potential concern are used, emitted, and released at various points in the plastic life cycle, from the extraction of oil and gas and the production of polymers and chemicals to the manufacturing, packaging, transportation, usage, and end-of-life management of plastics, posing risks to ecosystems and humans.

These substances have been linked with a broad spectrum of acute, chronic, and multigenerational toxic effects, such as toxicity to specific organs, genetic mutations, different cancer types, reproductive and developmental toxicity, endocrine disruption, and ecotoxicity. In addition, microplastics (MPs) act as vectors for hazardous contaminants due to their propensity to adsorb xenobiotic chemicals and pathogenic organisms [4]. It is estimated that the fossil-fuel-based plastics that were produced in 2015 emitted 1.7 gigatons (Gt) of carbon dioxide equivalent (CO_2e) greenhouse gases (GHGs) during their life cycle, 3.8% of the total GHGs (47 $GtCO_2e$) emitted globally that year. If things do not change, the GHG emissions from plastics are projected to reach nearly 6.5 $GtCO_2e$ by 2050, 15% of the total carbon budget [5–7].

MPs refer to plastic particles measuring less than 5 mm in diameter, while nanoplastics (NPs) measure less than 0.001 mm. These tiny plastic particles mostly result from the abrasion of larger plastic particles dumped into the environment. Cox and his colleagues analyzed the data from 26 published studies representing more than 3,600 processed samples to determine the levels of MPs in air and in commonly consumed items. They estimated that, based on 15% of caloric intake, a person consumes around 52,000 MPs particles annually, varying by age and sex. Inhalation increases these numbers to 74,000 to 121,000, and individuals who consume bottled water may be consuming an additional 90,000 MPs annually, compared with 4,000 for tap water [8]. According to Li and his colleagues, polypropylene infant feeding bottles, representing two-thirds of the global bottle market, can release up to 16,200,000 MPs particles per liter. They found that depending upon region, polypropylene-bottle-fed babies (up to 12 months old) are exposed to 14,600–4,550,000 particles per capita per day. Scientists determined that the process of heating bottles for sanitizing and mixing formula was causing millions of MPs to be shed from inside the plastic bottles [9, 10].

Efforts are being made to reduce the spread of plastic pollution by phasing out its use in consumer goods such as ear buds with plastic sticks, drinking straws, candy and ice-cream sticks, plastic plates, cups, packing films around sweet boxes, plastic and PVC banners less than 100 microns, and so on. However, new applications of plastic are still emerging in the industry. For instance, some tea producers have switched to utilizing plastic teabags as an alternative to the conventional paper teabags. This is worrisome because water used for brewing tea is often at or above 95 °C, and even plastics labeled food grade can break down or release harmful substances when exposed to temperatures above 40 °C. Hernandez and his colleagues found that at a brewing temperature of 95 °C, a single plastic teabag can release about 11,600 million MPs and 3,100 million NPs into a single cup of the beverage. Furthermore, the composition of discharged particles was similar to that of the original teabags (i.e., nylon and polyethylene terephthalate) [11].

As MPs are present almost everywhere, it is important to comprehend the growing concern over their exposure. However, isolating and characterizing these minuscule fragments pose significant challenges. MPs can infiltrate the human body through inhalation, ingestion, and skin absorption. Recent studies have revealed the extensive presence of NPs and MPs in vital tissues of the human body, including the placenta. These findings are concerning, particularly for newborns, whose developing immune systems make them more vulnerable to these substances. In 2021, Ragusa and colleagues documented the initial detection of MPs in the human placenta. In this study, out of six human placentas gathered from women with uncomplicated pregnancies, four were discovered to have a total of 12 different MPs of sizes ranging from 5 to 10 μm that had either spherical or irregular morphology [12]. The placenta is a transient organ that connects the baby to the mother via the tubelike umbilical cord. It supplies nutrients and oxygen to the developing baby and protects the fetus from infections and viruses. Therefore, the presence of any contaminant or xenobiotics in the placenta raises serious concerns regarding the likelihood of these particles entering the fetus and their long-term adverse effects.

In another study, Braun et al. observed the presence of MPs in the human placenta and meconium samples (newborn's first fecal matter), indicating a possibility that MPs are entering the fetus' body through placenta. However, the study also emphasized the necessity for a more comprehensive evaluation of potential contamination sources during sampling and analyses [13]. In one recent study, Garcia and his team unveiled a concerning discovery about the presence of MPs in all 62 human placenta samples they examined. The NPs and MPs concentrations in placental tissue ranged from 6.5 to 685 μg/g, with an average of 126.8 ± 147.5 μg/g [14].

Regarding long-term consequences and health impacts, scientific studies have revealed that MPs exposure can induce particle toxicity, leading to oxidative stress and inflammation. However, our comprehension of MP toxicity remains limited and is strongly influenced by factors such as particle properties, exposure concentration, absorbed contaminants, affected organs, and individual sensitivity. Further investigation is crucial for broadening our understanding in this area.

According to a recent conducted study by Marfella and his colleagues, patients who had tiny plastic particles lodged in their arteries had a 4.5-fold increased risk of having a heart attack or stroke or passing away in the 34 months following surgery compared with those whose arteries were clear of plastic particles. In a study of more than 250 people undergoing surgery, nearly 150 patients (58.4%) had polyethylene MPs (21.7 ± 24.5 μg/mg of plaque) or even smaller NPs in a main artery. Thirty-one patients (12.1%) also had measurable amounts of PVC, a mean of 5.2 ± 2.4 μg/mg of plaque [15, 16].

In summary, the widespread occurrence of MPs in diverse environments poses a significant and multifaceted threat to ecosystems and human health. Addressing the influx of plastics into the environment requires a comprehensive approach beyond single solutions. Implementing multiple synergistic interventions across systems is essential for effectively combating this issue. This entails reducing plastic consumption, implementing comprehensive recycling programs, gradually eliminating unnecessary and hazardous monomers, additives and processing aid, promoting alternate biocompatible materials, improving waste management practices, fostering innovative removal technologies, raising public awareness, and enforcing more stringent policies and regulations [17]. Emerging interventions are already on the horizon. In

March 2022, 175 nations reached a consensus to forge a legally binding international agreement aimed at combating plastic pollution. Through shared responsibilities and innovative solutions, we can create a sustainable planet for future generations.

REFERENCES

1. Geyer, R., 2020. Chapter 2—production, use and fate of synthetic polymers in plastic waste and recycling. In *Plastic Waste and Recycling: Environmental Impact, Societal Issues, Prevention, and Solutions*. Letcher, T.M. (ed.). Cambridge, MA: Academic Press, pp. 13–32. ISBN 9780128178805.
2. Avio, C.G., Gorbi, S. and Regoli, F., 2017. Plastics and microplastics in the oceans: From emerging pollutants to emerged threat. *Marine Environmental Research*, 128, pp. 2–11.
3. Geyer, R., 2020. Chapter 2—production, use and fate of synthetic polymers in plastic waste and recycling. In *Plastic Waste and Recycling: Environmental Impact, Societal Issues, Prevention, and Solutions*. Letcher, T.M. (ed.). Cambridge, MA: Academic Press, pp. 13–32.
4. Lim, C., Kim, N., Lee, J. and Yoon, Y., 2022. Potential of adsorption of diverse environmental contaminants onto microplastics. *Water*, 14(24), p. 4086.
5. Olivier, J.G.J., Schure, K.M. and Peters, J.A.H.W., 2017. *Trends in Global CO_2 and Total Greenhouse Gas Emissions: Summary of the 2017 Report*. The Hague: PBL Netherlands Environmental Assessment Agency.
6. Zheng, J. and Suh, S., 2019. Strategies to reduce the global carbon footprint of plastics. *Nature Climate Change*, 9, pp. 374–378.
7. United Nations Environment Programme, 2021. *From Pollution to Solution: A global assessment of marine litter and plastic pollution*. Nairobi: UNEP.
8. Cox, K.D., Covernton, G.A., Davies, H.L., Dower, J.F., Juanes, F. and Dudas, S.E., 2019. Human consumption of microplastics. *Environmental Science & Technology*, 53(12), pp. 7068–7074.
9. https://www.foodpackagingforum.org/news/study-investigates-microplastics-release-from-baby-bottles accessed on 30/06/2024.
10. Li, D., Shi, Y., Yang, L., Xiao, L., Kehoe, D.K., Gun'ko, Y.K., Boland, J.J. and Wang, J.J., 2020. Microplastic release from the degradation of polypropylene feeding bottles during infant formula preparation. *Nature Food*, 1(11), pp. 746–754.
11. Hernandez, L.M., Xu, E.G., Larsson, H.C., Tahara, R., Maisuria, V.B. and Tufenkji, N., 2019. Plastic teabags release billions of microparticles and nanoparticles into tea. *Environmental Science & Technology*, 53(21), pp. 12300–12310.
12. Ragusa, A., Svelato, A., Santacroce, C., Catalano, P., Notarstefano, V., Carnevali, O., Papa, F., Rongioletti, M.C.A., Baiocco, F., Draghi, S. and D'Amore, E., 2021. Plasticenta: First evidence of microplastics in human placenta. *Environment International*, 146, p. 106274.
13. Braun, T., Ehrlich, L., Henrich, W., Koeppel, S., Lomako, I., Schwabl, P. and Liebmann, B., 2021. Detection of microplastic in human placenta and meconium in a clinical setting. *Pharmaceutics*, 13(7), p. 921.
14. Garcia, M.A., Liu, R., Nihart, A., El Hayek, E., Castillo, E., Barrozo, E.R., Suter, M.A., Bleske, B., Scott, J., Forsythe, K., Gonzalez-Estrella, J., Aagaard, K.M. and Campen, M.J. 2024, Apr 29. Quantitation and identification of microplastics accumulation in human placental specimens using pyrolysis gas chromatography mass spectrometry. *Toxicological Sciences*, 199(1), pp. 81–88.
15. Kozlov, M., 2024 Mar 6. Landmark study links microplastics to serious health problems. *Nature*.

16. Marfella, R., Prattichizzo, F., Sardu, C., Fulgenzi, G., Graciotti, L., Spadoni, T., D'Onofrio, N., Scisciola, L., La Grotta, R., Frigé, C. and Pellegrini, V., 2024. Microplastics and nanoplastics in atheromas and cardiovascular events. *New England Journal of Medicine*, 390(10), pp. 900–910.
17. Ghosh, S., Sinha, J.K., Ghosh, S., Vashisth, K., Han, S. and Bhaskar, R., 2023. Microplastics as an emerging threat to the global environment and human health. *Sustainability*, 15(14), p. 10821.

2 The Emergence, Chemical Nature, Classification, and Properties of Microplastics

Neeraj Chauhan

2.1 INTRODUCTION: EMERGENCE

Because of the huge demand for and rapid development in plastic-based industries, numerous tiny pieces of plastic are used and discarded worldwide. Plastics are synthetic and semi-synthetic polymers found widely dispersed in soil and water bodies because of their low cost and durability. The most common form of plastics includes polyethylene (PE), polypropylene (PP), polystyrene (PS), polyvinyl chloride (PVC), and polyethylene terephthalate (PET). Microplastics (MPs) have emerged due to the breakdown of larger plastic polymers through various physical and chemical processes like weathering, exposure to UV radiation, hydrolysis, biodegradation, and mechanical activities and plastic debris is fragmented into smaller particles. MPs are defined as plastic particles smaller than five millimeters [1, 2]. These tiny particles arise from diverse origins, presenting a peril to ecosystems, fauna, and conceivably human well-being [3]. MPs have attracted significant interest owing to their widespread presence, easy manipulation, and economic characteristics. In recent years, plastic pollution has emerged as one of the most serious threats to the environment. In aquatic ecosystems, MPs can adsorb and transport chemical pollutants, which raises concerns about bioaccumulation [4]. Moreover, the consumption of MPs by marine organisms and their subsequent transmission within the ecological network presents a potential threat to biodiversity [5, 6].

This book chapter provides a comprehensive analysis of the emergence, chemical nature, classification, and properties of MPs. The scientific community's ongoing efforts to understand and address this issue are critical to developing effective mitigation strategies and preserving global ecosystems. Recent multidisciplinary research has demonstrated the distribution of MPs in the hydrosphere, aerosphere, and terrestrial environment, which can alter physiological functions, interfere with commensal micro-organisms, and potentially promote the emergence of infectious diseases [7]. In aquatic ecosystems, MPs can sink into the sediment and present a potential hazard to organisms residing in sediments [8]. Exposure to concentrations of MPs representative of

DOI: 10.1201/9781003468141-2

environmental conditions detrimentally impacts the growth and emergence of freshwater organisms, with observed effects on the size of the particles [9]. Therefore, it is important to develop new strategies for mitigating input of MPs into freshwater ecosystems and to further understand the long-term effects of continuous exposure to them [10, 11].

Previously restricted to water, MPs have recently been detected in soil, which is another vital environment. MP contamination in soil is a growing environmental concern; MPs can enter the soil through various sources such as plastic mulch, sewage sludge, and atmospheric fallout [12]; they negatively impact soil health and fertility by altering soil pH, porosity, soil microbial enzyme activity, and water-holding capacity [13]. The ingestion of MPs affects earthworms, which subsequently affects the dynamics of soil organic matter (SOM) [14, 15].

MPs also interact with pollutants in the soil, delaying their degradation and affecting soil nutrient cycling. Studies have shown that increasing concentrations of MPs alter soil compaction, soil structure, and water retention [16]. Additionally, MPs travel through the soil layers via water infiltration, with smaller particles having higher mobility [17]. Understanding the transport fate and effects of MPs in agricultural soil is crucial for assessing the potential risks and developing strategies for their control and mitigation.

Airborne MPs are ubiquitous in metropolitan areas, natural environments, and distant locales. The decomposition of bigger plastic objects, the deterioration of synthetic textiles, and the emission of plastic particles from industrial processes and road surfaces are some of the sources of atmospheric MPs [18]. The existence of MPs in air debris underscores the possibility of human exposure via breathing [19]. Moreover, research like that conducted by Bergmann et al. (2019) has demonstrated the worldwide distribution of MPs in the atmosphere by finding them in Arctic snow. The ongoing exploration of the potential health risks associated with inhaling MPs underscores the need for further research in this area [20]. Figure 2.1 illustrates the different air-, water-, and soilborne sources of microplastic pollution.

FIGURE 2.1 Sources of microplastic pollution: airborne, waterborne and soilborne microplastics.

2.2 CHEMICAL NATURE OF MICROPLASTICS

The chemical nature of MPs, comprising a spectrum of synthetic polymers, underscores their resilience and longevity. These particles not only resist natural degradation but also serve as vectors for various environmental contaminants. Consequently, MPs have become integral components of ecosystems, infiltrating marine habitats, freshwater systems, and even terrestrial landscapes. Depending on their size, shape, density, polymer type, and surface characteristics, MPs can be classified into a wide variety of types [21, 22]. These entities can absorb chemicals, metals, and pharmaceuticals and subsequently transmit them through the food chain [23].

The sorption property of MPs is affected by various mechanisms including hydrogen bonds, pore filling, Van der Waals forces, hydrophobic interactions, electrostatic interactions, and π-π interactions [24]. Degradable MPs such as polylactic acid or polybutylene succinate have a higher affinity for hydrophobic contaminants than conventional synthetic nondegradable MPs. Exposure to environmental stress can cause MPs to undergo chemical and physical changes that can potentially pose a health risk to humans [25]. The mechanisms of chemical adsorption into MPs have further been analyzed using molecular dynamics simulations.

Being faced with environmental challenges, various thermoplastics were found responsible for microplastic contamination. The most detected polymers in global water systems are PE followed by PP, PS, PVC, and PET. These plastic polymers can be degraded by different processes such as chemical, thermal degradation, radiolytic, and mechanical degradation; biodegradation; and photodegradation. Figure 2.2 illustrates the distinct types of polymers depending upon their chemical nature.

2.2.1 Polyethylene

Polyethylene is a major source of MPs and is present extensively in the environment. It is a synthetic polymer with $-CH_2-CH_2$ monomer units that is further divided as high-density polyethylene (HDPE) and low-density polyethylene (LDPE). PE shows high adsorption capacity and is also immune to photooxidation because it lacks the unsaturated chromophore groups that allow light absorption [26].

The beginning of photodegradation is heavily reliant on external contaminants or structural flaws. Biodegradation is restricted in the absence of photo-initiated oxidation because of PE's large molecular mass and lack of functional groups. Furthermore, the introduction of carbonyl functional groups by abiotic degradation makes PE more hydrophilic and hence more available for biodegradation. As a result, photooxidation turns out to be an important mechanism that facilitates the production of MPs in the environment and acts as a first step in the breakdown of PE's lengthy molecular chains [27].

2.2.2 Polypropylene

Polypropylene is a crystalline polymer of the propene ($-C_3H_6$) monomer units. Because of the lengthy polyolefin chains, it is resistant to biodegradation. However,

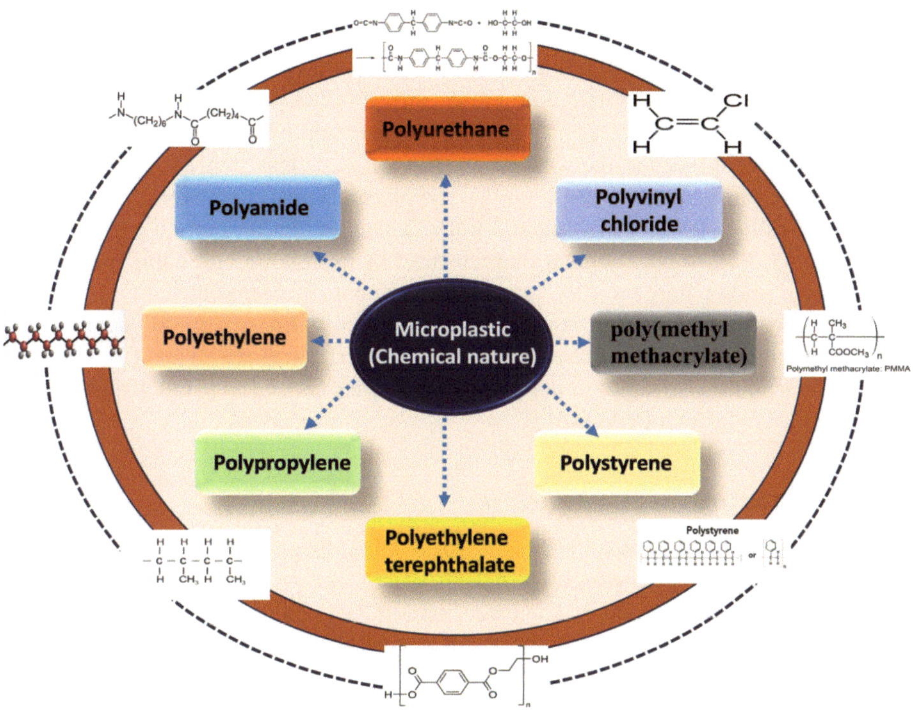

FIGURE 2.2 Schematic illustration of distinct types of polymers depending upon their chemical nature.

photo-oxidative intermediates are biodegradable; therefore, the microplastic from PP is primarily formed by photodegradation [28]. In contrast to other polymers, the material exhibits good resistance to chemicals, heat, etc. and shows low density, which makes it flow over the surface of water bodies. PP's excessive use and low recycling efficacy have led to environmental pollution.

2.2.3 Polystyrene

Polystyrene is a synthetic polymer consisting of the aromatic hydrocarbon styrene monomers ($-C_6H_5CH = CH_2$). PS disintegration predominantly arises from ultraviolet (UV) irradiation rather than biodegradation, as the inclusion of a phenyl ring facilitates the transference of energy from UV light to the adjacent C-H bond. This phenomenon results in the cleavage of hydrogen bonds and the initiation of radical formation, ultimately contributing to the fragmentation [29]. Due to its economic nature and prominent physical properties, PS is utilized in the inner walls of refrigerators, in food packaging, and in lamp coverings. However, its lower density makes PS disproportionately present in sewage water, pipelines, etc.

2.2.4 Polyvinyl Chloride

Polyvinyl chloride (PVC) is a synthetic resin that is derived from the polymerization of vinyl chloride ($-CH_2 = CHCl$) monomer units. Pollution from PVC MPs alters soil bacterial communities as well as available nitrogen and phosphorus levels, changing the nutrient availability in the soil [30]. Further, PVC is less stable for photodegradation because of its high sensitivity to UV radiation. Exposure to sunlight initiates the dechlorination of PVC, which results in the formation of HCl and a carbon-carbon double bond. This double bond is characterized by low photodegradability and serves as the focal point where the PVC backbone fragments into smaller segments. Despite this susceptibility to photodegradation, the inherent resistance to biodegradation in PVC is attributed to the presence of halogens, particularly chlorine. Consequently, pre-dechlorination through photooxidation becomes crucial in promoting the subsequent biodegradation of PVC [31].

2.2.5 Polyethylene Terephthalate

Polyethylene terephthalate (PET) consists of repeating ($-C_{10}H_8O_4$) units. The fundamental components consist of ethylene glycol and terephthalic acid, which are both sourced from petroleum. The ester bond in PET undergoes photooxidation and hydrolytic oxidation and generates carboxylic acids and vinyl end groups [31]. It is easier for PET to fracture into MPs because the maximum fracture stress reduces linearly with molar mass [32]. However, because of its high crystallinity and compact structure, PET is highly stable against biodegradation [33], so it should be pretreated abiotically before biodegradation. Additionally, biodegradation in nature has been shown to take decades [28].

In addition to traditional plastics, newer materials such as poly (methyl methacrylate), polyurethane (PUR), and polyamide (PA) also contribute to microplastic pollution. Bioplastics such as polylactide and polybutylene adipate co-terephthalate are becoming increasingly popular in areas such as food packaging and agriculture. In addition to tire wear particles, paint particles, and surface coatings made from synthetic polymers such as styrene-butadiene rubber are also MPs. Synthetic polymers are often used in curing coating systems (e.g., polyester, alkyd resins, epoxy resins, and urethane resins) and physical drying systems (e.g., acryl and vinyl (co)polymers). This diversity of microplastic sources underlines the complexity and multilayered nature of the problem of microplastic pollution [34].

MPs are characterized using various techniques including visual microscopy, stereomicroscopy, optical microscopy, scanning electron microscopy, and spectroscopic methodologies such as attenuated total reflection Fourier transform infrared spectroscopy, focal plane array and micro-FTIR, near infrared spectroscopy, Raman spectroscopy, and nuclear magnetic resonance. Additionally, chromatographic techniques, specifically liquid chromatography-tandem mass spectrometry and pyrolysis-gas chromatography-mass spectrometry are utilized in the analysis of MPs [35, 36]. To identify the polymer composition of MPs, it is important to conduct spectroscopic analysis [37].

2.3 CLASSIFICATION OF MICROPLASTICS

According to their origins, MPs can be categorized into primary and secondary particles. Primary particles refer to those manufactured within the size range characteristic of MPs, while secondary particles are generated through processes like the fragmentation and degradation of larger plastic components and released into the environment.

2.3.1 Primary Microplastics

Primary MPs are purposely manufactured for specific applications and represent a noteworthy source of MP pollution. The prevalence of MPs in personal care products emphasizes the need for regulatory measures to mitigate their release into the environment [38]. The intentional use of MPs in industrial processes such as abrasives and blasting media adds another layer to the intentional introduction of these particles. Primary MPs are manufactured by fragmentation, the breakdown of larger plastic items into smaller particles through mechanical processes.

2.3.2 Secondary Microplastics

The unintentional emergence of MP particles results from the disintegration of larger plastic items through various processes. Weathering, exposure to UV radiation, and mechanical abrasion contribute to the fragmentation of plastic debris into smaller particles [39]. The abundant presence of MPs in aquatic ecosystems including secondary MPs generated through the degradation of larger plastic objects, a prevailing phenomenon [40]. Secondary MPs are produced from weathering, fragmentation, and degradation, which includes the breakdown of plastics due to environmental factors like UV radiation, mechanical abrasion, and temperature fluctuations.

2.4 PROPERTIES OF MICROPLASTICS

MPs have different characteristics, including size, shape, and chemical composition. Understanding the variability in size and shape is essential for assessing the transport and fate of MPs in different environmental matrices.

2.4.1 Size Range of Microplastics

Plastic contamination can be divided into five main categories based on size (Figure 2.3).

i) **Megaplastics**

Megaplastics is the largest category of plastic debris; they comprise fragments exceeding 1 meter in diameter. These are easily visible to the naked eye and easily countable as well. These large plastic fragments are often found littering coastlines and terrestrial environments.

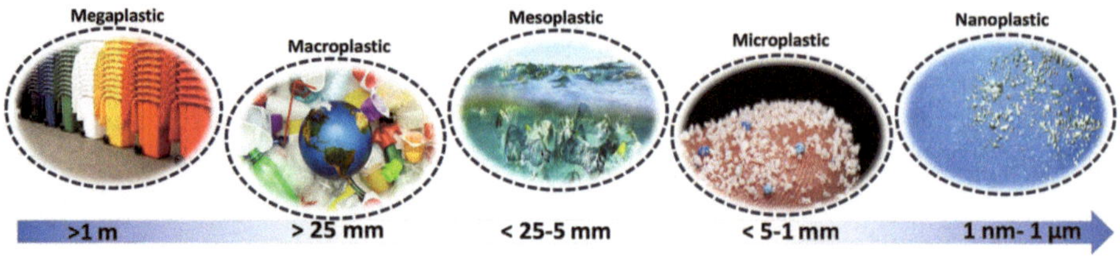

FIGURE 2.3 Classification of plastics based on their size distribution.

ii) Macroplastics

Macroplastics are more than 25 mm in diameter and can be visualized with the naked eye and counted. They play a significant role in the visual pollution of ecosystems.

iii) Mesoplastics

Mesoplastics are 5–25 mm in diameter and can be seen with the naked eye or an optical microscope. The smaller size of mesoplastics poses unique challenges in monitoring and managing their impacts on the environment.

iv) Microplastics

MPs are typically defined as particles with a size less than 5 millimeters in diameter, and these can be visualized with an optical microscope.

v) Nanoplastics

Nanoplastics are less than 1 micrometer in diameter and are gaining attention. These can be observed with an electron microscope [41, 42].

2.4.2 Shape

MPs of diverse shapes contribute to the complexity of their impacts on ecosystems. The range of forms that MPs can take is influenced by both their sources and the processes that lead to their fragmentation. MPs can take the form of asymmetric fragments, films, fibers, foams, pellets, balls, flakes, foils, and spheres, depending on their source and degradation process. Source-dependent variation in MP shapes is intricately linked to their sources. Whether originating from textiles, packaging materials, industrial processes, or intentional manufacturing, the diversity in shapes provides clues to their origins and the pathways through which they enter ecosystems. MPs degrade influenced by environmental factors. Sunlight, mechanical forces, and biological activities contribute to the diverse shapes observed in MPs, emphasizing the importance of understanding the degradation pathways [42–44].

2.4.3 Density

MPs can float or sink in water depending on their density. This property influences their distribution and transport in aquatic environments. The density of MPs varies based on their composition, size, and shape. Dense materials like PE and PP may sink in water, affecting their distribution in aquatic environments. Conversely, lighter particles such as expanded PS may remain suspended in water.

It is important to consider these factors when assessing MP density and distribution. Understanding the density of MPs is essential for predicting their transport mechanisms [45]. Buoyancy, sedimentation, and surface currents influence the movement of MPs within aquatic ecosystems [46]. The density of MPs contributes to their interaction with biota. Buoyant MPs in surface waters may be ingested by organisms at the ocean's surface, while sedimenting MPs can accumulate in benthic environments. Accurately measuring the density of MPs poses challenges due to their small size and diverse compositions. Techniques such as density gradient centrifugation and microscopic analysis are employed, but standardization and validation are still ongoing concerns. Preventing plastic pollution at its source, improving waste management, and looking for sustainable alternatives are crucial steps in mitigating the environmental impact of MPs [47].

2.4.4 Surface Area

The surface area of MPs is pivotal in influencing their interactions with various particles and organisms. Augmented specific surface area, larger porosity, and rough surface topography collectively enhance the adsorption capacity of MPs by creating potential sites for adsorption [48]. The rapid aging of MPs increases their surface area due to cracks and fragmentation, which facilitate light and oxygen access to their internal layers [49]. MPs have a critical influence on bacterial colonization as a result of their surface roughness and physicochemical properties, with eroded MPs showing a higher abundance of bacteria as a result of the lower contact angles [50]. Geometric shapes of MPs can also be influenced by their surface features, such as convex structures, which influence wear modes and production. Overall, the surface area of MPs affects their adsorption capacity, adhesion, and potential for carrying contaminants which highlights the importance of considering surface characteristics in understanding the behavior and impacts of MPs in the environment [51, 52].

2.4.5 Adsorption Capacity

MPs can adsorb various pollutants including pharmaceuticals, personal care products, and environmental contaminants. The adsorption capacity of MPs depends on factors such as the type of plastic, surface functional groups, the specific pollutant being adsorbed, the total energy, the energy of the highest occupied molecular orbital, and the energy gap of the pollutants [53, 54]. The adsorption behavior of MPs can be described using models such as the Freundlich and Langmuir isotherm models. MP adsorption mechanisms involve interactions such as π-π and hydrophobic interactions, hydrogen bonding, and electrostatic forces. Further, the adsorption capacity of MPs can also be influenced by factors such as pH, ionic strength, temperature, salinity, and the presence of other substances like humic acid. Overall, understanding the adsorption capacity of MPs is crucial for assessing their impacts on the environment and developing strategies to mitigate their effects [55, 56].

2.4.6 Hydrophobicity

The hydrophobicity properties of MPs play an important role in their environmental fate and effects. Hydrophobic MPs tend to repel water and have a high affinity for hydrophobic contaminants, leading to their accumulation in aquatic environments [57]. UV weathering can alter their hydrophobicity of MPs, reducing their sorption capacity to hydrophobic contaminants [58]. Additionally, weathered MPs exhibit increased settling velocity, which can enhance their deposition in sediments. Sludge treatments in wastewater treatment plants can also affect the hydrophobicity of MPs, potentially influencing their environmental risk when sludge is applied to agricultural soils. Understanding the hydrophobicity of MPs is crucial for assessing their transport potential, accumulation, and interactions with contaminants in different environmental compartments [59, 60].

2.4.7 Color

The original color of the plastic material is often dictated by additives and pigments during manufacturing, and this becomes a distinctive feature of resulting MPs. This inherent coloration not only aids in the identification and characterization of MPs but also contributes to their environmental behavior. The color of MPs affects their interaction with sunlight, influencing the rate of photodegradation and weathering [61, 62]. Additionally, the visual appearance of colored MPs may attract or repel certain organisms, impacting their ingestion patterns and potential ecological consequences [63]. Furthermore, the diverse colors of MPs contribute to their spatial distribution, as different colors exhibit varying buoyancies, affecting their transport in aquatic environments. Understanding the color-related dynamics of MPs is crucial for comprehending their environmental impact and designing effective strategies for mitigation. MPs can adsorb and accumulate harmful chemicals from the surrounding environment, leading to potential ecological and health risks.

2.5 CONCLUSION

Overall, conventional research and significant progress in the field of MPs have generated considerable attention in the past few years. At present, most MPs can be identified based on their shape, size, color, and other characteristics. This segregation allows differentiation and sampling of MPs along with their treatment using standardized techniques. Moreover, this approach could be beneficial for differentiating different forms of microplastic based on their bio-accumulative nature. However, the efficiency of these methods relies on combating microplastic pollution. The most prevalent forms of microplastic include polyethylene, polystyrene, polyamide, polypropylene, and polyethylene terephthalate. The potential adverse effects on human health warrant comprehensive exploration, emphasizing the need for continued research to establish a more nuanced understanding of the consequences of microplastic exposure. This approach offers an economic way to reduce microplastic pollution, thus producing a barrier between microplastic and freshwater sources.

REFERENCES

1. Andrady, A.L., **2011**. Microplastics in the marine environment. *Marine Pollution Bulletin*, *62*(8), pp. 1596–1605.
2. Cole, M., Lindeque, P., Halsband, C. and Galloway, T.S., **2011**. Microplastics as contaminants in the marine environment: a review. *Marine Pollution Bulletin*, *62*(12), pp. 2588–2597.
3. Krause, S., Schneidewind, U., Wazne, M., Kukkola, A., Lynch, I., Haverson, L., Kelleher, L., Davies, G., Dendievel, A.M., Mourier, B. and Mermillod-Blondin, F., **2023**. *The plastic underground: Are microplastics in the subsurface a ticking time bomb for soil and groundwater ecosystems?* (No. EGU23-4405). Copernicus Meetings.
4. Wright, S.L. and Kelly, F.J., **2017**. Plastic and human health: a micro issue? *Environmental Science & Technology*, *51*(12), pp. 6634–6647.
5. Lusher, A.L., Tirelli, V., O'Connor, I. and Officer, R., **2015**. Microplastics in Arctic polar waters: the first reported values of particles in surface and sub-surface samples. *Scientific Reports*, *5*(1), p. 14947.
6. Krause, S., Baranov, V., Nel, H.A., Drummond, J.D., Kukkola, A., Hoellein, T., Smith, G.H.S., Lewandowski, J., Bonet, B., Packman, A.I. and Sadler, J., **2021**. Gathering at the top? Environmental controls of microplastic uptake and biomagnification in freshwater food webs. *Environmental Pollution*, *268*, p. 115750.
7. Loiseau, C. and Sorci, G., **2022**. Can microplastics facilitate the emergence of infectious diseases? *Science of the Total Environment*, *823*, p. 153694.
8. Kurki-Fox, J.J., Doll, B.A., Monteleone, B., West, K., Putnam, G., Kelleher, L., Krause, S. and Schneidewind, U., **2023**. Microplastic distribution and characteristics across a large river basin: insights from the Neuse River in North Carolina, USA. *Science of the Total Environment*, *878*, p. 162940.
9. Ziajahromi, S., Kumar, A., Neale, P.A. and Leusch, F.D., **2018**. Environmentally relevant concentrations of polyethylene microplastics negatively impact the survival, growth and emergence of sediment-dwelling invertebrates. *Environmental Pollution*, *236*, pp. 425–431.
10. Wagner, M. and Lambert, S., **2018**. *Freshwater microplastics: emerging environmental contaminants?* (p. 303). Springer Nature.
11. Tibbetts, J., Krause, S., Lynch, I. and Sambrook Smith, G.H., **2018**. Abundance, distribution, and drivers of microplastic contamination in urban river environments. *Water*, *10*(11), p. 1597.
12. Lei, J., Zhang, X., Yan, W., Chen, X., Li, Z., Dan, P., Dan, Q., Jiang, W., Liu, Q. and Li, Y., **2023**. Urban microplastic pollution revealed by a large-scale wetland soil survey. *Environmental Science & Technology*, *57*(21), pp. 8035–8043.
13. Rai, M., Pant, G., Pant, K., Aloo, B.N., Kumar, G., Singh, H.B. and Tripathi, V., **2023**. Microplastic pollution in terrestrial ecosystems and its interaction with other soil pollutants: a potential threat to soil ecosystem sustainability. *Resources*, *12*(6), p. 67.

14. Guo, J.J., Huang, X.P., Xiang, L., Wang, Y.Z., Li, Y.W., Li, H., Cai, Q.Y., Mo, C.H. and Wong, M.H., **2020**. Source, migration and toxicology of microplastics in soil. *Environment International, 137*, p. 105263.
15. Xiang, Y., Jiang, L., Zhou, Y., Luo, Z., Zhi, D., Yang, J. and Lam, S.S., **2022**. Microplastics and environmental pollutants: key interaction and toxicology in aquatic and soil environments. *Journal of Hazardous Materials, 422*, p. 126843.
16. Chang, J., Liang, J., Fang, W., Zhang, H., Zhang, Y., Zhao, H., Zhang, R., Zhang, P. and Zhang, G., **2023**. Adsorption behaviours and bioavailability of tetrabromobisphenol A in the presence of polystyrene microplastic in soil: effect of microplastics aging. *Environmental Pollution, 334*, p. 122156.
17. Maqbool, A. and Gomez, J.A., **2023**. *Does plastic and microplastic change the soil physical properties? A review* (No. EGU23-1626). Copernicus Meetings.
18. Jahandari, A., **2023**. Microplastics in the urban atmosphere: Sources, occurrences, distribution, and potential health implications. *Journal of Hazardous Materials Advances, 12*, p. 100346.
19. Dris, R., Gasperi, J., Mirande, C., Mandin, C., Guerrouache, M., Langlois, V. and Tassin, B., **2017**. A first overview of textile fibers, including microplastics, in indoor and outdoor environments. *Environmental Pollution, 221*, pp. 453–458.
20. Bergmann, M., Mutzel, S., Primpke, S., Tekman, M.B., Trachsel, J. and Gerdts, G., **2019**. White and wonderful? Microplastics prevail in snow from the Alps to the Arctic. *Science Advances, 5*(8), p. eaax1157.
21. Godoy, V., Blazquez, G., Calero, M., Quesada, L. and Martín-Lara, M.A., **2019**. The potential of microplastics as carriers of metals. *Environmental Pollution, 255*, p. 113363.
22. Leng, Y., Wang, W., Cai, H., Chang, F., Xiong, W. and Wang, J., **2023**. Sorption kinetics, isotherms and molecular dynamics simulation of 17β-estradiol onto microplastics. *Science of the Total Environment, 858*, p. 159803.
23. Torres, F.G., Dioses-Salinas, D.C., Pizarro-Ortega, C.I. and De-la-Torre, G.E., **2021**. Sorption of chemical contaminants on degradable and non-degradable microplastics: recent progress and research trends. *Science of the Total Environment, 757*, p. 143875.
24. Ivleva, N.P., **2021**. Chemical analysis of microplastics and nanoplastics: challenges, advanced methods, and perspectives. *Chemical Reviews, 121*(19), pp. 11886–11936.
25. Kim, T., Park, K. and Hong, J., **2022**. Understanding the hazards induced by microplastics in different environmental conditions. *Journal of Hazardous Materials, 424*, p. 127630.
26. Hu, K., Yang, Y., Zuo, J., Tian, W., Wang, Y., Duan, X. and Wang, S., **2022**. Emerging microplastics in the environment: properties, distributions, and impacts. *Chemosphere, 297*, p. 134118.
27. Dumichen, E., Barthel, A.K., Braun, U., Bannick, C.G., Brand, K., Jekel, M. and Senz, R., **2015**. Analysis of polyethylene microplastics in environmental samples, using a thermal decomposition method. *Water Research, 85*, pp. 451–457.
28. Bond, T., Ferrandiz-Mas, V., Felipe-Sotelo, M. and Van Sebille, E., **2018**. The occurrence and degradation of aquatic plastic litter based on polymer physicochemical properties: a review. *Critical Reviews in Environmental Science and Technology, 48*(7–9), pp. 685–722.
29. Yousif, E. and Haddad, R., **2013**. Photodegradation and photostabilization of polymers, especially polystyrene. *SpringerPlus, 2*(1), pp. 1–32.
30. Yan, Y., Chen, Z., Zhu, F., Zhu, C., Wang, C. and Gu, C., **2021**. Effect of polyvinyl chloride microplastics on bacterial community and nutrient status in two agricultural soils. *Bulletin of Environmental Contamination and Toxicology, 107*, pp. 602–609.
31. Gewert, B., Plassmann, M.M. and MacLeod, M., **2015**. Pathways for degradation of plastic polymers floating in the marine environment. *Environmental Science: Processes & Impacts, 17*(9), pp. 1513–1521.

32. Arhant, M., Le Gall, M., Le Gac, P.Y. and Davies, P., **2019**. Impact of hydrolytic degradation on mechanical properties of PET-Towards an understanding of microplastics formation. *Polymer Degradation and Stability*, *161*, pp. 175–182.
33. Chen, Z., Wang, Y., Cheng, Y., Wang, X., Tong, S., Yang, H. and Wang, Z., **2020**. Efficient biodegradation of highly crystallized polyethylene terephthalate through cell surface display of bacterial PETase. *Science of the Total Environment*, *709*, p. 136138.
34. Ivleva, N.P., **2021**. Chemical analysis of microplastics and nanoplastics: challenges, advanced methods, and perspectives. *Chemical Reviews*, *121*(19), pp. 11886–11936.
35. Lamichhane, G., Acharya, A., Marahatha, R., Modi, B., Paudel, R., Adhikari, A., Raut, B.K., Aryal, S. and Parajuli, N., **2023**. Microplastics in environment: global concern, challenges, and controlling measures. *International Journal of Environmental Science and Technology*, *20*(4), pp. 4673–4694.
36. Ormaniec, P., **2022**. Occurrence and characteristics of microplastics in leachate at a large municipal wastewater treatment plant. *Civil and Environmental Engineering Reports*, *32*(4), pp. 105–115.
37. Tirkey, A. and Upadhyay, L.S.B., **2021**. Microplastics: an overview on separation, identification and characterization of microplastics. *Marine Pollution Bulletin*, *170*, p. 112604.
38. Law, K.L. and Thompson, R.C., 2014. Microplastics in the seas. *Science*, *345*(6193), pp. 144–145.
39. Gewert, B., Ogonowski, M., Barth, A. and MacLeod, M., **2017**. Abundance and composition of near surface microplastics and plastic debris in the Stockholm Archipelago, Baltic Sea. *Marine Pollution Bulletin*, *120*(1–2), pp. 292–302.
40. Eriksen, M., Lebreton, L.C., Carson, H.S., Thiel, M., Moore, C.J., Borerro, J.C., Galgani, F., Ryan, P.G. and Reisser, J., **2014**. Plastic pollution in the world's oceans: more than 5 trillion plastic pieces weighing over 250,000 tons afloat at sea. *PLOS One*, *9*(12), p. e111913.
41. Van Cauwenberghe, L., Devriese, L., Galgani, F., Robbens, J. and Janssen, C.R., **2015**. Microplastics in sediments: a review of techniques, occurrence and effects. *Marine Environmental Research*, *111*, pp. 5–17.
42. Kooi, M. and Koelmans, A.A., **2019**. Simplifying microplastic via continuous probability distributions for size, shape, and density. *Environmental Science & Technology Letters*, *6*(9), pp. 551–557.
43. Ricciardi, M., Pironti, C., Motta, O., Miele, Y., Proto, A. and Montano, L., **2021**. Microplastics in the aquatic environment: occurrence, persistence, analysis, and human exposure. *Water*, *13*(7), p. 973.
44. Lozano, Y.M., Lehnert, T., Linck, L.T., Lehmann, A. and Rillig, M.C., **2021**. Microplastic shape, polymer type, and concentration affect soil properties and plant biomass. *Frontiers in Plant Science*, *12*, p. 616645.
45. Liu, D., Guo, Z.F., Xu, Y.Y., Chan, F.K.S., Xu, Y.Y., Johnson, M. and Zhu, Y.G., **2022**. Widespread occurrence of microplastics in marine bays with diverse drivers and environmental risk. *Environment International*, *168*, p. 107483.
46. Cozar, A., Echevarria, F., Gonzalez-Gordillo, J.I., Irigoien, X., Ubeda, B., Hernandez-Leon, S., Palma, A.T., Navarro, S., Garcia-de-Lomas, J., Ruiz, A. and Fernandez-de-Puelles, M.L., **2014**. Plastic debris in the open ocean. *Proceedings of the National Academy of Sciences*, *111*(28), pp. 10239–10244.
47. Wright, L.S., Napper, I.E. and Thompson, R.C., **2021**. Potential microplastic release from beached fishing gear in Great Britain's region of highest fishing litter density. *Marine Pollution Bulletin*, *173*, p. 113115.
48. Yuan, Z. and Xu, X.R., **2023**. Surface characteristics and biotoxicity of airborne microplastics. In *Comprehensive analytical chemistry* (Vol. 100, pp. 117–164). Elsevier.
49. Luo, H., Zhao, Y., Li, Y., Xiang, Y., He, D. and Pan, X., **2020**. Aging of microplastics affects their surface properties, thermal decomposition, additives leaching and interactions in simulated fluids. *Science of the Total Environment*, *714*, p. 136862.

50. Hossain, M.R., Jiang, M., Wei, Q. and Leff, L.G., **2019**. Microplastic surface properties affect bacterial colonization in freshwater. *Journal of Basic Microbiology*, *59*(1), pp. 54–61.
51. Wang, T., Wang, L., Chen, Q., Kalogerakis, N., Ji, R. and Ma, Y., **2020**. Interactions between microplastics and organic pollutants: effects on toxicity, bioaccumulation, degradation, and transport. *Science of the Total Environment*, *748*, p. 142427.
52. Ricardo, I.A., Alberto, E.A., Junior, A.H.S., Macuvele, D.L.P., Padoin, N., Soares, C., Riella, H.G., Starling, M.C.V. and Trovo, A.G., **2021**. A critical review on microplastics, interaction with organic and inorganic pollutants, impacts and effectiveness of advanced oxidation processes applied for their removal from aqueous matrices. *Chemical Engineering Journal*, *424*, p. 130282.
53. Yao, J., Li, H. and Yang, H.Y., **2023**. Predicting adsorption capacity of pharmaceuticals and personal care products on long-term aged microplastics using machine learning. *Journal of Hazardous Materials*, *458*, p. 131963.
54. Yu, F., Yang, C., Zhu, Z., Bai, X. and Ma, J., **2019**. Adsorption behavior of organic pollutants and metals on micro/nanoplastics in the aquatic environment. *Science of the Total Environment*, *694*, p. 133643.
55. Torres, F.G., Dioses-Salinas, D.C., Pizarro-Ortega, C.I. and De-la-Torre, G.E., **2021**. Sorption of chemical contaminants on degradable and non-degradable microplastics: recent progress and research trends. *Science of the Total Environment*, *757*, p. 143875.
56. Fu, L., Li, J., Wang, G., Luan, Y. and Dai, W., **2021**. Adsorption behavior of organic pollutants on microplastics. *Ecotoxicology and Environmental Safety*, *217*, p. 112207.
57. Qiu, Y., Zhang, T. and Zhang, P., **2023**. Fate and environmental behaviors of microplastics through the lens of free radical. *Journal of Hazardous Materials*, *453*, p. 131401.
58. Hooge, A., Hauggaard-Nielsen, H., Heinze, W.M., Lyngsie, G., Ramos, T.M., Sandgaard, M.H., Vollertsen, J. and Syberg, K., **2023**. Fate of microplastics in sewage sludge and in agricultural soils. *TrAC Trends in Analytical Chemistry*, p. 117184.
59. Wang, C., Zhao, J. and Xing, B., **2021**. Environmental source, fate, and toxicity of microplastics. *Journal of Hazardous Materials*, *407*, p. 124357.
60. Wang, T., Wang, L., Chen, Q., Kalogerakis, N., Ji, R. and Ma, Y., **2020**. Interactions between microplastics and organic pollutants: effects on toxicity, bioaccumulation, degradation, and transport. *Science of the Total Environment*, *748*, p. 142427.
61. Llorca, M., Alvarez-Munoz, D., Abalos, M., Rodriguez-Mozaz, S., Santos, L.H., Leon, V.M., Campillo, J.A., Martinez-Gomez, C., Abad, E. and Farre, M., **2020**. Microplastics in Mediterranean coastal area: toxicity and impact for the environment and human health. *Trends in Environmental Analytical Chemistry*, *27*, p.e00090.
62. La Daana, K.K., Officer, R., Lyashevska, O., Thompson, R.C. and O'Connor, I., **2017**. Microplastic abundance, distribution and composition along a latitudinal gradient in the Atlantic Ocean. *Marine Pollution Bulletin*, *115*(1–2), pp. 307–314.
63. Hidalgo-Ruz, V., Gutow, L., Thompson, R.C. and Thiel, M., **2012**. Microplastics in the marine environment: a review of the methods used for identification and quantification. *Environmental Science & Technology*, *46*(6), pp. 3060–3075.

3 Identifying and Characterizing Microplastics

Jianchao Wang, Yanli Shi, Chunhui Wang, Ao Shi, Yuan Liu, Dongbei Yue, Jianbing Wang, Chunrong Wang, and Pin Hou

3.1 INTRODUCTION: MICROPLASTICS IN VARIOUS ENVIRONMENTAL MEDIA

Globally, massive plastic waste is produced annually; by 2050, the generation of waste plastics is projected to be 26,000 Mt (million tons), and about half will be discarded into the environment [1, 2] (Figure 3.1). In the environment, factors such as UV radiation, wind, and bacteria induce the erosion, abrasion, and degradation of large waste plastics, producing microplastics (MPs) and nanoplastics, with particle sizes of 100 nm–5 mm and <100 nm, respectively [3].

As a result, MPs can be found in various environmental media, including ocean, surface water, soil, underground water, and air. Revell et al. [4] found that in the West Pacific Ocean, the concentration of airborne MPs was within 0.01–5,650 particles per cubic meter. The presence of MPs in 81% of 159 tap water samples was also reported [5]. In another study, the concentrations of MPs in the underlying soil of landfill sites ranged from 590–103,080 items·kg^{-1} [6]. Soil has in fact been established as a significant long-term sink for MP pollution [7]. The common types of MPs found in various environments consist of polystyrene (PS), polyvinyl chloride (PVC), and other polymers, and their ubiquity causes severe human and environmental health concerns.

Recently, the annual estimated human inhalation and ingestion of MPs was reported to have reached 74,000–121,000 particles [8]. Remarkably, MPs particles have been detected in the human placenta, resulting in serious concerns for newborn health [9]. In human bodies, MPs are foreign matter that trigger severe immunoreactions in host organisms. Although the direct health risks of MPs to human bodies have not been established, the strong interactions of MPs (as vectors) with organic and inorganic pollutants have been widely reported.

For instance, in aqueous environments, micro-organisms rapidly colonize on the surface of MPs to form biofilms consisting of bacteria, fungus, and extracellular polymeric substances (Figure 3.2) [10]. These biofilms strongly facilitate the adsorption of contaminants, up to 98.6% of the dosed viruses on MPs in one study [11]. MPs have been proven as vectors of bisphenol A in aqueous environments, leading

DOI: 10.1201/9781003468141-3

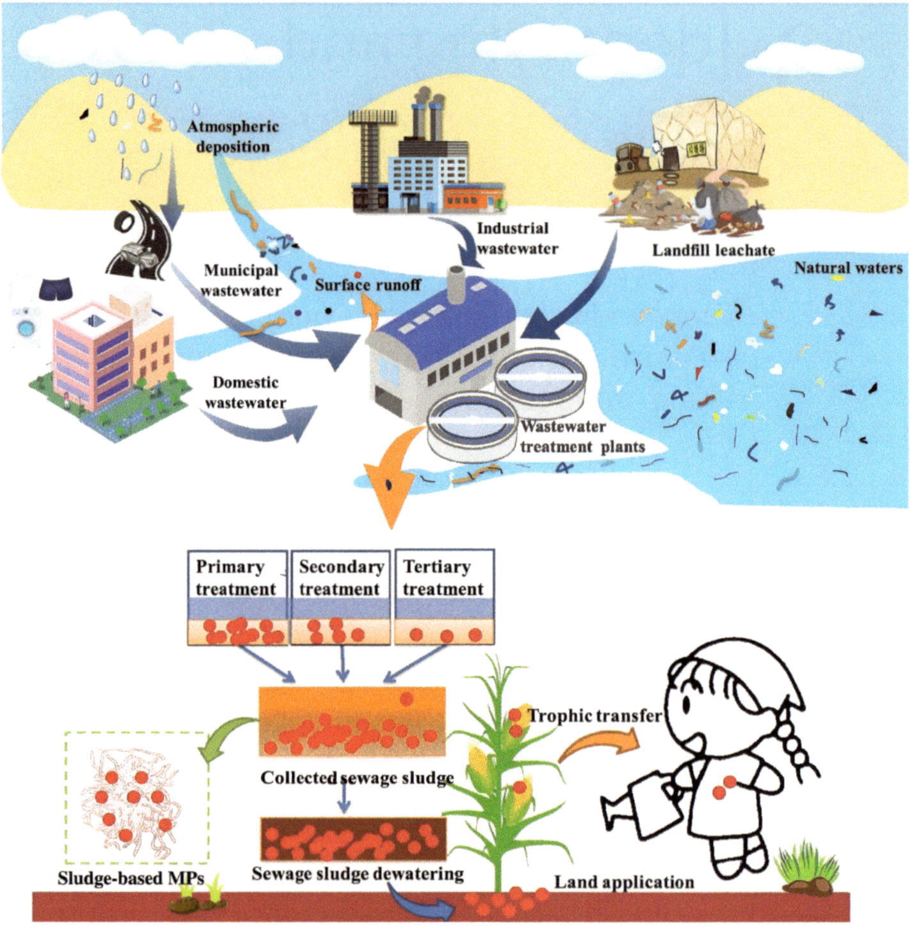

FIGURE 3.1 The MPs sources and distribution in current wastewater treatment plants [2].

to ecological risks [12]. Additionally, plastics leach hazardous additives, and most are carcinogenic, mutagenic, and endocrine-disrupting [13–15].

In the soil environment, the presence of mineral and organic matter in the matrix makes it very difficult to extract and analyze MPs, and researchers have not yet established a comprehensive protocol for identifying and quantifying MPs in complex soil environments [7]. In the aqueous environment, natural organic matter (NOM) and other organic pollutants adsorb on MPs, inhibiting their precise identification using spectroscopy. It is particularly difficult to estimate MPs in organisms because of the complexity in such interactions.

In nature, MPs show great heterogeneity in sizes, density, surface morphology, shapes, color, elemental composition, crystalline, and additives [16]. This and the other considerations reveal the great challenges in the identification, quantification, and characterization of MPs. Although the ubiquity of MPs pollution in various

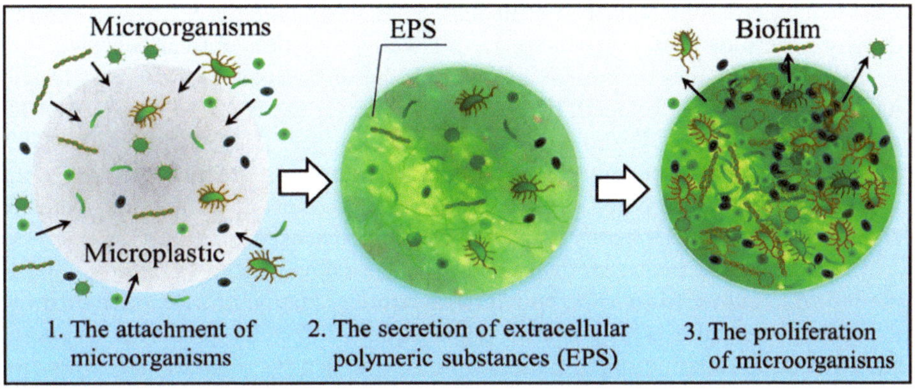

FIGURE 3.2 The formation of biofilm observed on MPs [10].

environments has been widely demonstrated, the pollution extent and comprehensive characteristics are still unknown, leading to great challenges in eliminating MPs. Therefore, it is essential to thoroughly introduce and summarize the analysis protocols of MPs in various environmental media, to provide perspectives on eliminating MPs pollution in different environments.

3.2 MPs EXTRACTION METHODS

In the environment, many organic and inorganic colloids coexist with MPs; therefore, after sampling, extracting MPs from the coexisting substances is necessary for identifying MPs. Methods such as separation based on differences in density, oil extraction, separation based on electrostatic interactions, and magnetic extraction have been developed to extract MPs.

3.2.1 Density Separation

Density separation is a frequently used method of extracting MPs from other substances in different environmental media (i.e., marine, aqueous, soil, and sediments). During extraction, the MPs float due to low density, and the higher-density substances sink [7, 17]. MPs density depends on the types of polymers and the additives but ranges from 0.8–1.4 $g{\cdot}cm^{-3}$. For extraction, scholars have used fixed-density solutions as separation media such as NaI, $ZnCl_2$, $ZnBr_2$, and KI.

For example, Vermeiren et al. applied $ZnCl_2$ as a separation medium and extracted more than 90% of the MPs from estuarine sediment in Japan, and the extracted samples had particle sizes of 62.5–125 μg [18]. Similarly, Coppock et al. reported using $ZnCl_2$ (1.5 $g{\cdot}cm^{-3}$) as a medium to effectively extract 70% to 100% of MPs from various sediments from fine silt/clay (10.25 μm) to coarse sand (149.3 μm) [19]. Researchers reported the successful application of density separation using NaCl solutions for extracting MPs from sediments and organisms collected from the Western Pacific Ocean [20]. In general, compounds adsorb on MPs in aqueous and sedimentary environments, restricting the accuracy of qualitative analysis. Therefore, density separation is often performed along with a digestion pretreatment step.

To remove organic compounds adsorbed on the surface of MPs, different digestion methods have been developed, including oxidation (H_2O_2, $H_2O_2 + Fe^{2+}$, and NaClO), alkaline treatment (KOH and NaOH), and acid treatment (HNO_3 and HNO_3:$HClO_4$) [7]. Since H_2O_2 is a frequently used oxidant for destroying organic matter in soil, it was proposed for destroying organic matter adsorbed on MPs, facilitating the efficient separation of MPs from complex sediment matrices.

Under oxidation treatment using H_2O_2 for seven days, scientists removed 50% of the biogenic matter from MPs, which triggered degradation of MPs after treatment using this procedure [21]. Digestion using $H_2O_2 + Fe^{2+}$ (i.e., Fenton) was reported effective in eliminating organic matter during the density separation of MPs from agriculture surface soils and sewage sludge [22]. Alkaline treatment using 10% KOH has also been used to destroy organic matter that interacted with MPs, although this procedure might cause the melding or discoloration of some specific MPs [23]. Particularly, alkaline digestion was used to decompose fish intestines, while the process is time-consuming (for about 2–3 weeks) [24]. Acidic digestion using HNO_3 or HNO_3:$HClO_4$ has been widely utilized to destroy marine animal tissue and identify MPs in the tissues [7, 25].

3.2.2 Oil Extraction

Based on the hydrophobic nature of MPs, oil extraction is regarded as a highly cost-effective and easy method for separating MPs from various environmental media [26]. The procedures of this method are adding a small amount of oil to samples consisting of MPs and sediments and shaking, removing and filtering the oil layer using funnel extraction, and treating the filter to remove oil residues. In a previous study, three extraction media (NaCl, oil in H_2O, and oil in NaCl) were systematically compared for separating MPs from mussels via digestion; the authors found that oil in H_2O and oil in NaCl extracted significantly more of three MPs (95.6%–100%) than did NaCl (51.1%–67.8%) [27]. Similarly, authors separated six types of MPs as separation targets into three groups of low, medium, or high density and olive oil as the medium to separate them from soil and compost samples [28] (Figure 3.3). The authors found the following MPs recovery rates: 90% for the low-density MPs, 97% for medium density, and 95% for high-density MPs.

3.2.3 Electrostatic Separation

Given the surface electrostatic properties of MPs, a novel electrostatic separation method was proposed for separating MPs from other solid matter. This method uses significantly lower amounts of chemicals, and as reported by Debraj and Lavanya, electrostatic extraction reduced almost 90% of mass when tested on MPs of various shapes and types [29]. In a previous report, the electrostatic separator was proposed to concentrate MPs, and MPs with the particle size of 63 μm–5 mm were completely recovered [30]. However, the performance of electrostatic separation is highly affected by the surface properties of MPs, such as density, organic matter

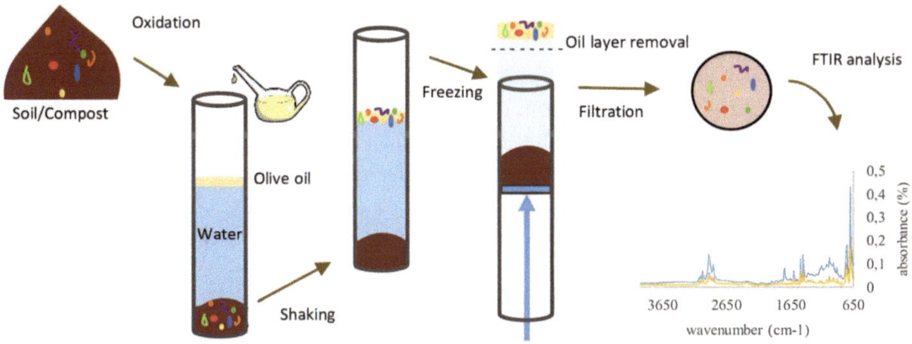

FIGURE 3.3 The oil extraction of microplastics from soil or compost [28].

adsorbed, shape, age, and biofouling. Accordingly, electrostatic separation to extract MPs needs to be further verified in real environmental samples.

3.2.4 Other Methods

Other methods of extracting MPs from various environments include magnetic separation, microrobots capture, and visual sorting. Magnetic separation is based on the hydrophobicity of polymers; magnetic iron nanoparticles are fabricated with a hydrophobic tail that interacts with hydrophobic MPs. In a previous study, magnetic porous nanocomposite was fabricated to extract MPs, and almost 98% of PS microspheres were removed from aqueous solutions within 5 min [31]. Magnetic carbon nanotubes were also utilized to remove MPs from aqueous solutions, and the results indicated that within 300 min, polyethylene, polyethylene terephthalate, and polyamide were completely removed [32]. Nevertheless, the practical application of these methods needs more comprehensive studies due to the great complexity of environmental media.

3.3 MPs IDENTIFICATION AND CHARACTERIZATION METHODS

To accurately measure MPs pollution requires identifying MPs in various environmental media. MPs are derived from synthetic polymers and a variety of additives, and these polymers have different chemical structures and functional groups. The differences in the chemical structures of synthetic polymers are a solid basis for identifying MPs. Researchers have developed several methods for identifying MPs based on this such as visual inspection, Fourier-transform infrared spectroscopy (FT–IR), and Raman spectroscopy [16, 33].

3.3.1 Visual Inspection

Visual identification with the naked eye is a very easy and straightforward method of identifying MPs sizes, shapes, colors, and morphologies [34]. Heo et al. [35] reported using visual identification with the naked eye to identify MPs of particle size

2–5 mm. Researchers in most early studies depended on visual identification to characterize MPs, but this method is error prone and limited to larger particles. It is difficult to visually detect MPs in complicated environmental samples, particularly as particle size decreases. In addition, PS and polymethyl methacrylate MPs are transparent and are therefore very challenging to characterize by visual identification. Researchers found a 70% error rate using visual identification [16].

3.3.2 FT–IR Spectroscopy

The vibrational spectroscopy of polymers can produce characteristic spectra, and FT–IR using a reference spectrum library can be used to characterize MPs. FT–IR can be used to confirm the types, functional groups, and additives of polymers in different modes depending on the targeted MPs size; two frequent modes are attenuated total reflectance (ATR) and focal plane array, the former for MPs greater than 500 μm and the latter mode, combined with an optical microscope, for MPs within 10–500 μm.

In landfill leachates, ATR–FT–IR confirmed the presence of MPs, and most were within 100 to 1,000 μm [36]. Napper et al. [37] combined FT–IR with microscopy to detect MPs of minimum particle size >30 μm after separation with filter paper. Although FT–IR is successful at identifying MPs, its use has some complications. In the environment, NOM is ubiquitous and interacts with MPs and can strongly interfere with FT–IR characterization; it forms foreign chemical bonds that reduce the precision of FT–IR for matching to reference spectra. Additionally, due to the high adsorption of infrared radiation, blank MPs particles are difficult to detect by FT–IR, and using micro-FT–IR to characterize these small MPs is highly expensive and time-consuming. More importantly, MPs smaller than 20 μm cannot be characterized by FT–IR, which is a major challenge.

3.3.3 Raman Spectroscopy

Similar to FT–IR spectroscopy, using the Raman spectra of polymers is another method of characterizing MPs. In measurements of Raman spectra, a monochromatic laser strongly interacts with the molecular structure of polymers to produce the characteristic peaks that identify different chemical structures [38, 39], and analyzing chemical structures facilitates analyzing MPs.

In contrast to the FT–IR method, Raman shows obvious advantages, including better spatial resolution (for MPs with particle size >1 μm), lower water and atmospheric CO_2 interference, and higher sensitivity to nonpolar functional groups. For example, Werbowski et al. [40] combined Raman and FT–IR to detect the MPs present in urban stormwater runoff; the former was used for smaller particles and the latter for particles larger than 500 μm. Schymanski et al. [41] utilized micro-Raman spectroscopy to detect MPs sized 5–20 μm.

Although Raman spectra take much longer to measure than FT–IR spectra, Raman spectra are not strongly affected by MPs shape, size, or thickness. However, the biggest obstacle to the Raman method is that Raman spectra strongly interfere with background fluorescence. Therefore, using Raman spectroscopy to identify MPs in environmental media with high concentrations of NOM or soil organic matter might be especially problematic.

3.3.4 Other Methods

Scanning electron microscopy (SEM), laser direct infrared (LDIR), pressurized fluid extraction (PFE), short-wave infrared (SWIR), and other techniques have been utilized to identify MPs. SEM using a focused electron beam provides high-resolution images of MP samples. SEM combined with energy-dispersive X-ray spectroscopy provides information on the chemical compositions of MPs [42].

However, SEM can provide morphological features of MPs but cannot determine MP polymer types. In addition, pretreatment for SEM sampling is complicated, including cleaning, drying, conductive metal spraying, and other processes. LDIR shows obvious advantages such as rapid and automatic measurements, and it can particularly provide MPs composition, diameter, and thickness. Importantly, LDIR doesn't need visual pre-selection of MPs particles or enhanced spectrum quality. For example, Ourgaud et al. [43] successfully applied LDIR to identify more than five types of MPs (such as polypropylene (PP), PS, and PVC) from marine samples.

PFE uses proper organic solvents under subcritical temperature and pressure to extract and identify MPs from marine and terrestrial samples. Fuller and Gautam [44] reported using PFE to extract 101%–111% of plastics from glass beads and 85%–94% from a composted municipal waste sample. Other authors reported on using SWIR to analyze MPs from surface water samples [45]. Before measurement by SWIR imaging spectrometer, the samples were purified using a hydrogen peroxide solution and filtered via several glass fiber filters. The measuring speed of SWIR was one sample per 2 minutes. The true detection efficiency of SWIR was reported 75%.

3.4 MPs QUANTIFICATION METHODS

The quantification of MPs in various environments is necessary for revealing the pollution problems. However, it is difficult to quantify MPs; different measures are used for different media. For instance, for MPs in sediment, the concentration is grams of MPs per m^2 or particle items per m^2. For samples collected from sea and surface water, grams per m^2 or particle items per m^2 are the most used units, but grams or items per volume (grams or items per m^3) is also used [16].

Optical microscopes are used to both characterize and quantify MPs. For example, in a previous report, only optical microscope was used to quantify MP particles in *Anomalocardia flexuosa* after filtering through a glass fiber filter (1 μm) [46]. Nevertheless, the optical method suffers from the major disadvantage of poor accuracy because it is highly dependent on the MPs handlers, color, shape, and size. Therefore, optical microscope is generally combined with other analytical techniques to quantify MPs.

Pyrolysis gas chromatography mass spectrometry (Py-GC-MS) is also widely applied for quantifying MPs in various environmental media. Py-GC-MS consists of pyrolysis and detection using gas chromatography mass spectrometry. For pyrolysis, the MPs are reduced to their constituents by heating to high temperature without oxygen; for the GC-MS, the pyrolysis products are separated and further analyzed. The obvious advantage of Py-GC-MS is its sensitivity; even at low concentration, it can identify MPs. Researchers confirmed the effectiveness of Py-GC-MS for quantifying and characterizing MPs in previous studies [47, 48]. However, Py-GC-MS is destructive, and therefore, its characterization results cannot be replicated. Additionally, it cannot produce the information – number, size, color, morphology – that is critical for assessing MP pollution.

Proton nuclear magnetic resonance spectroscopy (^{1}H NMR) has been developed to qualitatively and quantitatively analyze MPs using a calibration curve [49]; researchers showed the successful qualitative and quantitative analysis of some MPs, including PE, PS, and PET. During NMR measurement, different types of MPs are dissolved by different solvents. For instance, PE was dissolved using deuterated toluene, PS was dissolved using deuterated chloroform, and PET was dissolved using trifluoracetic acid. However, the method has some drawbacks. For example, the adsorptive interaction between MPs with NOM or SOM result in signal overlapping; therefore, NMR cannot be used to extract MPs in soil and wastewater because these sources contain high concentrations of organic matter.

3.5 CONCLUSION AND FUTURE PERSPECTIVES

Because of the massive production of waste plastics, MPs pollution is omnipresent in every environments. MPs identification and quantification are important for assessing the extent of MPs pollution, and in this chapter, we summarize and discuss the technological advances for identifying, quantifying, and characterizing MPs.

Several methods have been proposed for extracting MPs from both organic and inorganic colloids in aqueous solution and soil matrix, such as density separation, oil extraction, and electrostatic separation. Density separation is a frequently used method, but its feasibility for small MPs has not been tested, and although oil extraction shows promising performance, the oil residues can restrict MPs identification and characterization. Electrostatic separation exhibits prominent efficacy for MPs extraction, but its efficacy is highly dependent on the MPs density, organic matter adsorbed, shape, aging, and biofouling. In this regard, novel methods, a standardized framework of extraction procedures, or automating extraction protocols would substantially facilitate understanding of MP pollution.

Researchers have proposed some novel methods for MP identification and characterization, including visual inspection and FT–IR and Raman spectroscopy. Visual identification is the simplest method and can characterize MP size, shape, and color; however, visual identification is highly error prone and limited by MP, and the transparency of MPs also restricts its application. FT–IR and Raman spectra are highly effective for analyzing large- and small-particle MPs, respectively, but the adsorptive adsorption of NOM or SOM greatly interferes with their effectiveness.

Other new methods such as LDIR, PFE, and SWIR have been utilized for MPs identification, but they need testing in practical application. In the near future, research on identifying MPs should focus on establishing standard validation processes and reference data. In addition, novel methods for detecting nanoparticles in different environments are needed.

For quantifying MPs in various environmental media, optical microscope, Py-GC-MS, ^{1}H NMR, and other methods have been utilized. The optical microscope is a popular approach that has the advantage of being simple; however, its accuracy for analysis is strongly affected by the MP handlers, shapes, and sizes. As a sensitive method, Py-GC-MS can detect MPs even at low sample concentrations. However, Py-GC-MS is a destructive technique that strongly restricts its practical application since the results cannot be replicated, and it does not provide MPs properties such as size, color, and shape. ^{1}H NMR is reported to be effective for quantifying microplastics, but the adsorptive adsorption of organic matter in the environmental media can interfere with the signal. In summary, efforts are needed to develop novel quantification methods and standard processes; more combinations of quantification methods and their practical applications should also be examined.

ACKNOWLEDGMENTS

This work is supported by National Natural Science Foundation of China (52300071) and Young Elite Scientist Sponsorship Program By BAST (BYESS2023097).

REFERENCES

1. S. Castan, C. Henkel, T. Hüffer, T. Hofmann, 2021. Microplastics and nanoplastics barely enhance contaminant mobility in agricultural soils, *Commun. Earth Environ.*, 2, pp. 1–9.
2. X. Wu, X. Zhao, R. Chen, P. Liu, W. Liang, J. Wang, M. Teng, X. Wang, S. Gao, 2022. Wastewater treatment plants act as essential sources of microplastic formation in aquatic environments: A critical review, *Water Res.*, 221, p. 118825.
3. K. Min, J.D. Cuiffi, R.T. Mathers, 2020. Ranking environmental degradation trends of plastic marine debris based on physical properties and molecular structure, *Nat. Commun.*, 11, p. 727.
4. L.E. Revell, P. Kuma, E.C. Le Ru, W.R.C. Somerville, S. Gaw, 2021. Direct radiative effects of airborne microplastics, *Nature*, 598, pp. 462–467.
5. Y. Zhang, A. Diehl, A. Lewandowski, K. Gopalakrishnan, T. Baker, 2020. Removal efficiency of micro- and nanoplastics (180 nm–125μm) during drinking water treatment, *Sci. Total. Environ.*, 720, p. 137383.
6. Y. Wan, X. Chen, Q. Liu, H. Hu, C. Wu, Q. Xue, 2022. Informal landfill contributes to the pollution of microplastics in the surrounding environment, *Environ Pollut.*, 293, p. 118586.
7. J.N. Moller, M.G.J. Loder, C. Laforsch, 2020. Finding microplastics in soils: A review of analytical methods, *Environ. Sci. Technol.*, 54, pp. 2078–2090.
8. K.D. Cox, G.A. Covernton, H.L. Davies, J.F. Dower, F. Juanes, S.E. Dudas, 2019. Human consumption of microplastics, *Environ. Sci. Technol.*, 53, pp. 7068–7074.
9. A. Ragusa, A. Svelato, C. Santacroce, P. Catalano, V. Notarstefano, O. Carnevali, F. Papa, M.C.A. Rongioletti, F. Baiocco, S. Draghi, E. D'Amore, D. Rinaldo, M. Matta, E. Giorgini, 2021. Plasticenta: First evidence of microplastics in human placenta, *Environ. Int.*, 146, p. 106274.
10. J. Wang, X. Guo, J. Xue, 2021. Biofilm-developed microplastics as vectors of pollutants in aquatic environments, *Environ. Sci. Technol.*, 55, pp. 12780–12790.
11. J. Lu, Z. Yu, L. Ngiam, J. Guo, 2022. Microplastics as potential carriers of viruses could prolong virus survival and infectivity, *Water Res.*, 225, p. 119115.

12. X. Liu, H. Shi, B. Xie, D.D. Dionysiou, Y. Zhao, 2019. Microplastics as both a sink and a source of bisphenol A in the marine environment, *Environ. Sci. Technol.*, 53, pp. 10188–10196.
13. H. Cheng, H. Luo, Y. Hu, S. Tao, 2020. Release kinetics as a key linkage between the occurrence of flame retardants in microplastics and their risk to the environment and ecosystem: A critical review, *Water Res.*, 185, p. 116253.
14. C. Chen, L. Chen, Y. Yao, F. Artigas, Q. Huang, W. Zhang, 2019. Organotin release from polyvinyl chloride microplastics and concurrent photodegradation in water: Impacts from salinity, dissolved organic matter, and light exposure, *Environ. Sci. Technol.*, 53, pp. 10741–10752.
15. A.T.N. Do, Y. Ha, J.H. Kwon, 2022. Leaching of microplastic-associated additives in aquatic environments: A critical review, *Environ. Pollut.*, 305, p. 119258.
16. V. Hidalgo-Ruz, L. Gutow, R.C. Thompson, M. Thiel, 2012. Microplastics in the marine environment: A review of the methods used for identification and quantification, *Environ. Sci. Technol.*, 46, pp. 3060–3075.
17. S.M. Patil, N.R. Rane, P.O. Bankole, P. Krishnaiah, Y. Ahn, Y.-K. Park, K.K. Yadav, M.A. Amin, B.-H. Jeon, 2022. An assessment of micro- and nanoplastics in the biosphere: A review of detection, monitoring, and remediation technology, *Chem. Eng. J.*, 430, p. 132913.
18. P. Vermeiren, C. Muñoz, K. Ikejima, 2020. Microplastic identification and quantification from organic rich sediments: A validated laboratory protocol, *Environ. Pollut.*, 262, p. 114298.
19. R.L. Coppock, M. Cole, P.K. Lindeque, A.M. Queirós, T.S. Galloway, 2017. A small-scale, portable method for extracting microplastics from marine sediments, *Environ. Pollut.*, 230, pp. 829–837.
20. D. Zhang, X. Liu, W. Huang, J. Li, C. Wang, D. Zhang, C. Zhang, 2020. Microplastic pollution in deep-sea sediments and organisms of the Western Pacific Ocean, *Environ. Pollut.*, 259, p. 113948.
21. M.T. Nuelle, J.H. Dekiff, D. Remy, E. Fries, 2014. A new analytical approach for monitoring microplastics in marine sediments, *Environ. Pollut.*, 184, pp. 161–169.
22. Q. Li, J. Wu, X. Zhao, X. Gu, R. Ji, 2019. Separation and identification of microplastics from soil and sewage sludge, *Environ. Pollut.*, 254, p. 113076.
23. A. Herrera, P. Garrido-Amador, I. Martínez, M.D. Samper, J. López-Martínez, M. Gómez, T.T. Packard, 2018. Novel methodology to isolate microplastics from vegetal-rich samples, *Mar. Pollut. Bull.*, 129, pp. 61–69.
24. E.M. Foekema, C. De Gruijter, M.T. Mergia, J.A. van Franeker, A.J. Murk, A.A. Koelmans, 2013. Plastic in North sea fish, *Environ. Sci. Technol.*, 47, pp. 8818–8824.
25. M. Claessens, L. Van Cauwenberghe, M.B. Vandegehuchte, C.R. Janssen, 2013. New techniques for the detection of microplastics in sediments and field collected organisms, *Mar. Pollut. Bull.*, 70, pp. 227–233.
26. A. Crew, I. Gregory-Eaves, A. Ricciardi, 2020. Distribution, abundance, and diversity of microplastics in the upper St. Lawrence River, *Environ. Pollut.*, 260, p. 113994.
27. X. Song, X. Wu, X. Song, Z. Zhang, 2022. Oil extraction following digestion to separate microplastics from mussels, *Chemosphere*, 289, p. 133187.
28. C. Scopetani, D. Chelazzi, J. Mikola, V. Leiniö, R. Heikkinen, A. Cincinelli, J. Pellinen, 2020. Olive oil-based method for the extraction, quantification and identification of microplastics in soil and compost samples, *Sci. Total. Environ.*, 733, p. 139338.
29. D. Debraj, M. Lavanya, 2023. Microplastics everywhere: A review on existing methods of extraction, *The Science of the Total Environment*, 893, p. 164878.
30. S. Felsing, C. Kochleus, S. Buchinger, N. Brennholt, F. Stock, G. Reifferscheid, 2018. A new approach in separating microplastics from environmental samples based on their electrostatic behavior, *Environ. Pollut.*, 234, pp. 20–28.

31. F. Pasanen, R.O. Fuller, F. Maya, 2023. Fast and simultaneous removal of microplastics and plastic-derived endocrine disruptors using a magnetic ZIF-8 nanocomposite, *Chem. Eng. J.*, 455, p. 140405.
32. Y. Tang, S. Zhang, Y. Su, D. Wu, Y. Zhao, B. Xie, 2021. Removal of microplastics from aqueous solutions by magnetic carbon nanotubes, *Chem. Eng. J.*, 406, p. 126804.
33. D. Elkhatib, V. Oyanedel-Craver, 2020. A critical review of extraction and identification methods of microplastics in wastewater and drinking water, *Environ. Sci. Technol.*, 54, pp. 7037–7049.
34. M.S. Kabir, H. Wang, S. Luster-Teasley, L. Zhang, R. Zhao, 2023. Microplastics in landfill leachate: Sources, detection, occurrence, and removal, *Environ. Sci. Ecotechnol.*, 16, p. 100256.
35. N.W. Heo, S.H. Hong, G.M. Han, S. Hong, J. Lee, Y.K. Song, M. Jang, W.J. Shim, 2013. Distribution of small plastic debris in cross-section and high strandline on Heungnam beach, South Korea, *Ocean Sci. J.*, 48, pp. 225–233.
36. P. He, L. Chen, L. Shao, H. Zhang, F. Lu, 2019. Municipal solid waste (MSW) landfill: A source of microplastics? Evidence of microplastics in landfill leachate, *Water Res.*, 159, pp. 38–45.
37. I.E. Napper, B.F.R. Davies, H. Clifford, S. Elvin, H.J. Koldewey, P.A. Mayewski, K.R. Miner, M. Potocki, A.C. Elmore, A.P. Gajurel, R.C. Thompson, 2020. Reaching new heights in plastic pollution—preliminary findings of microplastics on Mount Everest, *One Earth*, 3, pp. 621–630.
38. C.F. Araujo, M.M. Nolasco, A.M.P. Ribeiro, P.J.A. Ribeiro-Claro, 2018. Identification of microplastics using Raman spectroscopy: Latest developments and future prospects, *Water Res.*, 142, pp. 426–440.
39. A. Käppler, D. Fischer, S. Oberbeckmann, G. Schernewski, M. Labrenz, K.-J. Eichhorn, B. Voit, 2016. Analysis of environmental microplastics by vibrational microspectroscopy: FTIR, Raman or both?, *Anal. Bioanal. Chem.*, 408, pp. 8377–8391.
40. L.M. Werbowski, A.N. Gilbreath, K. Munno, X. Zhu, J. Grbic, T. Wu, R. Sutton, M.D. Sedlak, A.D. Deshpande, C.M. Rochman, 2021. Urban stormwater runoff: A major pathway for anthropogenic particles, black rubbery fragments, and other types of microplastics to urban receiving waters, *ACS ES&T Water*, 1, pp. 1420–1428.
41. D. Schymanski, C. Goldbeck, H.U. Humpf, P. Fürst, 2018. Analysis of microplastics in water by micro-Raman spectroscopy: Release of plastic particles from different packaging into mineral water, *Water Res.*, 129, pp. 154–162.
42. G. Kalčíková, B. Alič, T. Skalar, M. Bundschuh, A.Ž. Gotvajn, 2017. Wastewater treatment plant effluents as source of cosmetic polyethylene microbeads to freshwater, *Chemosphere*, 188, pp. 25–31.
43. M. Ourgaud, N.N. Phuong, L. Papillon, C. Panagiotopoulos, F. Galgani, N. Schmidt, V. Fauvelle, C. Brach-Papa, R. Sempere, 2022. Identification and quantification of microplastics in the marine environment using the laser direct infrared (LDIR) technique, *Environ. Sci. Technol.*, 56, pp. 9999–10009.
44. S. Fuller, A. Gautam, 2016. A procedure for measuring microplastics using pressurized fluid extraction, *Environ. Sci. Technol.*, 50, pp. 5774–5780.
45. L.K. Schmidt, M. Bochow, H.K. Imhof, S.E. Oswald, 2018. Multi-temporal surveys for microplastic particles enabled by a novel and fast application of SWIR imaging spectroscopy—study of an urban watercourse traversing the city of Berlin, Germany, *Environ. Pollut.*, 239, pp. 579–589.
46. D.N.A. Bruzaca, A.K.S. Justino, G.C.P. Mota, G.A. Costa, F. Lucena-Frédou, A.O. Gálvez, 2022. Occurrence of microplastics in bivalve molluscs Anomalocardia flexuosa captured in Pernambuco, Northeast Brazil, *Mar. Pollut. Bull.*, 179, p. 113659.
47. K. Harata, S. Kitagawa, Y. Iiguni, H. Ohtani, 2020. Identification of polymer species in a complex mixture by pyrolysis-gas chromatography-atmospheric pressure chemical

ionization-high resolution time-of-flight mass spectrometry as a basis for environmental microplastic analysis, *J. Anal. Appl. Pyrol.*, 148, p. 104828.

48. E. Fries, J.H. Dekiff, J. Willmeyer, M.T. Nuelle, M. Ebert, D. Remy, 2013. Identification of polymer types and additives in marine microplastic particles using pyrolysis-GC/MS and scanning electron microscopy, *Environ. Sci-Proc. Imp.*, 15, pp. 1949–1956.
49. N. Peez, M.C. Janiska, W. Imhof, 2019. The first application of quantitative ^{1}H NMR spectroscopy as a simple and fast method of identification and quantification of microplastic particles (PE, PET, and PS), *Anal. Bioanal. Chem.*, 411, pp. 823–833.

4 Unraveling the Microplastics Mystery

Computational Insights, Real-World Cases, and the Road Ahead

Ravi Kumar

4.1 INTRODUCTION

In contemporary times, pervasive microplastics (MP) contamination has evolved as a noteworthy environmental challenge with extensive implications for ecosystems and human well-being. The fast evolution of the plastic industry has resulted in a substantial surge in production: from 1.5 million tons (Mt) to 390.7 Mt from in the span of 70 years from 1950 to 2021 [1]. Borelle et al. projected that there will be increases of 20 to 50 Mt plastic waste in the aquatic environment by 2030 because of inadequate disposal practices [2]. Under the influences of mechanical friction [3] and photo-aging [4], this plastic waste degrades into MPs of diameter <5 mm [5], ubiquitously present in oceans, freshwater bodies, the atmosphere, and soil [6].

The origins of MPs are intricately tied to human activities, notably processes such as laundering clothes and vehicular activities but also other means. A significant portion enters sewage treatment facilities through municipal network pipes, becoming sequestered in the soil. Furthermore, the extensive utilization of plastic mulches in agriculture significantly contributes to soil MP pollution [7]. Atmospheric MPs primarily result from the aging of fibrous fabrics and engineered plastics. Through wind dispersion, atmospheric MPs infiltrate soil and water environments propelled by precipitation and natural deposition, subsequently accumulating extensively in ponds, seas, lakes, and other aquatic systems as well, following intricate movement patterns [8]. MP contamination has already inflicted harm upon biological health and instigated considerable ecological damage.

Owing to their limited specific surface areas, MPs possess the capacity to adsorb organic pollutants, heavy metals, and pathogens, leading to the accumulation of toxicity [9]. Additionally, the detrimental impacts of MPs are worsened by the migration of the additives in them [10]. MPs obstruct plant pores, thereby affecting crucial processes such as gas exchange and photosynthesis. Additionally, the exposure of

DOI: 10.1201/9781003468141-4

humans and other animals to water and air contaminated by MPs poses a significant risk, with ingested MPs potentially leading to oxidative stress, apoptosis, and metabolic disorders.

The increasing ubiquity of MPs presents a complex and pressing concern that demands innovative solutions. In response to this environmental crisis, the intersection of computational studies, sensor technologies, and remediation strategies has become a focal point for researchers seeking effective and sustainable approaches [11]. This chapter delves into the realm of computational studies dedicated to the sensing and remediation of MPs in the environment. Through the lens of advanced computational models, I explore the intricate dynamics of MP dispersion, optimal sensing techniques, and novel remediation methodologies.

Computational tools not only enhance our understanding of the intricate pathways of MP pollution but also offer pathways toward informed decision-making in environmental management. As I embark on this exploration, I navigate through the intricacies of molecular simulations, machine learning (ML) algorithms, and data-driven insights to unravel the behavior of MPs at both macro and micro scales. Additionally, we scrutinize emerging sensor technologies designed to detect and quantify MPs present in diverse environmental matrices. These advancements in sensing lay the foundation for precise monitoring, enabling timely interventions and policy formulations.

Furthermore, the chapter addresses the imperative aspect of remediation strategies, emphasizing the role of computational simulations in the design and optimization of innovative techniques for removing MPs. From engineered nanomaterials to biological remediation, the computational framework serves as a virtual laboratory, accelerating the advancement of sustainable solutions to minimize the effects of MP pollution. Through this comprehensive exploration, we aim to contribute to the evolving discourse on MP pollution, offering valuable insights and a roadmap for the integration of computational studies in the quest for a cleaner and healthier environment. As I delve into the intricate web of sensing and remediation, I uncover the potential of computational approaches for revolutionizing our response to the MPs crisis, paving the way for a more sustainable coexistence with the planet we call home.

4.2 COMPUTATIONAL APPROACHES IN ENVIRONMENTAL SCIENCE

In this section, I delve into the intricate world of computational techniques for sensing and remediating MPs. From molecular simulations elucidating particle dynamics to ML algorithms revolutionizing detection precision, and from quantum mechanical modeling guiding nanomaterial design to meta-genomic analysis exploring biological remediation strategies, the marriage of computation and environmental science opens new frontiers in our battle against MP pollution. The microscopic nature of MPs poses a unique challenge for traditional sensing methods. Computational techniques, particularly molecular simulations, offer a microscopic lens for understanding the behaviors and interactions of individual MP particles.

These simulations delve into the molecular structure of plastics, predicting their movement patterns, aggregation tendencies, and potential environmental transformations. Molecular dynamics (MD) simulations, for instance, enable researchers to track the trajectories of individual MP particles in aquatic environments. By considering factors like water currents, temperature, and the presence of other substances, MD simulations provide an enhanced understanding of how MPs disperse and accumulate, guiding the design of effective sensing strategies.

ML is also a useful approach that is receiving attention for its precise detection of MPs. The sheer volume and variety of environmental data make traditional analysis methods insufficient for robust MP detection. ML algorithms, fueled by vast data sets generated from sensors and imaging technologies, have emerged as powerful tools for precise detection and classification. Image recognition algorithms trained on diverse sets of MP images can identify and classify particles in environmental samples. The integration of ML enhances the accuracy of MP sensing, allowing for real-time monitoring and rapid identification of pollution hotspots.

ML constitutes a crucial subset of artificial intelligence, demonstrating the ability to dynamically and swiftly process intricate and voluminous linear and nonlinear data sets. Leveraging algorithms, ML extracts features from data sets to unveil possible relationships among relevant factors, thereby achieving automatic categorization or regression [12]. Due to its notable advantages in terms of speed, precision, and automation, ML has found widespread application in advancing scientific research, encompassing domains such as water treatment [13] and catalysis [14].

Through training with MP spectroscopy or other images, ML can achieve highly precise classification and identification depending upon various MPs properties, including characteristic peaks, shape, and RGB [15]. Additionally, integrating with ML with hyperspectral imaging technology, capable of acquiring MPs spectral information via multiple channels and visualizing corresponding position information, is anticipated to expedite advancements in identifying MPs [16].

ML plays a pivotal role not only in predicting MPs' toxicity [17] but also in assessing the tendency of MPs to adsorb pollutants [18]. The deployment of ML models represents an effective approach to extracting pertinent features defining the characteristics of MPs within the burgeoning data set landscape of relevant studies. Furthermore, ML models contribute to predicting environmental MP pollution and identifying critical variables affecting toxicity, adsorption capacity, and abundance of microbes, thereby mitigating research load. Consequently, ML stands as a valuable tool with considerable prospects for recognizing and understanding the complexities associated with MPs [19]. By extracting the internal insights and discerning patterns from huge data sets, ML models can handle nonlinear regression as well as classification [20, 21].

A range of ML methods has been created to categorize or detect MPs within environmental contexts [22, 23]. For instance, integrating a deep UV Raman spectrometer with convolutional neural networks very effectively detected MPs in aquatic environments [23]. ML models showcase their capability to automatically tally and categorize MP particles from images [24]. Extant ML models have predominantly leaned upon data reminiscent of images when addressing MPs. The burgeoning reservoir of literature pertaining to anaerobic digestion, replete with a myriad of MPs, holds promise for the evolution of ML models. These models adeptly simulate the intricate ramifications of MPs on anaerobic digestion processes, while duly considering their concentrations and physicochemical attributes [25, 26].

Within the realm of traditional ML model development, practitioners are intricately engaged in every facet of the design journey. This encompasses the meticulous crafting of features, judicious selection of models and algorithms, fine tuning of hyperparameters, and beyond. However, traditional ML models demand a significant investment of time and expertise, rendering them less accessible and adaptable for individuals lacking specialized knowledge in the field [27, 28]. Addressing this, several automated machine learning (AutoML) frameworks, like Auto-Sklearn [29], tree-based pipeline optimization tool [30], genetic automated ML assistant [31], AutoGluon [32], and H_2O AutoML [33], have been proposed. These frameworks are designed to reduce the need for human involvement in ML applications by automatically comparing and deploying ML models that achieve high performance.

4.2.1 Optimizing Nanomaterials for Remediation Efficiency

The quest for effective MP remediation involves the design and deployment of innovative materials. Computational simulations, such as quantum mechanical calculations and molecular docking studies, play a pivotal role in optimizing the properties of nanomaterials tailored for MP capture. Quantum mechanical modeling provides insights into the electronic properties of nanomaterials, guiding the selection of materials with enhanced adsorption capabilities. Coupled with molecular docking simulations, researchers can predict the affinity of these materials for different types of MPs, facilitating the engineering of highly efficient remediation agents.

4.2.2 Biological Remediation through Computational Biology

Harnessing the power of nature for MP remediation requires a nuanced understanding of biological processes. Computational biology techniques, including bioinformatics and metabolic pathway analysis, aid in exploring the potential of microorganisms for biodegradation. Metagenomic analysis unveils the microbial communities capable of degrading plastics, guiding the identification and engineering of enzymes involved in the biodegradation process. These computational insights contribute to the development of environmentally friendly, biology-driven remediation strategies.

4.3 SENSING MPs

Sensing and detecting MPs in the environment using computational methodologies involve advanced technologies and data analysis techniques (Figure 4.1) that I discuss in this section.

4.3.1 Machine Learning and Image Processing

Researchers conducted a scholarly investigation with the aim of bridging the conspicuous void in critical literature reviews by synthesizing the advancements achieved over the past decade in leveraging ML algorithms for the discernment and measurement of MPs. The authors identified various amalgamations of ML methodologies for MP identification, such as Fourier-transform infrared, Raman, and near-infrared

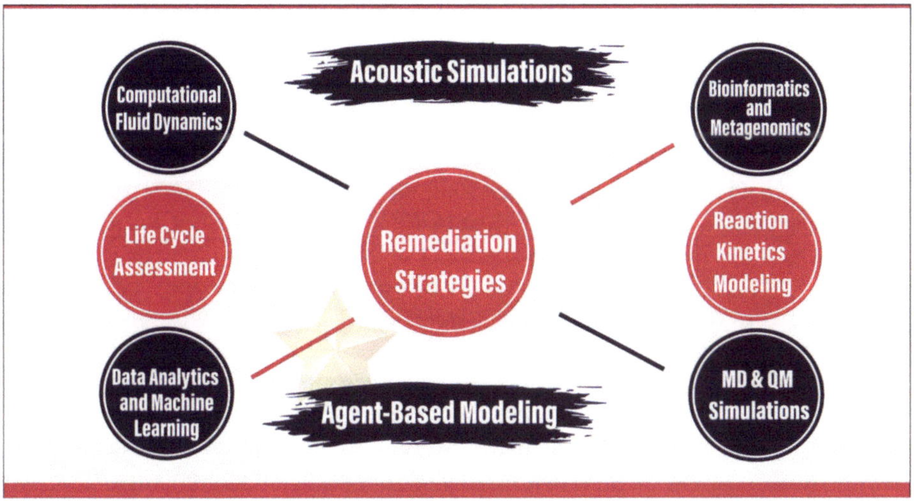

FIGURE 4.1 Various computational approaches for sensing MPs in the environment.

(NIR) spectroscopy. Notably, support vector machine emerged as the most frequently employed ML model, markedly augmenting traditional analytical approaches for detecting spectral irregularities and enhancing detection precision. Furthermore, artificial neural network models exhibit heightened recognition capabilities, offering expedited detection times and enhanced efficacy in MP identification. Cumulatively, the review underscored the promising potential of ML in propelling the field of MP identification and quantification forward [15].

4.3.2 Automated Image Analysis

ML algorithms can be trained to analyze images captured through microscopes or other imaging devices. These algorithms can identify and quantify MPs based on their size, shape, and color. A scholarly inquiry introduces a semi-automated image processing methodology devised for the quantification and measurement of MPs on filter membrane substrates [34]. Its objective is to alleviate the arduous and error-prone nature inherent in manual visual counting protocols. The innovative algorithm, integrated into a user-friendly Python script, underwent validation through comparison with visual counting conducted by a skilled operator. The evaluation encompassed both synthetically generated MP images and specimens procured from natural settings, including cave sediment and river water. Findings revealed notable accuracy and expeditiousness, with an average processing time of approximately 10 seconds per image. Moreover, the average deviation remained below 10%, indicative of the method's precision. Notably, when the samples were treated with Nile Red staining, this deviation was further diminished to nearly 8%, underscoring the efficacy of the proposed image analysis approach [35].

4.3.3 Computer Vision

Computer vision techniques can be employed to recognize MP patterns and features in images, making the process of detection more efficient. Researchers built a computer vision and ML model to swiftly and automatically count and categorize MPs into four morphology and size groups, eliminating manual intervention. Initially, the authors devised the ML algorithm to count and categorize the MPs; subsequently, they implemented both supervised K-nearest neighbor and unsupervised classification to assess numbers and characteristics of MPs, uncovering latent insights [19].

While the ML algorithm exhibited good results in counting and size categorization, it requires refinement for visual class categorization. Likewise, the supervised classification consistently attained satisfactory accuracy surpassing 0.9. Conversely, the unsupervised classification underestimated certain MP shapes due to sampling methodology, serving as a valuable tool for revealing imperceptible information missed by traditional research methods. To sum up, the anticipated system provides a trustworthy automatic approach to quantifying MPs, relying upon particle numbers, size distribution, and morphology. This method shows considerable potential for standardizing procedures in the field [19].

Another suggested approach melded computer vision with multivariate calibration methodologies, notably leveraging partial least squares in tandem with interval selection and successive algorithm projection. The quantities varied for blue, black, yellow, green, red, and colorless MPs, with ranges spanning from 7 to 134, 0 to 50, 1 to 212, 0 to 113, 6 to 290, and 20 to 392, respectively. Upon application to an independent sample set, the models yielded root mean square errors of prediction of 9.8 for yellow, 6.4 for blue, 3.5 for black, 8.1 for colorless, 7.5 for green, and 19.3 for red. These findings underscore the potential of image processing for quantifying MPs according to their color, presenting a time-efficient and objective alternative for forthcoming analyses [36].

4.3.4 Modeling and Simulation

Modeling and simulation are important in MP detection in the environment, for instance with environment fate modeling and numerical simulations. Researchers introduced Canadian MPs Simulation (CaMPSim-3D), a new model for predicting MP movement in water. It integrates a 3D particle-tracking model with a hydrodynamic system. The model accurately simulates particle behavior and integrates well with various advection schemes. Testing in the Saint John Rever Estuary showed significant differences in results between advection schemes, with TVD3 Runge–Kutta proving optimal with lower computational costs [37].

In another study, those authors assessed four Lagrangian particle-tracking models, namely, the Delft3D Water Quality Particle tracking module, Ichthyoplankton, Track Marine Plastic Debris, and CaMPSim-3D, in terms of their effectiveness in simulating the fate and transport of MPs. They examined how each model handles key processes like diffusion, advection, beaching, windage, and degradation and considered the effects of MPs' physical properties. The authors identified strengths, weaknesses, and areas for improvement in simulating MPs' behavior, providing valuable insights for future model enhancements [38].

4.3.5 Environmental Fate Modeling [39]

Computational models can simulate the transport and fate of MPs across various environmental compartments, aiding in the projection of their distribution patterns over time.

4.3.6 Numeric Simulations

Computational fluid dynamics (CFD) simulations play a pivotal role by giving valuable insights into the dispersion of MPs inside aquatic bodies, facilitating the identification of zones where accumulation occurs [37]. Numeric models confer a distinct advantage by allowing for scenario analyses that enable assessing the relative contributions of various mechanisms of transport and predicting the potential pathways under the altered conditions of climate [40]. Sherman and Van Sebille, for example, employed a global floating MP model to simulate diverse cleanup scenarios, pinpointing the coasts of Indonesia and China as the best removal locations [41]. Numerical transport models also play a critical role in identifying potential sources of ocean plastics [42]. Isobe et al. devised a two-way particle-tracking methodology encompassing both backward and forward tracking in time to statistically locate the sources of drifting objects in the East China Sea [42]. Liubartseva et al. employed a comparable methodology to systematically generate a source-receptor matrix spanning various regions within the Adriatic Sea. This method serves as a valuable tool for stakeholders and policymakers involved in tackling issues related to plastic pollution [43].

Numeric models can also predict future distribution patterns amid evolving climate and circulation dynamics. Variables such as sea level rise, melting sea ice, shifts in precipitation patterns, changes in storm frequency and intensity, and fluctuations in thermo-haline conditions could potentially influence the dispersion and accumulation of plastics within the marine ecosystem. Therefore, the evaluation of these factors using numeric models is imperative [44]. The advancement and utilization of numeric models significantly enhance our ability to precisely evaluate environmental risks and implement management strategies effectively, thereby mitigating the impacts of MP pollution.

4.4 REMEDIATION STRATEGIES

Remediating MPs from the environment involves employing various strategies to either remove or degrade these particles. Computational techniques can be instrumental in designing, optimizing, and monitoring these remediation strategies. Figure 4.2 displays some remediation approaches using computational methodologies.

4.4.1 Filtration and Physical Removal

Authors proposed electrokinetic-assisted filtration as a novel approach to address concerns associated with membrane filtration in removing MPs from aquatic environments. Integrating electrokinetic assistance with physical filtration, the hybrid system achieved outstanding filtration performance (>99.9% removal efficiency and

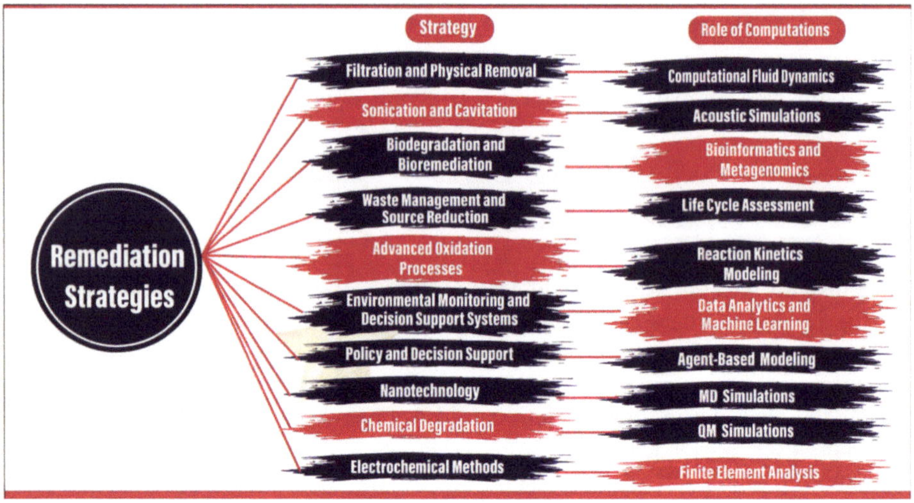

FIGURE 4.2 Computational approaches to remediating MPs in the environment.

a flow of 10,000 $Lm^{-2}h^{-1}$) using just a coarse physical filter. Electrokinetic assistance reliably halted the MP particles as they passed through a filter lattice, ensuring consistent filtration performance regardless of variations in MPs types, sizes, concentrations, or chemical compositions. This scalable electrokinetic system offers a practical solution to combat MP pollution in aquatic environments, providing valuable insights for future applications [45].

4.4.2 Computational Fluid Dynamics

Simulating fluid flow patterns to optimize the design of filtration systems helps in understanding how water currents and other factors influence the efficiency of physical barriers and filters [46]. Numerous researchers have employed CFD in their studies related to bio-electrochemistry applications [47–49]. The primary focus of these CFD investigations has been on mass transport within microbial fuel cells (MFCs). Various computational packages, such as Fluent, GAMBIT, COMSOL, and SimuLink, have been utilized for these inquiries.

For instance, Kim et al. [47] utilized CFD to examine mass transport of MFCs with two electrodes, noting non-uniform flow with Lagrangian velocity structure function. Additionally, they explored the interaction between biological organisms and electrodes in MFC flow, employing CFD simulations through the Fluent package [48]. Karimi and colleagues made a recent contribution involving the development of a CFD model tailored for simulating mass transfer within bio-electrochemical systems [49]. It is important to highlight that, unlike methods such as density functional therapy, MD, and quantum mechanics/molecular mechanics approaches described later, CFD does not delve into individual molecular interactions with electrodes. Nevertheless, CFD provides insights into the overall flow of particles, allowing for understanding their average behavior.

4.4.3 Biodegradation and Bioremediation

Plastics undergo degradation through biodegradation, a process facilitated by microorganisms such as bacteria and fungi. These microorganisms employ enzymatic activities to degrade plastics into metabolic byproducts, including methane, carbon dioxide, and water [50]. The biological breakdown of plastic pollutants is influenced by a multitude of factors, including surface area, molecular weight, hydrophilicity or hydrophobicity, presence of functional groups, crystallinity, and chemical structure. Typically, an increase in molecular weight correlates with a decrease in the degradation rate due to reduced solubility, thereby rendering plastics less susceptible to microbial attacks. This difficulty arises from the challenge microbes face in assimilating plastics through the cell membrane [51].

Another critical determinant of biodegradability is crystallinity, where polymers containing amorphous structure are more susceptible to enzymatic action by microbes [52]. Plastics with high hydrophobicity are also less prone to attack by microbes by impeding water adsorption. Nevertheless, this challenge can be mitigated through biofilm formation on plastics [53].

4.4.4 Bioinformatics and Metagenomics

Computational tools play a pivotal role in the examination of microbial communities with the potential to degrade or assimilate MPs. Identifying and understanding the enzymes and genes implicated in plastic degradation serve as a foundation for the development of targeted bioremediation strategies [54]. Several software tools, including Vina, YASARA, Autodock, Glide, GEMDOCK, GOLD, DARWIN, PatchDOCK web server, and Schrödinger, can be utilized for molecular docking (MD) analysis. This analytical approach enables exploring the capacity of an enzyme to bind with a pollutant, potentially revealing its role in biodegradation [55, 56].

Skariyachan et al. conducted molecular interaction analysis between polythene, polystyrene, and lipase from *Pseudomonas spp.* and revealed robust interactions between lipases and both polyethylene and polystyrene. Bioinformatic analysis emerged as a valuable tool for hypothesizing mechanisms of degradation [57]. In MD studies, proteins are typically kept static; however, this constraint can be overcome through simulations. MD simulations enable the evaluation of the stability of a receptor–ligand complex by allowing for conformational changes in both the enzyme and ligand complexes during biological processes. Several programs, such as Amsterdam density functional (ADF), Abalone, and Desmond, are available for conducting molecular simulations [58].

The multiomics approach integrates transcriptomics, genomics, metagenomics, proteomics, and metabolomics, providing a comprehensive strategy for identifying uncharacterized biosynthetic gene clusters within the genomes of sequenced species. Metagenomics in particular has commanded significant acclaim for its capacity to degrade petroleum, oil, plastics, and other hydrocarbons via the direct scrutiny of the genomes of nonculturable microorganisms. Fang et al. utilized metagenomics to explore biodegradation genes involved in breaking down dichloro-diphenyl-trichloroethane, atrazine, and hexachlorocyclohexane. Through genome annotation, they successfully identified the genes responsible for biodegradation [59].

4.4.5 Chemical Degradation

Recent strides in research regarding the chemical eradication and degradation of MPs in water have been outlined, shedding light on the efficacy and merits of various techniques. Authors undertook a meticulous analysis of the intricate chemical mechanisms governing the elimination and degradation of MPs, identifying key factors that impact the effectiveness of treatment. They systematically assessed the performance of MP treatment methodologies, taking into account MP attributes, operational parameters, and other pertinent variables [60]. Quantum chemistry simulations can be utilized to address the kinetic behavior of the reactions involved.

4.4.6 Quantum Chemistry Simulations

Computational chemistry can be used to predict the reactivity of different chemicals with MPs. This information helps in selecting suitable chemicals for degradation and in understanding reaction mechanisms [61]. In a recent study, PET MPs were degraded using HCl, HNO_3, and H_2SO_4 under varying conditions. Weight loss, variable pressure SEM, and ATR-FTIR were used to monitor degradation, using response surface methodology (RSM) to optimize. Results showed increased weight loss with longer contact times and higher temperatures and shaking speeds. ATR-FTIR analysis indicated that breakdown primarily occurred in PET MP structures containing oxygen atoms and aromatic rings. RSM optimization revealed optimal weight loss of 33% for HCl, 30% for H_2SO_4, and 22% for HNO_3, reflecting the efficacy of these acids in PET MP degradation, with higher rates requiring longer durations, higher temperatures, and faster shaking speeds [62].

4.4.7 Nanotechnology

Nanotechnology assumes a paramount position in the realm of water and wastewater treatment, especially when confronting formidable pollutants like magnetic nanoparticles (NPs). Authors of a comprehensive review explored nanomaterial-based strategies that enhance water treatment processes, with a specific focus on the removal of micro- and nanoplastics through adsorption, photocatalysis, and membrane filtration [63].

Nanomaterials provide numerous benefits, encompassing a notable surface-to-volume ratio, customizable surface charges, versatile functionalities, and reduced dosage requirements. These attributes render them exceedingly potent in endeavors aimed at remediation. Through a critical analysis of property–performance relationships drawn from recent literature, we identify key challenges and gaps in the field, such as quantification methods, the influence of coexisting contaminants, and considerations regarding the size and shape of micro- and nanoplastics. Furthermore, we discuss process-specific limitations to provide insights for further research and development in this domain [63].

4.4.8 Molecular Dynamics Simulations

MD simulations emulate the behavior of NPs engineered for MP remediation at the molecular scale, facilitating the optimization of their size, shape, and composition to enhance efficacy [64]. Authors of one inquiry meticulously examined the adsorption of phenol onto various types of PET MPs—pristine (Pr-PET), modified (Mod-PET), and aged (Ag-PET)— through both experimental and theoretical means. They employed a range of models encompassing kinetics, isotherms, and thermodynamics to elucidate the complexities inherent in the adsorption process [65].

Findings revealed that Ag-PET MPs exhibited superior removal efficiency attributable to their larger surface area. Adsorption adhered to pseudo-second-order kinetics, with intraparticle diffusion dictating the rate-limiting step. However, Mod-PET showcased the highest maximum monolayer adsorption capacity, trailed by Ag-PET and Pr-PET. Thermodynamic scrutiny underscored the spontaneous and favorable nature of adsorption. Molecular simulations employing grand canonical Monte Carlo coupled with MD revealed that PET MPs possess the capability to effectively adsorb phenol molecules through Van der Waals and electrostatic interactions. This discovery positions them as promising and innovative adsorbents for the removal of phenol in aqueous environments [65].

4.4.9 Electrochemical Methods

Authors have comprehensively examined the latest electrochemical techniques for identifying and removing MPs and NPs from urban waters. They covered electrochemical sensing principles, featured devices/electrodes, and applications like electro-coagulation and electro-adsorption and analyzed critical parameters influencing electrochemical identification and removal. The review concludes with discussions on current challenges and future prospects, aiming to enhance electrochemical technologies for managing urban water pollution by MPs and NPs [66].

4.4.10 Finite Element Analysis

Computational simulations can be employed to model the distribution of electric fields and optimize electrode configurations for electrochemical methods aimed at degrading or removing MPs. A scholarly inquiry delved into the dynamics of MP migration within a saturated porous medium, employing sophisticated mathematical formulations and simulations via the finite element method [67].

Through meticulous analysis, the authors elucidated the influence of various factors on MP transport, highlighting the critical role of reducing MP diffusivity while increasing water flow velocity, porosity, and the first-order attachment coefficient in mitigating this environmental hazard. The simulation findings underscored the effectiveness of an integrated approach, emphasizing the importance of orchestrating a harmonious interplay between reduced diffusivity, increased water flow velocity, and enhanced soil porosity to effectively mitigate the infiltration of MPs into groundwater ecosystems. These modeling and simulation findings enhance our understanding of MP transport and offer insights for combating water pollution [67].

4.4.11 Sonication and Cavitations

Assessing MP concentrations in landfill samples is crucial, but extracting MPs from these samples effectively is challenging. Ultrasonication (US) with proper dispersing agents helps release MPs from solid aggregates, but complete extraction is difficult due to adhesion between MPs. Researchers investigated US parameters' influence on surface cleaning of native MPs (NMPs) and release of adhered MPs (AMPs). Using 500 kJ/L of energy during US indirectly resulted in thorough cleaning of NMP surfaces and complete release of AMPs without structural damage. This study lays groundwork for guidelines on US-assisted MP extraction and provides a relationship between energy and AMP concentrations, useful for studying the impact of landfill type on MP release during US [68].

4.4.12 Acoustic Simulations

Computational simulations can model the propagation of acoustic waves and cavitation effects in water, helping to optimize sonication methods for breaking down MPs [69]. A groundbreaking acoustofluidics model harnessing traveling surface acoustic waves was innovated and seamlessly integrated into COMSOL Multiphysics software. This cutting-edge model demonstrated precise predictive capabilities in tracing microparticle trajectories, enabling the capture of MPs on the microchannel ceiling for subsequent quantification. By meticulously optimizing design parameters such as microchannel flow rate, voltage, and frequency, this model achieved the adept entrapment of MPs with diameters exceeding 7 μm and a density of 950 kgm^{-3} [70].

4.4.13 Advanced Oxidation Processes

AOPs show promise for addressing MPs and microfiber pollution [71]. However, conclusive findings are premature due to limited research. Authors assessed the impacts of AOPs on MP microfibers, focusing on degradation efficiency, toxicity, and sustainability. Polyamide microfibers lost over 90% of their mass with TiO_2 photocatalysis. Conventional Fenton processes had limited MP oxidation (<30%), but alternative methods achieved up to 96% MP weight loss under specific conditions. AOPs' full-scale applicability requires further optimization. Ozone treatment increases MP fragmentation. Standardized analysis methods and research on AOP sustainability are crucial for effective MP treatment in water [72].

4.4.14 Reaction Kinetics Modeling

Computational models can predict the rates of chemical reactions involved in advanced oxidation processes, such as ozonation and photo-catalysis. This information helps in optimizing reaction conditions for efficient MP degradation. In the domain of chemical engineering, simulations serve as indispensable instruments for modeling and refining wastewater treatment processes, thereby bolstering their efficiency, effectiveness, and sustainability. Initially, these simulations forecast and scrutinize the mechanisms involved in pollutant removal, empowering engineers to fine-tune treatment systems for heightened efficiency while concurrently mitigating resource utilization and waste production [73].

Subsequently, simulations facilitate the conception of innovative treatment methodologies, including membrane filtration and AOPs, enabling engineers to assess performance and feasibility across a spectrum of scenarios. Additionally, simulations address emerging challenges like the removal of pharmaceuticals, MPs, and endocrine-disrupting chemicals, helping devise targeted removal strategies. In summary, the utilization of chemical engineering simulations spearheads advancements in wastewater treatment, propelling us toward elevated levels of environmental sustainability. These simulations provide efficient, cost-effective solutions for wastewater management, safeguarding ecosystems in the process. Embracing simulation-driven approaches is imperative for realizing sustainable development objectives and securing a cleaner, healthier future [74].

4.5 ENVIRONMENTAL MONITORING AND DECISION SUPPORT SYSTEMS

Data analysis of environmental monitoring system data using ML algorithms can help in identifying hotspots of MP pollution, optimizing the deployment of remediation technologies and predicting future trends [15].

4.6 WASTE MANAGEMENT AND SOURCE REDUCTION

In life cycle assessment, computational methodologies can be employed to evaluate the environmental footprint of diverse plastic products across their entire life cycle [75]. This information is crucial for designing effective waste management and source reduction strategies.

4.7 POLICY AND DECISION SUPPORT

In agent-based modeling, models simulate the interactions between various stakeholders, including policymakers, industries, and the public, to assess the impacts of different policy interventions on MP pollution and remediation efforts. Interdisciplinary collaboration between environmental scientists, engineers, chemists, and computational experts is essential for developing and implementing effective computational strategies for MP remediation validated through experiments.

4.8 CASE STUDIES

The scientific community is currently directing considerable attention toward investigating the effect of MPs on anaerobic oxidation, which forms methane in the environment. However, information regarding this relationship is scattered across various sources. One research group systematically collected pertinent data from the literature and identified key variables [76].

Instead of employing manual ML implementation, the authors employed four AutoML algorithms from an available library to forecast methane production based on variables like digestion time, MPs, and substrates and select the optimal ML model. They determined that MP type was more important than size or concentration. Partial dependence plots (PDPs) indicated the highest production of methane in the presence of the polystyrene and an inverse relationship between MP diameter and methane production. Their PDP findings enabled identifying the optimal operational variables, like MP size, composition, and digestion duration and substrate properties, to optimize methane production. To improve our knowledge of the effects of MPs on anaerobic digestion, it is advisable to comprehensively characterize and incorporate a wider array of MPs variables into the data set [76].

In another investigation, authors used an innovative numeric modeling approach to examine MP transport and retention. The model served validate the outcomes of the concurrent flume experiment simulations and was highly effective at capturing particle transport even in a chaotic flow of water. Additionally, in the hyporheic zone (HZ), the model could represent the movement of MP particles as small as 1 μm into sandy stream. This underscores the applicability of the advection–dispersion equation in adequately depicting particle transport at the pore scale within the HZ [77].

To test the method with larger MPs, the authors replicated the experiment with MPs of 10 μm, and these larger MPs exhibited not only delayed infiltration but also slow transport behaviors. Although the model was successful in depicting the spatial limit of the MPs' migration via HZ, it could not fully replicate the hyporheic transit times. The study was significant as the first to show validation along with the experimental results in not only qualitative observations of MP concentration patterns but also quantitative flux measurements. The findings significantly enhance our comprehension of MP transport in fluvial systems, emphasizing the merits and demerits of using fully integrated modeling to study the behavior of MPs in streams and rivers [77].

Some investigators studied applying deep knockoff, a contemporary statistical method for variable selection, to reveal the spectral characteristics of marine debris. Utilizing a generative model driven by deep neural networks, deep knockoff comprehensively grasps the intricate, high dimensionality of reflectance distribution spanning from the visible to the NIR spectrum. Drawing from a publicly accessible data set derived from ground-based annotations of observations captured by the Multi-Spectral Instrument aboard the Sentinel-2 (S2) satellite, the authors revealed the pivotal significance of NIR (band 8, 833 nm) and red (band 4, 665 nm) wavelengths in distinguishing marine (plastic) debris from the surrounding water. Furthermore, deep knockoff effectively managed the false discovery rate in accordance with established physical causality principles. Moreover, it identified the green (band 3, 560 nm) and the narrow NIR (band 8a, 864.7 nm) spectra as supplementary spectral bands, particularly pertinent for analysis in the presence of dense Sargassum macroalgae [78].

Researchers examined the applicability of kernel weight histograms for identifying an appropriate neural network for segmenting minute substances in photographs depicting sand containing a few MP particles. The authors investigated MultiResUNet (MRU-Net) and U-Net as potential target networks. However, upon scrutinizing kernel weight histograms through Tensor Board visualization, they observed that the initial encoder stages of U-Net and MRU-Net excelled in detecting small features, in contrast with the later encoder stages [79].

Consequently, the authors modified U-Net and MRU-Net to Half U-Net, Half MRU-Net, and Quarter MRU-Net. Experimental findings demonstrate that the highest average recall-weighted F1 (40%) and recall-weighted mean Intersection over Union (mIoU) (26%) for the MP data sets were achieved using Half MRU-Net. Quarter MRU-Net emerged as the second-best performer in terms of average recall-weighted F1 as well as recall-weighted mIoU. Importantly, these streamlined models necessitate only 1/5 or fewer floating-point operations and 1/50 or fewer parameters, in contrast with U-net and MRU-Net, rendering them more efficient for the specified segmentation task [79].

In a comprehensive study, researchers employed MD simulations to investigate the structural characteristics of MPs by analyzing atom movements over a specified period, offering insights into the dynamic changes in the system. Individual polymer chains were folded in MD simulations lasting 500 ps at 500 K in the NVT ensemble to generate the initial nano PET structures, followed by energy minimization via conformational search using force field parameters derived from AMBER using Gabedit. The study provides valuable conclusions into how the PET-derived MPs (PET-aMPs or nanoPET) interact with the primary air pollutants [80].

PET-aMPs exhibited high selectivity to air pollutants, with adsorption energies ranging from 6 to 20 kcal/mol, showing the nature of the adsorption to be physisorption. These values were nearly equal to and in some cases more than those observed with metals/metalloids, metal oxides, and carbon-based materials. The surface adsorption of air pollutants on PET-aMPs occurs in two distinct manners, outer or inner, depending upon their molecular polarity, which is decided by their dipole moment, and their atomic composition (influenced by electronegativity). Notably, CO, SO_2, NO, and NH_3 preferred outer surface adsorption, while N_2O and CO_2 favored inner adsorption [80].

Additionally, there was less competition between molecules in inert atmosphere, whereas ozone may increase adsorption by oxidizing the surface and breaking down the structure. The adsorption mechanisms were governed by the interplay between the dispersion (21%–28%) and electrostatic (46%–61%) forces. Air pollutants with lower polarity demonstrated nearly equal contributions of electrostatic and dispersion forces, while in case of molecules with high polarity, electrostatic forces dominated [80].

From a chemical point of view, strong dispersion and electrostatic interactions like hydrogen bonding and dipole-dipole-induced pairs dominated in the outer adsorption, while inner adsorption was characterized by the weak forces like lone-pair π interactions. PET-aMPs thus served as carriers for co-transporting coexisting air pollutants. This knowledge of adsorption mechanisms enhances our understanding of the harmfulness and the risks associated with atmospheric and airborne plastic trash, providing insights into their effect, co-transportability, and environmental interactions [80].

4.9 CHALLENGES AND FUTURE DIRECTIONS

4.9.1 Computational Approaches for Complex Environmental Systems

The dynamic and complex nature of environmental systems presents a significant challenge for computational models. Simulating MP behavior in real-world environments, with their myriad interacting factors like water currents, temperature fluctuations, and the presence of diverse contaminants, requires increasingly sophisticated models. Integrating these complexities while maintaining computational efficiency remains a formidable task.

4.9.2 Data Variability and Validation

Computational approaches rely heavily on input data, and the variability in environmental conditions and MP characteristics poses challenges. Acquiring accurate and representative data for model training and validation can be difficult due to the spatial and temporal heterogeneity of MP distribution. Additionally, establishing standardized protocols for data collection and validation is crucial for ensuring the reliability and robustness of computational models.

4.9.3 Interdisciplinary Integration

Addressing the multifaceted issue of MP pollution demands interdisciplinary collaboration. Computational scientists must work closely with environmental scientists, engineers, and biologists to integrate diverse knowledge domains into comprehensive models. Bridging the gap between disciplines requires effective communication and shared frameworks, presenting a challenge in fostering collaboration among experts with different backgrounds.

4.9.4 ML for Enhanced Sensing

The evolution of ML holds great promise for the future of MP sensing. Continued advancements in deep learning algorithms and diverse data sources can lead to more accurate and adaptable sensing technologies. The development of autonomous sensing systems, capable of continuously learning and adapting to changing environmental conditions, represents a frontier in the application of computational approaches.

4.9.5 Sensor Networks and IoT Technologies

The future of MP surveillance lies in the harmonious amalgamation of computational models with sensor networks and Internet of Things (IoT) technologies. The deployment of sensor arrays across aquatic environments, intricately linked with real-time computational models, promises to elevate the precision and timeliness of monitoring initiatives. This synergy enables swift detection and response to nascent MP concentrations, fostering a proactive stance in environmental stewardship.

4.9.6 In Silico Design of Novel Remediation Strategies

Computational approaches will play a pivotal role in the in silico design of novel remediation strategies. Quantum computing, with its capacity to model complex molecular interactions at unprecedented speeds, holds the potential to accelerate the discovery of highly efficient nanomaterials for MP capture. This frontier in computational science opens avenues for precision engineering of remediation agents with enhanced environmental impact.

4.9.7 Global Collaboration and Standardization

The future success of computational approaches in addressing MP pollution necessitates global collaboration and standardization. Establishing a shared framework for data collection, model validation, and information exchange will enhance the reliability and comparability of computational models worldwide. International partnerships can leverage collective expertise to address regional variations in MP pollution and promote a unified approach to environmental conservation.

In conclusion, while challenges persist in the complex landscape of MP research, the future of computational approaches is marked by exciting opportunities. Advancements in ML, integration with sensor networks, in silico design of remediation strategies, and global collaboration hold the key to more effective and sustainable solutions in the ongoing battle against MP pollution.

4.10 CONCLUSION: NAVIGATING THE DIGITAL FRONTIERS OF MP SOLUTIONS

In the relentless pursuit of mitigating the pervasive impact of MP pollution, computational approaches have emerged as indispensable allies, reshaping the landscape of environmental science. As we navigate the digital frontiers of MP solutions, the synergy between advanced modeling techniques, innovative sensor technologies, and precision remediation strategies has unveiled unprecedented opportunities and formidable challenges.

The complexity of environmental systems, with their intricate interplay of variables, demands ever-evolving computational models. From molecular simulations illuminating the microscopic behavior of individual MP particles to ML algorithms revolutionizing the precision of detection, these computational tools enable us to unravel the mysteries of MP dynamics at scales previously inaccessible. Challenges abound, though, from the need for standardized data collection and validation to the imperative of interdisciplinary collaboration. The journey toward effective computational solutions necessitates the convergence of expertise from diverse fields, emphasizing the interconnectedness of environmental science, computational science, and engineering.

Looking forward, the future holds promising trajectories. The evolution of ML promises enhanced sensing capabilities capable of adapting to the dynamic nature of MP environments. The integration of computational models with sensor networks and IoT technologies opens new avenues for real-time monitoring and proactive intervention. Quantum computing, with its unparalleled computational power, charts a course toward the in silico design of nanomaterials poised to revolutionize MP remediation.

Yet as we embrace these opportunities, a call for global collaboration and standardization echoes. The challenge of MP pollution transcends geographical boundaries, necessitating a unified approach. Only through international cooperation, shared frameworks, and the pooling of collective expertise can we hope to address the global variations in MP distribution and formulate holistic solutions.

In the grand tapestry of environmental science, computational approaches for MP sensing and remediation marks a crucial chapter in development. It signifies not just a technological advance but a paradigm shift in our approach to environmental challenges. The digital tools at our disposal empower us to explore, understand, and ultimately address the intricate challenges posed by MPs in our ecosystems.

As I conclude the exploration in this chapter, I observe that the path ahead is both challenging and promising. The integration of computational approaches into environmental science is not merely a scientific endeavor; it is a commitment to a sustainable future. By harnessing the power of computation, we stand on the precipice of transformative change, armed with the knowledge and tools to safeguard our environment for generations to come.

ACKNOWLEDGMENTS

Ravi Kumar is highly grateful to Ms. Hemjot for proofreading this chapter and Mr. Hariom for helping with the pictures.

REFERENCES

1. Lase, I.S., Tonini, D., Caro, D., Albizzati, P.F., Cristóbal, J., Roosen, M., Kusenberg, M., Ragaert, K., Van Geem, K.M., Dewulf, J. and De Meester, S., 2023. How much can chemical recycling contribute to plastic waste recycling in Europe? An assessment using material flow analysis modeling. *Resources, Conservation and Recycling*, *192*, p. 106916.
2. Borrelle, S.B., Ringma, J., Law, K.L., Monnahan, C.C., Lebreton, L., McGivern, A., Murphy, E., Jambeck, J., Leonard, G.H., Hilleary, M.A. and Eriksen, M., 2020. Predicted growth in plastic waste exceeds efforts to mitigate plastic pollution. *Science*, *369*(6510), pp. 1515–1518.
3. Ren, S.Y., Sun, Q., Ni, H.G. and Wang, J., 2020. A minimalist approach to quantify emission factor of MPs by mechanical abrasion. *Chemosphere*, *245*, p. 125630.
4. Wu, X., Liu, P., Wang, H., Huang, H., Shi, Y., Yang, C. and Gao, S., 2021. Photo aging of polypropylene MPs in estuary water and coastal seawater: Important role of chlorine ion. *Water Research*, *202*, p. 117396.
5. Auta, H.S., Abioye, O.P., Aransiola, S.A., Bala, J.D., Chukwuemeka, V.I., Hassan, A., Aziz, A. and Fauziah, S.H., 2022. Enhanced microbial degradation of PET and PS MPs under natural conditions in mangrove environment. *Journal of Environmental Management*, *304*, p. 114273.
6. Jiang, H., Bu, J., Bian, K., Su, J., Wang, Z., Sun, H., Wang, H., Zhang, Y. and Wang, C., 2023. Surface change of MPs in aquatic environment and the removal by froth flotation assisted with cationic and anionic surfactants. *Water Research*, *233*, p. 119794.
7. Sarkar, B., Dissanayake, P.D., Bolan, N.S., Dar, J.Y., Kumar, M., Haque, M.N., Mukhopadhyay, R., Ramanayaka, S., Biswas, J.K., Tsang, D.C. and Rinklebe, J., 2022. Challenges and opportunities in sustainable management of MPs and nanoplastics in the environment. *Environmental Research*, *207*, p. 112179.
8. Jiang, H., Zhang, Y., Bian, K., Wang, C., Xie, X., Wang, H. and Zhao, H., 2022. Is it possible to efficiently and sustainably remove MPs from sediments using froth flotation? *Chemical Engineering Journal*, *448*, p. 137692.
9. Singh, S., Naik, T.S.K., Anil, A.G., Dhiman, J., Kumar, V., Dhanjal, D.S., Aguilar-Marcelino, L., Singh, J. and Ramamurthy, P.C., 2022. Micro (nano) plastics in

wastewater: A critical review on toxicity risk assessment, behaviour, environmental impact and challenges. *Chemosphere*, *290*, p. 133169.
10. Wang, L., Wu, W.M., Bolan, N.S., Tsang, D.C., Li, Y., Qin, M. and Hou, D., 2021. Environmental fate, toxicity and risk management strategies of nanoplastics in the environment: Current status and future perspectives. *Journal of Hazardous Materials*, *401*, p. 123415.
11. Al Mamun, A., Prasetya, T.A.E., Dewi, I.R. and Ahmad, M., 2023. MPs in human food chains: Food becoming a threat to health safety. *Science of the Total Environment*, *858*, p. 159834.
12. Zhu, M., Wang, J., Yang, X., Zhang, Y., Zhang, L., Ren, H., Wu, B. and Ye, L., 2022. A review of the application of machine learning in water quality evaluation. *Eco-Environment & Health*, *1*(2), pp. 107–116.
13. Viet, N.D. and Jang, A., 2023. Machine learning-based real-time prediction of micropollutant behaviour in forward osmosis membrane (waste) water treatment. *Journal of Cleaner Production*, *389*, p. 136023.
14. Keith, J.A., McKone, J.R., Snyder, J.D. and Tang, M.H., 2022. Deeper learning in electrocatalysis: Realizing opportunities and addressing challenges. *Current Opinion in Chemical Engineering*, *36*, p. 100824.
15. Xie, L., Luo, S., Liu, Y., Ruan, X., Gong, K., Ge, Q., Li, K., Valev, V.K., Liu, G. and Zhang, L., 2023. Automatic identification of individual nanoplastics by Raman spectroscopy based on machine learning. *Environmental Science & Technology*, *57*(46), pp. 18203–18214.
16. Adão, T., Hruška, J., Pádua, L., Bessa, J., Peres, E., Morais, R. and Sousa, J.J., 2017. Hyperspectral imaging: A review on UAV-based sensors, data processing and applications for agriculture and forestry. *Remote Sensing*, *9*(11), p. 1110.
17. Xu, J.L., Lin, X. and Gowen, A.A., 2022. Combining machine learning with meta-analysis for predicting cytotoxicity of micro-and nanoplastics. *Journal of Hazardous Materials Advances*, *8*, p. 100175.
18. Guo, X. and Wang, J., 2021. Projecting the sorption capacity of heavy metal ions onto MPs in global aquatic environments using artificial neural networks. *Journal of Hazardous Materials*, *402*, p. 123709.
19. Massarelli, C., Campanale, C. and Uricchio, V.F., 2021. A handy open-source application based on computer vision and machine learning algorithms to count and classify MPs. *Water*, *13*(15), p. 2104.
20. Huang, R., Ma, C., Ma, J., Huangfu, X. and He, Q., 2021. Machine learning in natural and engineered water systems. *Water Research*, *205*, p. 117666.
21. Xu, R.Z., Cao, J.S., Feng, G., Luo, J.Y., Feng, Q., Ni, B.J. and Fang, F., 2022. Fast identification of fluorescent components in three-dimensional excitation-emission matrix fluorescence spectra via deep learning. *Chemical Engineering Journal*, *430*, p. 132893.
22. Cowger, W., Gray, A., Christiansen, S.H., DeFrond, H., Deshpande, A.D., Hemabessiere, L., Lee, E., Mill, L., Munno, K., Ossmann, B.E. and Pittroff, M., 2020. Critical review of processing and classification techniques for images and spectra in MPs research. *Applied Spectroscopy*, *74*(9), pp. 989–1010.
23. Post, C., Brülisauer, S., Waldschläger, K., Hug, W., Grüneis, L., Heyden, N., Schmor, S., Förderer, A., Reid, R., Reid, M. and Bhartia, R., 2021. Application of laser-induced, deep UV Raman spectroscopy and artificial intelligence in real-time environmental monitoring—solutions and first results. *Sensors*, *21*(11), p. 3911.
24. Lorenzo-Navarro, J., Castrillón-Santana, M., Sánchez-Nielsen, E., Zarco, B., Herrera, A., Martínez, I. and Gómez, M., 2021. Deep learning approach for automatic MPs counting and classification. *Science of the Total Environment*, *765*, p. 142728.
25. Chen, H., Tang, M., Yang, X., Tsang, Y.F., Wu, Y., Wang, D. and Zhou, Y., 2021. Polyamide 6 MPs facilitate methane production during anaerobic digestion of waste activated sludge. *Chemical Engineering Journal*, *408*, p. 127251.

26. Wei, W., Huang, Q.S., Sun, J., Wang, J.Y., Wu, S.L. and Ni, B.J., 2019. Polyvinyl chloride MPs affect methane production from the anaerobic digestion of waste activated sludge through leaching toxic bisphenol-A. *Environmental Science & Technology*, *53*(5), pp. 2509–2517.
27. Abd Elrahman, A., El Helw, M., Elshawi, R. and Sakr, S., 2020, November. D-smartml: A distributed automated machine learning framework. In *2020 IEEE 40th International Conference on Distributed Computing Systems (ICDCS)* (pp. 1215–1218). IEEE.
28. Ferreira, L., Pilastri, A., Martins, C.M., Pires, P.M. and Cortez, P., 2021, July. A comparison of AutoML tools for machine learning, deep learning and XGBoost. In *2021 International Joint Conference on Neural Networks (IJCNN)* (pp. 1–8). IEEE.
29. Feurer, M., Eggensperger, K., Falkner, S., Lindauer, M. and Hutter, F., 2022. Auto-sklearn 2.0: Hands-free automl via meta-learning. *Journal of Machine Learning Research*, *23*(261), pp. 1–61.
30. Le, T.T., Fu, W. and Moore, J.H., 2020. Scaling tree-based automated machine learning to biomedical big data with a feature set selector. *Bioinformatics*, *36*(1), pp. 250–256.
31. Gijsbers, P. and Vanschoren, J., 2019. GAMA: Genetic automated machine learning assistant. *Journal of Open Source Software*, *4*(33), p. 1132.
32. Erickson, N., Mueller, J., Shirkov, A., Zhang, H., Larroy, P., Li, M. and Smola, A., 2020. Autogluon-tabular: Robust and accurate automl for structured data. *arXiv preprint arXiv:2003.06505.*
33. LeDell, E. and Poirier, S., 2020, July. H2o automl: Scalable automatic machine learning. In *Proceedings of the AutoML Workshop at ICML* (Vol. 2020). ICML.
34. Gauci, A., Deidun, A., Montebello, J., Abela, J. and Galgani, F., 2019. Automating the characterisation of beach MPs through the application of image analyses. *Ocean & Coastal Management*, *182*, p. 104950.
35. Giardino, M., Balestra, V., Janner, D. and Bellopede, R., 2023. Automated method for routine MPs detection and quantification. *Science of the Total Environment*, *859*, p. 160036.
36. Bertoldi, C., Lara, L.Z., Gomes, A.A. and Fernandes, A.N., 2021. MPs abundance quantification via a computer-vision-based chemometrics-assisted approach. *Microchemical Journal*, *160*, p. 105690.
37. Pilechi, A., Mohammadian, A. and Murphy, E., 2022. A numerical framework for modeling fate and transport of MPs in inland and coastal waters. *Marine Pollution Bulletin*, *184*, p. 114119.
38. Bigdeli, M., Mohammadian, A., Pilechi, A. and Taheri, M., 2022. Lagrangian modeling of marine MPs fate and transport: The state of the science. *Journal of Marine Science and Engineering*, *10*(4), p. 481.
39. Venel, Z., Tabuteau, H., Pradel, A., Pascal, P.Y., Grassl, B., El Hadri, H., Baudrimont, M. and Gigault, J., 2021. Environmental fate modeling of nanoplastics in a salinity gradient using a lab-on-a-chip: Where does the nanoscale fraction of plastic debris accumulate? *Environmental Science & Technology*, *55*(5), pp. 3001–3008.
40. Hardesty, B.D., Harari, J., Isobe, A., Lebreton, L., Maximenko, N., Potemra, J., Van Sebille, E., Vethaak, A.D. and Wilcox, C., 2017. Using numerical model simulations to improve the understanding of micro-plastic distribution and pathways in the marine environment. *Frontiers in Marine Science*, *4*, p. 30.
41. Sherman, P. and Van Sebille, E., 2016. Modeling marine surface MPs transport to assess optimal removal locations. *Environmental Research Letters*, *11*(1), p. 014006.
42. Isobe, A., Kako, S.I., Chang, P.H. and Matsuno, T., 2009. Two-way particle-tracking model for specifying sources of drifting objects: Application to the East China sea shelf. *Journal of Atmospheric and Oceanic Technology*, *26*(8), pp. 1672–1682.
43. Liubartseva, S., Coppini, G., Lecci, R. and Creti, S., 2016. Regional approach to modeling the transport of floating plastic debris in the Adriatic Sea. *Marine Pollution Bulletin*, *103*(1–2), pp. 115–127.

44. Welden, N.A. and Lusher, A.L., 2017. Impacts of changing ocean circulation on the distribution of marine MPs litter. *Integrated Environmental Assessment and Management*, *13*(3), pp. 483–487.
45. Lee, M., Choi, W. and Lim, G., 2023. Electrokinetic-assisted filtration for fast and highly efficient removal of MPs from water. *Chemical Engineering Journal*, *452*, p. 139152.
46. Fatahi, M., Akdogan, G., Dorfling, C. and Van Wyk, P., 2021. Numerical study of MPs dispersal in simulated coastal waters using cfd approach. *Water*, *13*(23), p. 3432.
47. Kim, J.R., Boghani, H.C., Amini, N., Aguey-Zinsou, K.F., Michie, I., Dinsdale, R.M., Guwy, A.J., Guo, Z.X. and Premier, G.C., 2012. Porous anodes with helical flow pathways in bioelectrochemical systems: The effects of fluid dynamics and operating regimes. *Journal of Power Sources*, *213*, pp. 382–390.
48. Vilà-Rovira, A., Puig, S., Balaguer, M.D. and Colprim, J., 2015. Anode hydrodynamics in bioelectrochemical systems. *RSC Advances*, *5*(96), pp. 78994–79000.
49. Karimi, M., Widén, T., Nygård, Y., Olsson, L. and Ström, H., 2021. Towards enhancement of gas–liquid mass transfer in bioelectrochemical systems: Validation of a robust CFD model. *Biotechnology and Bioengineering*, *118*(10), pp. 3953–3961.
50. Mohan, K., 2011. Microbial deterioration and degradation of polymeric materials. *Journal of Biochemical Technology*, *2*(4), pp. 210–215.
51. Gumienna, M. and Górna, B., 2021. Antimicrobial food packaging with biodegradable polymers and bacteriocins. *Molecules*, *26*(12), p. 3735.
52. Slor, G., Papo, N., Hananel, U. and Amir, R.J., 2018. Tuning the molecular weight of polymeric amphiphiles as a tool to access micelles with a wide range of enzymatic degradation rates. *Chemical Communications*, *54*(50), pp. 6875–6878.
53. Syranidou, E., Karkanorachaki, K., Amorotti, F., Franchini, M., Repouskou, E., Kaliva, M., Vamvakaki, M., Kolvenbach, B., Fava, F., Corvini, P.F.X. and Kalogerakis, N., 2017. Biodegradation of weathered polystyrene films in seawater microcosms. *Scientific Reports*, *7*(1), p. 17991.
54. Sharma, P., Singh, S.P., Iqbal, H.M. and Tong, Y.W., 2022. Omics approaches in bioremediation of environmental contaminants: An integrated approach for environmental safety and sustainability. *Environmental Research*, *211*, p. 113102.
55. Saar, A., Ghahremanpour, M.M., Tirado-Rives, J. and Jorgensen, W.L., 2023. Assessing metadynamics and docking for absolute binding free energy calculations using severe acute respiratory syndrome Coronavirus 2 main protease inhibitors. *Journal of Chemical Information and Modeling*, *63*(22), pp. 7210–7218.
56. Daina, A. and Zoete, V., 2019. Application of the SwissDrugDesign online resources in virtual screening. *International Journal of Molecular Sciences*, *20*(18), p. 4612.
57. Skariyachan, S., Megha, M., Kini, M.N., Mukund, K.M., Rizvi, A. and Vasist, K., 2015. Selection and screening of microbial consortia for efficient and ecofriendly degradation of plastic garbage collected from urban and rural areas of Bangalore, India. *Environmental Monitoring and Assessment*, *187*, pp. 1–14.
58. Lindorff-Larsen, K., Piana, S., Palmo, K., Maragakis, P., Klepeis, J.L., Dror, R.O. and Shaw, D.E., 2010. Improved side-chain torsion potentials for the Amber ff99SB protein force field. *Proteins: Structure, Function, and Bioinformatics*, *78*(8), pp. 1950–1958.
59. Fang, H., Cai, L., Yang, Y., Ju, F., Li, X., Yu, Y. and Zhang, T., 2014. Metagenomic analysis reveals potential biodegradation pathways of persistent pesticides in freshwater and marine sediments. *Science of the Total Environment*, *470*, pp. 983–992.
60. Zhou, T., Song, S., Min, R., Liu, X. and Zhang, G., 2024. Advances in chemical removal and degradation technologies for MPs in the aquatic environment: A review. *Marine Pollution Bulletin*, *201*, p. 116202.
61. Cortés-Arriagada, D., Miranda-Rojas, S., Camarada, M.B., Ortega, D.E. and Alarcón-Palacio, V.B., 2023. The interaction mechanism of polystyrene MPs with pharmaceuticals and personal care products. *Science of the Total Environment*, *861*, p. 160632.

62. Chowdhury, T., Wang, Q. and Enyoh, C.E., 2022. Degradation of polyethylene terephthalate MPs by mineral acids: Experimental, molecular modelling and optimization studies. *Journal of Polymers and the Environment*, *30*(12), pp. 5211–5227.
63. Ouda, M., Banat, F., Hasan, S.W. and Karanikolos, G.N., 2023. Recent advances on nanotechnology-driven strategies for remediation of MPs and nanoplastics from aqueous environments. *Journal of Water Process Engineering*, *52*, p. 103543.
64. Li, H., Wang, F., Li, J., Deng, S. and Zhang, S., 2021. Adsorption of three pesticides on polyethylene MPs in aqueous solutions: Kinetics, isotherms, thermodynamics, and molecular dynamics simulation. *Chemosphere*, *264*, p. 128556.
65. Enyoh, C.E. and Wang, Q., 2024. Combined experimental and molecular dynamics removal processes of contaminant phenol from simulated wastewater by polyethylene terephthalate MPs. *Environmental Technology*, *45*(6), pp. 1183–1202.
66. Chen, Z., Wei, W., Liu, X. and Ni, B.J., 2022. Emerging electrochemical techniques for identifying and removing micro/nanoplastics in urban waters. *Water Research*, *221*, p. 118846.
67. Ryu, H.S., Moon, J., Kim, H. and Lee, J.Y., 2021. Modeling and parametric simulation of MPs transport in groundwater environments. *Applied Sciences*, *11*(16), p. 7189.
68. Goli, V.S.N.S. and Singh, D.N., 2023. Effect of ultrasonication conditions on polyethylene MPs sourced from landfills: A precursor study to establish guidelines for their extraction from environmental matrices. *Journal of Hazardous Materials*, *459*, p. 132230.
69. Jonai, T., Ohori, Y., Fujii, T., Nakayama, A., Moriwaki, H. and Akiyama, Y., 2023. A collection device for various-sized microparticles that uses four serial acoustic separations: Working toward MPs emission prevention. *Separation and Purification Technology*, *315*, p. 123697.
70. Farahi Bolamiri, F., 2023. *Numerical simulation of MPs trapping in a microfluidic channel utilizing travelling acoustic waves* (Master's thesis, University of South-Eastern Norway).
71. Dewil, R., Mantzavinos, D., Poulios, I. and Rodrigo, M.A., 2017. New perspectives for advanced oxidation processes. *Journal of Environmental Management*, *195*, pp. 93–99.
72. Dos Santos, N.D.O., Busquets, R. and Campos, L.C., 2023. Insights into the removal of MPs and microfibres by advanced oxidation processes. *Science of the Total Environment*, *861*, p. 160665.
73. Yu, F., Li, Y., Huang, G., Yang, C., Chen, C., Zhou, T., Zhao, Y. and Ma, J., 2020. Adsorption behavior of the antibiotic levofloxacin on MPs in the presence of different heavy metals in an aqueous solution. *Chemosphere*, *260*, p. 127650.
74. Obiuto, N.C., Ugwuanyi, E.D., Ninduwezuor-Ehiobu, N., Ani, E.C. and Olu-lawal, K.A., 2024. Advancing wastewater treatment technologies: The role of chemical engineering simulations in environmental sustainability. *World Journal of Advanced Research and Reviews*, *21*(3), pp. 019–031.
75. Rigamonti, L. and Mancini, E., 2021. Life cycle assessment and circularity indicators. *The International Journal of Life Cycle Assessment*, *26*, pp. 1937–1942.
76. Xu, R.Z., Cao, J.S., Ye, T., Wang, S.N., Luo, J.Y., Ni, B.J. and Fang, F., 2022. Automated machine learning-based prediction of MPs induced impacts on methane production in anaerobic digestion. *Water Research*, *223*, p. 118975.
77. Dichgans, F., Boos, J.P., Ahmadi, P., Frei, S. and Fleckenstein, J.H., 2023. Integrated numerical modeling to quantify transport and fate of MPs in the hyporheic zone. *Water Research*, *243*, p. 120349.
78. Olyaei, M., Ebtehaj, A. and Hong, J., 2022. Optical detection of marine debris using deep knockoff. *IEEE Transactions on Geoscience and Remote Sensing*, *60*, pp. 1–12.
79. Lee, G. and Jhang, K., 2021. Neural network analysis for MPs segmentation. *Sensors*, *21*(21), p. 7030.
80. Ortega, D.E. and Cortés-Arriagada, D., 2023. Atmospheric MPs and nanoplastics as vectors of primary air pollutants-A theoretical study on the polyethylene terephthalate (PET) case. *Environmental Pollution*, *318*, p. 120860.

5 Monitoring Microplastics and Nanoplastics in a Marine Realm

Devlina Das Pramanik and Roudrashwa Nag

5.1 INTRODUCTION

Increases in plastic production worldwide from 1.5 million tons (in the 1950s) to 335 million tons in 2016 have led to an environmental build-up of plastic waste and toxicological effects on biota [1–3]. When Carpenter and Smith sampled the pelagic Sargassum population in the western Sargasso Sea in 1972, they made the first recorded observations of plastic waste in the marine environment. They expressed concern that given the increased rate of plastic manufacturing and inadequate trash disposal policies, plastic litter might eventually amass in the sea, and they ascribed their findings to the disposal of debris from towns or by cargo and passenger ships [4].

Since that report was published more than 40 years ago, plastic pollution in the marine environment has grown to be a serious worldwide environmental and public health issue [5–7]. Microplastics (MPs), which are particles of size less than 5 mm, constitute 92.4% of marine plastic debris [5, 8]. These microscopic particles have been found in a wide range of environmental matrices worldwide, such as the beach, sea ice, water columns, deep-sea sediment, marine creatures, and other marine sources. Surface waters in all major seas have also been found to contain these particles [9–11]. While the detrimental effects of macroplastics on marine life and ecosystems have been thoroughly studied, the increasing recognition of MPs, and nanoplastics (NPs) has added a new dimension to this ecological challenge. As these tiny plastic particles infiltrate marine environments, there is an urgent demand for innovative methods to monitor and evaluate their potential ecological and human health consequences.

A species, or collection of species, that provides information about the condition or health of an environment is called a bioindicator. These metrics are frequently used to assess how human actions—like pollution or habitat destruction—affect the natural world. They are chosen because they are easily monitored due to their sensitivity and capacity to detect changes in the surrounding environment. When assessing an environmental element directly or when data interpretation is difficult or impossible, bioindicators come in handy. These indicators provide insights on pollutant levels, which can be compared with measurements made using instruments.

DOI: 10.1201/9781003468141-5

Numerous species, including fish, seabirds, turtles, invertebrates, and marine mammals, have been shown to be negatively impacted by plastic pollution, and the list of affected species is still growing [1, 12]. Under the framework of the Marine Strategy Framework Directive (2008/56/EC) for evaluating the environmental status of European waters, the use of species indicators has been promoted as a long-term monitoring tool for plastic pollution. Nevertheless, monitoring the impacts across all species is inefficient. Ehlers et al. presented a methodology for imaging MP ingestion and egestion in aquatic invertebrates [13], and Figure 5.1 offers a beautiful image of the freshwater snail *Radix balthica* exposed to MPs and microfibers.

Selecting suitable sentinel species depends on their ability to provide essential information about pollution impacts, resource efficiency, and stakeholder engagement. Criteria for bioindicators include lower trophic level, nonselective feeding, ease of sampling and identification, abundance, defined biology, and global distribution. General criteria can guide species selection for assessing plastic pollution, with various organisms like zooplankton, crustaceans, mollusks, and fish serving as biomonitors. This comprehensive approach aims to highlight the role of diverse marine sentinel organisms in monitoring NPs and MPs in the marine environment, addressing the increasing need to understand the extent and consequences of these pollutants [14–19].

In this chapter, we extensively explore various organisms as biomonitors for MPs and NPs. We thoroughly examine MP ingestion, accumulation, uptake, and egestion in both low-trophic-level marine organisms (such as fish, bivalves, crustaceans, zooplanktons) and marine predators (including birds, mammals, and reptiles).

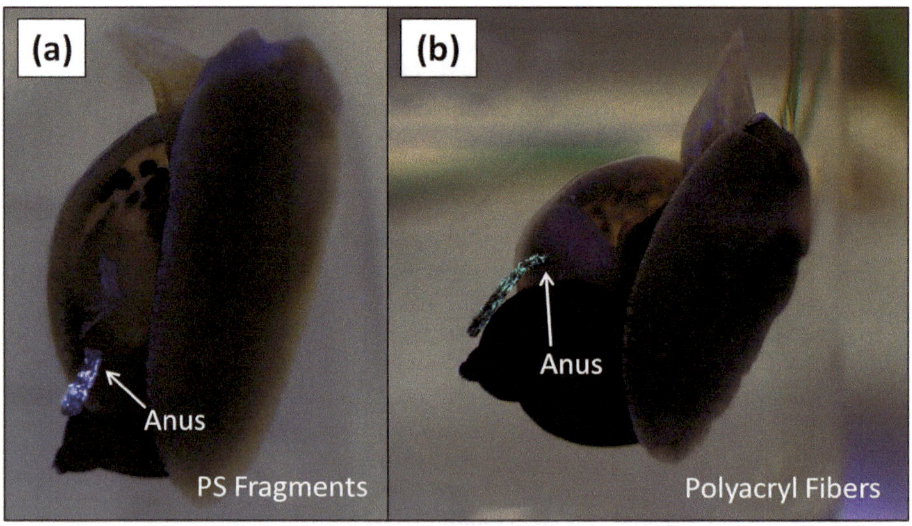

FIGURE 5.1 Accumulation of polystyrene fragments and fibers in the snail *Radix balthica* [13].

Subsequent sections provide in-depth insights into the biomonitoring potential of diverse organisms concerning MPs and NPs, aiming to enhance understanding for the selection of sentinel species and stimulate further research in the field of monitoring emerging pollutants.

5.2 MARINE FISH

While observations in the field might not immediately showcase the impact of MPs on fish, laboratory experiments have conclusively demonstrated the adverse effects of MP exposure. After 20 days of exposure to two concentrations of high-density polyethylene (PE) and polystyrene (PS) MPs, a transcriptome analysis of adult zebrafish revealed a downregulation in the expression of genes related to immune system function, epithelial integrity, and lipid metabolism [20]. Several recent investigations highlight the prevalence of microplastic ingestion in diverse fish species, including both bony fish and commercially important elasmobranchs known for their regular consumption of litter [21]. While previous researchers have successfully monitored MP ingestion by looking at a particular target species [22], it is acknowledged that differences in fish species' feeding patterns can affect rates of MP ingestion [23, 24]. Table 5.1 describes different marine fish species as effective MP biomonitoring agents.

Zhang et al. examined MP contamination in 11 wild fish species (193 individuals) and 8 wild crustacean species (136 individuals) that were caught in the East China Sea's Zhoushan fishing area [25]. For each individual, the average MP abundance in the gastrointestinal tract (GI tract) and the gill tissues was 0.52 ± 0.90 and 0.77 ± 1.25 items/individual. The most prevalent MP was fiber-shaped polyester (PES). These findings indicated the pervasive nature of MP pollution in coastal areas, with MPs accumulating in fish species occupying higher trophic levels. The outcomes of the research suggest that marine organisms at elevated trophic levels can serve as bioindicator species for the presence of MPs in aquatic ecosystems.

The team of Barboza et al. reported the accumulation of MPs in three commercially important marine fish species: *Trachurus trachurus*, *Dicentrarchus labrax*, and *Scomber colias* [26]. Figure 5.2(a) shows a schematic illustration of the different organs where they noted accumulation; this included gills, GI tract, and muscles. The ingested MPs were further quantified and graphically represented in Figure 5.2(b). Among all fish samples tested for microplastic ingestion, *S. colias* exhibited the maximum percentage of ingested MPs at 62%. Figure 5.3(a,b) represents the various shapes of MPs recovered from the respective fish species.

In 2022, Kılıç and Yücel's team discovered four different commercial fish species—*Mullus surmuletus*, *Mugil cephalus*, *Mullus barbatus*, and *Saurida undosquamis*—as suitable bioindicators of MP contamination in the Mediterranean region [27]. The frequencies of occurrence in the GI tract and gills ranged from 66% to 100% and 68% to 90%, respectively. When *Mugil cephalus* samples were taken from the Asi River estuary, their GI tracts showed the highest MP concentrations, predominantly black fibers, measuring less than 1 mm in size. Previous research on *M. surmuletus* showed MP ingestion rates of 27% in the Aegean Sea, 35% in the northeastern Mediterranean Sea, and 56% in the western Mediterranean [27, 28, 29, 30].

TABLE 5.1
Marine Fish Species as MP and NP Biomonitoring Agents

S. No	Species	Overview	Location	References
1	*Mullus barbatus*	*Mullus barbatus* was used as a bioindicator due to the occurrence of microplastics in its GI tract and its gills. MP bioavailability was attributed to their proximity to urban settlements. MP concentration is also dependent on the feeding habits of the species. MPs were also identified as vectors for microorganisms in this socioeconomically important species.	Northeastern Mediterranean Sea Adriatic Sea Tyrrhenian Sea Western Sardinia Northern Tyrrhenian Sea between Gulf of Oristano and Gulf of Cagliari	[27, 43, 57, 44, 256]
2	*Mullus surmuletus*	*M. surmuletus* is a bioindicator species that is used due to its high ability to concentrate MPs in its gills and GI tract. MP contamination remained the same irrespective of distance from land, which indicated the extent of MP pollution in the marine environment.	Northeastern Mediterranean Sea	[27, 28, 257]
3	*Mugil cephalus*	*M. cephalus* accumulates MPs in its gills and GI tract, which makes it a very useful bioindicator.	Northeastern Mediterranean Sea	[27]
4	*Saurida undosquamis*	*Saurida undosquamis* are used as bioindicators as they amass MPs in their gills and GI tract.	Northeastern Mediterranean Sea	[27]
5	*Merluccius merluccius*	The MP contamination in *M. merluccius* decreased with increasing distance from urban settlements. MP contamination is linked to the feeding habits of the species. MPs also acted as vectors for microorganisms in this socioeconomically important species.	Adriatic Sea Tyrrhenian Sea Western Sardinia Northern Tyrrhenian Sea between Gulf of Oristano and Gulf of Cagliari	[43, 57, 44, 256]
6	*Scomber*	MPs are directly relevant to *Scomber* feeding habits. Proximity to urban locations increased MP contamination as well. In a comparative study between multiple species, *Scomber* had among the highest occurrences.	Adriatic Sea Tyrrhenian Sea Western Sardinia between Gulf of Oristano and Gulf of Cagliari	[43, 256]
7	*Boops boops*	MP frequency, abundance, and proportions varied with respect to geographic areas. *B, boops* are a potential bioindicator for MP contamination because fish in poorer physical condition ingest more MPs. Males also ingest significantly more MPs than females.	Mediterranean Sea Western Mediterranean Sea NW Iberian shelf	[22, 256, 52, 258]

8	*Seriola dumerili*	Pelagic fish such as *Seriola dumerili* are considered good bioindicators for environmental pollution monitoring.	Balearic Islands	[259]
9	*Galeus melastomus*	*G. melastomus* did not exhibit significant differences in MP content when their GI tract was observed. The small differences might be attributed to individual differences in characteristics. It is considered a target species for microlitter monitoring in deep water habitats.	Tyrrhenian Sea	[43]
10	*Scyliorhinus canicula*	*S. canicular* is considered a target for microlitter monitoring in deep water settings because they do not display significant differences in their intestinal contents, marking them as suitable bioindicators. The ability of sharks to regurgitate their stomach contents also affects the extent of MP contamination.	Tyrrhenian Sea Southwest coast of the United Kingdom	[43, 260]
11	*Etmopterus spinax*	*E. spinax* is a deep-water elasmobranch species that is an amazing bioindicator for deep water habitats for monitoring microlitter. They show less difference in their stomach contents than similar species, with the minimal differences attributed to individual characteristics.	Tyrrhenian Sea	[43]
12	*Trachurus mediterraneus*	*T. mediterraneus* exhibited the highest MP concentrations among species that were recorded in a similar environment. The higher MP contamination is due to its benthic nature.	Mediterranean Sea NW Iberian shelf	[258]
13	*Sardinia pilchardus*	*S. pilchardus* exhibited similar contamination to that in other species recorded in similar environments. As a pelagic fish, it has lower MP concentrations than its benthic counterparts.	Mediterranean Sea NW Iberian shelf	[257, 258]
14	*Engraulis encrasicolus*	*E. encrasicolus* exhibited moderate MP contamination but less than that in benthic species due to its pelagic nature.	Mediterranean Sea NW Iberian shelf	[257, 258]
15	*Callionymus lyra*	*C. lyra* exhibited similar characteristics to other species that were tested with it, not displaying very high or very low MP contamination.	NW Iberian shelf	[257]
16	*Oblada melanural*	*O. melanural* displayed moderate MP contamination, neither very high nor very low occurrence.	Between Gulf of Oristano and Gulf of Cagliari	[256]
17	*Spicara smaris*	*S. smaris* had among the lowest MP contamination in a multispecies study and the lowest occurrence.	Between Gulf of Oristano and Gulf of Cagliari	[256]

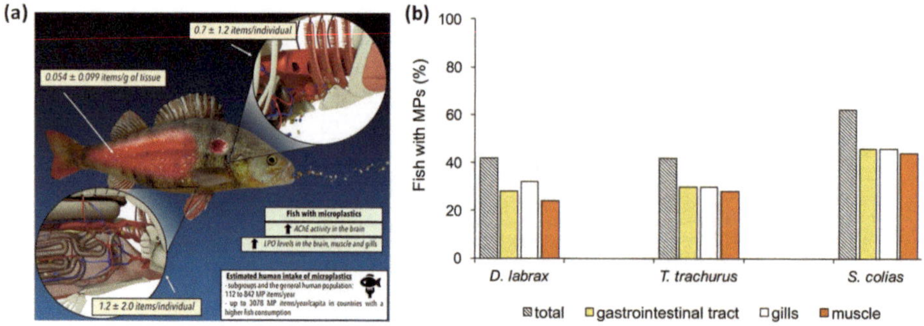

FIGURE 5.2 (a) Illustrative representation of microplastic accumulation in commercially important fish species obtained from the Northeast Atlantic Ocean; (b) Graphical representation showing comparative analyses of ingested microplastics in three commercially important marine fish species [26].

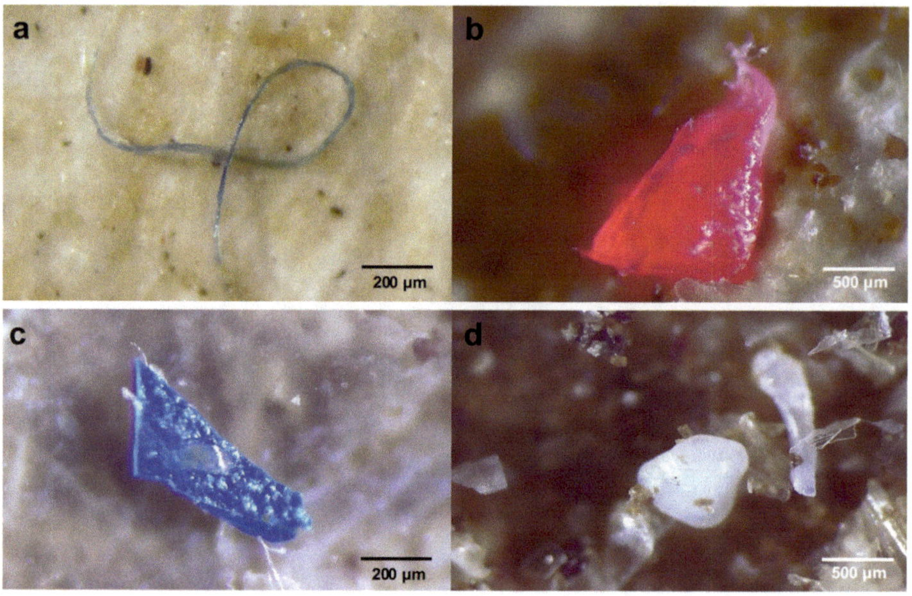

FIGURE 5.3 Microplastics recovered from three commercially important fish species: *Trachurus trachurus*, *Dicentrarchus labrax*, and *Scomber colias* [26].

In the northeastern Mediterranean Sea, the GI tract includes MPs in 36% of *S. undosquamis* [29]. Regarding *M. cephalus*, reported frequencies of MP occurrence in the GI tract were 60% in Hong Kong [31], 42% in Mexico [32], 64% in Sydney Harbour [33], 73% in South Africa [34], and 100% in China [35].

MPs can either be actively ingested when mistaken for food or passively ingested through foraging or as prey [36]. Trophic transfer of MPs has been observed in both marine environments and laboratory experiments [37, 38]. A significant inverse relationship was found between the trophic levels of *M. surmuletus* and MP dosage present in the GI tract. MPs adhere to gills in the presence of water flow. In the Persian Gulf the average MP content in the gills of various fish was 3.81 microparticles per fish [39]. Capillo et al. found fibers in the gills of *Trigla lyra* in the central Mediterranean Sea averaging 0.19 MPs per fish [40]. Similarly, in South China, the mean MP abundance was 0.16 MPs per fish across 24 different species [41]. Documenting a mean MP abundance of 0.62 MPs per fish in the Black Sea, Atamanal et al. focused on the gills of *M. barbatus* [42].

Valente et al. [43] evaluated the effectiveness of monitoring MP ingestion in fish species with different feeding behaviors. They studied three marine zones, each with unique circulation patterns and sources of microplastic pollution. The objective was to elucidate variations in MP ingestion among different fish species at specific locations. The chosen species included the pelagic feeder species from the genus *Scomber* and the species *Merluccius merluccius* and *Mullus barbatus*. They selected these species based on their known MP ingestion, availability in the targeted locations, commercial relevance, and variety of feeding patterns [44, 40, 45, 46].

As a benthivorous species, the red mullet (*M. barbatus*) eats on the sea floor by eating sediment with its prey and expelling the sediment through its gills [47]. As such, it may function as a marker for exclusively benthic MP consumption. However, being a demersal top predator, the European hake (*M. merluccius*) regularly swims between the sea bottom and mid- and surface waters, where it usually feeds on quickly moving food like fish and cephalopods. Pelagic fish species are thought to be useful markers for tracking environmental pollution in maritime environments because of their extensive dispersion.

In a 2022 study, researchers investigated MP presence in the digestive systems of the predatory pelagic fish *Seriola dumerili* in the Balearic Islands in the Mediterranean Sea. They also examined potential associations with oxidative stress by analyzing biomarkers in the liver tissue and found MPs in 98% of the samples that were tested, with an average of 12 particles per sample. Particles that resembled fibers were more common than pieces. Notably, there was no association between the fish's size and the prevalence of MPs in its GI tract. Increased activity of phase II detoxification enzyme glutathione-S-transferase and antioxidant enzymes like catalase and superoxide dismutase was linked to higher MP loads in fish. Specific types of polymers, such as PE, are known to persist in the water column, while others, like polyethylene terephthalate (PET), are more inclined to sink to the seabed [48], and fish habitat has demonstrated an association with rate of MP ingestion [49]. According to Anastasopoulou et al. [50], MPs are more common in demersal fish in the southern Adriatic Sea than they are in pelagic fish in the northeastern Ionian Sea. Comparably, research has indicated that MP ingestion occurs more frequently in pelagic fish [29, 51] than in demersal fish [30, 41].

Tsangaris et al. [22] evaluated MP consumption in *Boops boops* in various Mediterranean Sea locations. They caught 884 fish in all, from 20 coastal locations in France, Spain, Greece, and Italy, and subjected them to conventional technique analysis, and found MPs in 46.8% of the fish that were sampled had MPs in them, with an average of 1.17 particles per fish. Filaments were the most commonly found shape type, and polypropylene (PP) and PE were the most commonly found polymers.

Regarding MP consumption in *B. boops* from Italian beaches, Sbrana et al. documented similar results [52]. To investigate the relationship between biological characteristics and environmental conditions, the authors used generalized linear mixed models on a total of 379 bogues gathered from three diverse Italian locations that were under varied anthropogenic stressors. Most observed individuals, 56%, had MPs in their GI tracts, an average of 1.8 particles per individual. The authors emphasized that *B. boops* could serve as an indicator of MP ingestion, with variations based on local anthropization, and they identified Latium as the most polluted regions.

Sbrana et al. also discovered that fish with poorer physical characteristics were more likely to consume MPs, potentially indicating the degree of environmental contamination in the area. Furthermore, MP intake was much higher among male specimens than in the females, indicating a gender difference in MP intake. These results highlight how important it is to comprehend a bioindicator species' biology to create a strong monitoring plan. Geographically, MP occurrence, abundance, and proportions varied regarding different ingested MP properties (size, shape, color, and polymer). The authors found that the primary driver of geographic variations in the abundance of swallowed MPs was the level of anthropogenic pressure prevalent throughout the coastal areas where the sample was taken. These findings further support *B. boops* as a bioindicator of MP contamination in the Mediterranean Sea [52].

Zhang et al. [53] discovered MP pollution in surface water from 19 sampling locations in Guangdong's eastern coastal regions in south China and identified spatial variations in MP size, shape, and color in the water samples. The authors selected *Mugil cephalus* and *Konosirus punctatus*, two species of marine fish, as potential biomonitoring species for MPs. Due to its ability to tolerate a broad range of salinities, *M. cephalus* is widely distributed globally, and as an omnivorous fish, it is potentially exposed to pollutants including MP ingestion, making it a valuable subject for ecotoxicological studies, as demonstrated in previous research [31, 34, 53, 54]. *K. punctatus* is widely consumed as a popular variety of sliced raw fish (sashimi) because of its low cost, mouthwatering flavor, and elevated levels of taurine and omega fatty acids [55].

Zhang et al. sought connections between the MP quantities in the water samples and the corresponding fish samples and identified differences between MP amount in water samples and the amount the fish consumed. In one sample, they found no MPs in *K. punctatus* irrespective of MP abundance in water samples, but in *K. punctatus* as biomonitors, the fish ingested more MPs with respect to the MP abundance in the surrounding aqueous environment. Establishing a relationship between fish MP levels and those in the water is essential for the successful active biomonitoring of MPs. Though there is data available, the authors were unable to find clear quantitative evidence that *K. punctatus* and *M. cephalus* are viable candidate species for detecting marine MP ingestion [53].

Bellas et al. discovered that three commercially significant demersal fish species located around the Mediterranean and Atlantic coasts of Spain consumed MPs [56]: *Mullus barbatus* (red mullet), *Merluccius merluccius* (European hake), and *Scyliorhinus canicula* (dogfish). The authors examined a total of 212 fish samples, and with an average of 1.56 ± 0.5 items per fish, the observed percentage of fish containing MPs was 17.5% (dogfish, 15.3%; red mullet, 18.8%, hake, 16.7%). The MPs had a size range of 0.38–3.1 mm. These fish species should work well for tracking temporal and spatial changes associated with ingested litter because they are currently used as biomonitors in the Spanish Marine Pollution Monitoring Programme.

Following that study, Miccoli et al. [57] examined areas close to the coastal epipelagic northern Tyrrhenian Sea to determine the presence of anthropogenic debris in *M. barbatus* and *M. merluccius*. Additionally, to assess any potential harmful effects, they subjected primary cell cultures from mucosal and lymphoid organs to marine microlitter. Micro-FTIR examination identified polyamide (PA), high-density PE, and PP, and each individual from these species consumed an average of 5–15 microlitter pieces; the authors verified microlitter role as a microbe carrier.

The comprehensive exploration of microplastic impacts on fish reveals a multifaceted scenario. Diverse fish species, including commercially significant ones, exhibit varying degrees of microplastic ingestion, emphasizing the need for a nuanced understanding of their feeding habits. Rios et al. [58] published the first study on the interactions between MPs and aquatic organisms in a Patagonian marine setting; specifically, the team assessed MP types and abundance in the benthic habitat in the Puerto Madryn urban region. They sampled several representative species, including bottom water and three small benthic fish: juvenile *Ribeiroclinus eigenmanni*, *Eleginops maclovinus*, and *Aulacomya atra* (ribbed mussel), microcarnivores with a body size of less than 10 cm. The selection of the specific biomonitoring species may have been motivated by their vulnerability to the deleterious effects of MP exposure, possibly via the food chain [59, 60].

In brief, the effectiveness of monitoring approaches is showcased by studies evaluating MP ingestion efficiency in fish species with distinct feeding behaviors. Studies involving marine species such as *Konosirus punctatus* and *Mugil cephalus* demonstrate the complexity of the link between fish MP consumption and MP quantity in water. Discrepancies in MP ingestion compared with environmental abundance highlight the challenges in establishing a quantitative correlation. Despite these complexities, active biomonitoring remains crucial for assessing the presence of MPs in fish and corresponding water levels. In this context, conclusive quantitative evidence supporting the suitability of *M. cephalus* and *K. punctatus* as biomonitoring species remains elusive.

5.3 BIVALVES

Bivalves are probably the primary source of MPs from seafood that people ingest as they are typically consumed whole. This makes them ideal markers of MP contamination [61–63]. To determine amounts and features of MP contamination, Cho et al. [64] monitored MP pollution along Korean beaches using filter-feeding bivalves (mussel, oyster, and Manila clam) as bioindicators. Their results suggested the applicability of bivalves as microplastic biomonitoring agents.

Table 5.2 presents an overview of various bivalve species used as MP and NP biomonitoring agents. Ding et al. [65] confirmed four bivalve species—*Chlamys farreri, Crassostrea gigas* (Pacific oysters), *Mytilus galloprovincialis*, and *Ruditapes philippinarum*—as MP pollution bioindicators, finding MPs in 80% of samples, 233 out of 290 collected from Qingdao, China. The four species showed MP abundance

TABLE 5.2
Bivalve Species as MP and NPBiomonitoring Agents

S. No.	Species	Overview	Location	References
1	*Donax trunculus*	*D. trunculus* species has been used in urbanized beaches as a microplastic bioindicator.	Morocco	[108]
2	*Corbicula fluminea*	*C. fluminea* has been used as a bioindicator in freshwater environment as they reflected the microplastic pollution variability.	China	[261]
3	*Arca noae*	Microplastic as well as trace elements have been studied using *A. noae*. The species was used as an early warning bioindicator for various biomonitoring programs.	Egypt	[66]
4	*Mytilus edulis*	*M. edulis* was used due to their large-scale filter feeding nature which exposed them to higher levels of microplastics. It has been proposed as potential bioindicators by global studies. TGA-FTIR-GC/MS is a technique which can be used to quantify microplastic in mussels. However recent studies have indicated bivalves to be poor bioindicators due to their selective nature during ingestion and egestion.	Korea China	[64, 75, 100, 11, 93]
5	*Mytilus galloprovincialis*	Microplastics presence in edible mussels has established an exposure route to humans with highest concentration in children as compared to adults. In some cases, samples collected during summers contained smaller sized microplastics.	Morocco Tunisia China Atlantic coast	[95, 65]
6	*Unio stevenianus*	*U. stevenianus* can potentially be utilized as a bioindicator due to its nature to uptake pollution in high microplastic polluted rivers.	Turkey	[262]
7	*Crassostrea gigas*	*C. gigas* was kept in rocky shores and is used as a bioindicator for the water column. In some cases, samples collected during summers contained smaller microplastics.	Korea China	[64, 65]
8	*Ruditapes philippinarum*	*R. philippinarum* also known as Manilla clams reside in the sediment and are filter feeders. It contains more smaller microplastics than other bivalves. They reflect microplastic polymer variability in sediment better than other bivalve species. *R. philippinarum* is used as a bioindicator at the benthic water level. Some studies indicate that there is a low risk of microplastic toxicity in humans through consumption.	Korea China	[64, 65, 263]

9	*Perna perna*	*P. perna* was indicated as a bioindicator in an estuary to investigate microplastic contamination on edible mussels.	Brazil	[97]
10	*Chlamys farreri*	*C. farreri* also known as Farrer's scallop is a proposed bioindicator for detection of microplastics in marine environment. In some cases, samples collected during summers contained smaller sized microplastics.	China	[65]
11	*Tegillarca granosa*	Microplastics present in *T. granosa* also known as blood cockles were compared in markets and aquaculture farms. It was observed that bivalves in the market had significantly higher microplastics in comparison to their counterparts.	Thailand	[101]
12	*Perna viridis*	*P. viridis* also known as green mussels were compared in markets and aquaculture farms setting. It was noted that bivalves in the market had significantly higher microplastics in comparison to their counterparts.	Thailand	[101, 271]
13	*Donax cuneatus*	*D, cuneatus* are known for their ability to ingest microplastic pollutants. An allometric relationship between the body size of the organism and microplastic concentration was drawn. Microplastics were observed. They can act as bioindicators for assessing microplastic pollution.	India	[109]
14	*Crassostrea virginica*	Studies conducted on *C. virginica* indicates that this species is a poor bioindicator due to differential ingestion and egestion of microplastics. Low recorded microplastics along with dissimilarity between microplastics present in the oyster and its surrounding environment further stressed bivalves to be poor indicators.	Long Island	[111, 93, 122]

ranging from 0.5–3.3 items/individual or 0.3–20.1 items/g wet weight digestive tract, with microfibers being the most common shape. Ding et al. identified 18 polymer types with sizes from 7 to 5,000 μm.

In another study, on *Arca noae* in the southeastern Mediterranean, authors found an average MP abundance of 1.65 particles/individual and 58 items/gram wet tissue weight, with a 48% incidence rate of MP ingestion [66]. Commonly identified polymers were PE, PS, and PP. This research and the findings in the table establish crucial baseline data for future environmental assessments, highlighting the potential of *A. noae* as an early warning biomonitoring agent.

5.3.1 Mussels

Given their unique properties, mussels are among the best biological indicators available for monitoring trends in anthropogenic pollution along coastal waterways [67, 68]. Mussels are among the first animals to be used to assess the environmental quality of saltwater because they fulfill nearly every prerequisite for an effective bioindicator species [69]. They are especially beneficial because of their widespread distribution, ease of access, and remarkable resistance to salt, temperature, oxygen saturation, and food supply.

Mussels are benthic filter feeders, which means they effectively gather chemical pollutants from saltwater and provide an accurate measure of the concentration and bioavailability of pollutants in the environment [68]. Because mussels have been a part of human seafood consumption for millennia, it is important to consider the potential dangers to human health that marine pollution poses [70]. In many regional environmental monitoring programs, including the Assessment and Control of Pollution in the Mediterranean region, OSPAR's Coordinated Environmental Monitoring Program, and U.S. Mussel Watch Project, mussels have been used extensively [68].

Based on geographic coverage, scientists have reported on a variety of different mussel species including *M. edulis* (blue mussel), *M. galloprovincialis* (Mediterranean mussel), *M. trossulus*, *Mytilus spp.*, *Modiolus modiolus*, *P. perna*, and *P. viridis*, and biomonitoring; among the most commonly reported polymers are PP, cellophane, PE, PES, PET, PS, PA, and polyvinyl chloride (PVC). Data collected from diverse investigations indicate a noteworthy alignment between the polymeric composition and primary morphotype found in mussels and surrounding aqueous environments [48, 71–74]. Qu et al. observed a consistent proportionality between these factors in mussels and seawater. These results indicate MP presence and properties in mussels as proxies for true pollution levels in the environment [73].

The prevalence of MPs in mussels exhibits a close correlation with human activities, as evidenced by significantly elevated quantities in areas with intense human influence [75]. This pattern is seen not just in areas where human activity is concentrated but also in areas where MPs may accumulate, like the Barents Sea [63]. Findings showing much higher concentrations in mussels than in the saltwater and surrounding soil highlight the possibility of microplastic buildup [76].

Numerous study findings revealed a robust association between the amount of MP pollution in mussels and the pollution levels in coastal environments, suggesting that mussels can precisely represent the true abundance of MPs in the environment within particular size ranges. Laboratory exposure studies have been conducted on *M. edulis*, *M. galloprovincialis*, *Mytilus spp.*, *Dreissena polymorpha*, *Geukensia demissa*, and *P. perna* [77], although *M. edulis* has been predominantly used as a bioindicator species. Li et al. measured the efficacy of the respective species for active MP biomonitoring by studying MP pollution in *M. edulis* at 22 sites along China's 12,400 miles of coastline [75]. Wild *M. edulis* exhibited more MPs than farmed specimens. Mussels from areas with intensive human activities showed significantly elevated MP abundance, ranging from 1.6 to 3.3 items/g. Among various MP shapes, fibers were the most prevalent, followed by fragments. MPs smaller than 250 μm constituted 17%–79% of the total MP content.

As a novel and innovative approach, diatoms could be successfully distinguished from MPs using SEM. Researchers studied the effects of aging PE MPs in *M. edulis* by exposing the mussels to different MP exposures and levels of breakdown, and they identified different expression levels according to kind of exposure and state of breakdown (aged vs. non-aged) [78]. The authors examined changes in gene expression levels linked to the cytoskeleton, immune system, detoxification, and cell cycle regulation using quantitative real-time quantitative polymerase chain reaction, in contrast with traditional biochemical techniques that indicated relatively small changes between examined circumstances. Additionally, the authors produced a great deal of information about the toxicological effects of MPs via in vitro studies.

Various polymer kinds and shapes have been the subject of experimentation. High-density (PE) [79], low-density PE [80], PS, PVC [81], PE [82], and PP [73, 82–92] are the common MP polymers observed. Particles, threads, spheres, powder, and other shapes are highlighted. The effects of different MP polymer types, sizes, and forms on *Crassostrea virginica* (eastern oysters) and *M. edulis* were studied, and the mussels and oysters rejected the PES microfibers based on their size; oysters consumed more shorter fibers than long [93].

M. galloprovincialis has been a valuable bioindicator in numerous toxicological and environmental studies for monitoring MP presence in marine environments and is widely distributed throughout the Mediterranean Sea. Specifically, *M. galloprovincialis* demonstrates C-reactive protein (CRP), an innate immune system acute phase protein, as a biomarker. Franco-Martinez et al. discovered and isolated a CRP-like protein from *M. galloprovincialis*'s digestive gland mussels subjected to NPs and hypoxia had higher levels of CRP-like protein than the control group (Figure 5.4) [94]. The authors established that a high-throughput automated approach could accurately and consistently screen CRP-like protein and that CRP-like protein is a promising biomarker for any environmental element that may directly or indirectly stimulate mussels. Therefore, measuring the CRP-like protein in mussels could be very beneficial for biomonitoring initiatives.

Microplastic contamination in the species *M. galloprovincialis* was reported by the team of Abelouah et al. [95]. The authors conducted studies in Morocco and Tunisia, and the frequency of available MPs ranged from 79%–100% in all the analyzed samples. Abelouah et al. detected higher-density MPs in mussel samples obtained from Morocco than in Tunisia. They observed maximum accumulations up to 1.5 MPs/g and 1.8 MPs/g in the gills and digestive glands of mussel samples obtained from Tunisia and Morocco, respectively. MPs were identified to be major

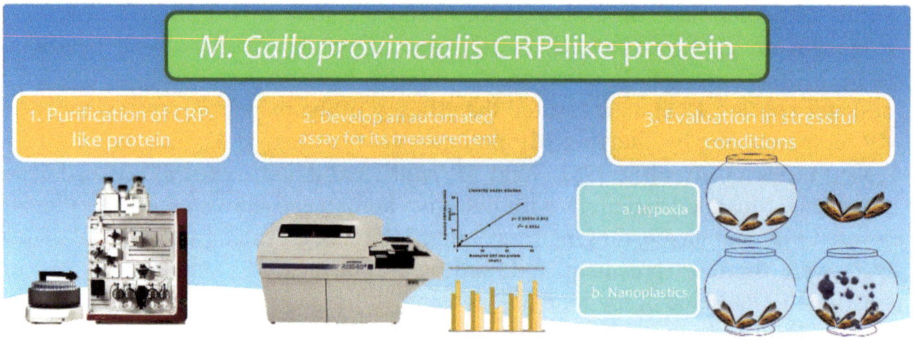

FIGURE 5.4 Schematic illustration of CRP-like protein isolation and evaluation from the Mediterranean mussel (*Mytilus galloprovincialis*) [94].

fibers of sizes less than 1,000 μm. Among different polymeric types, PET, PP, and PE were identified to be the predominant ones (Figure 5.5).

The team of Ferriera et al. monitored the Douro estuary (NE Atlantic coast) for water physicochemical parameters and the seasonal levels of MPs and biomarkers (neurotoxicity, condition index, energy, oxidative stress) in *M. galloprovincialis* mussels [96]. They also used integrated biomarker response to measure mussel stress and, based on consumer behavior and mussel MP concentrations, calculated the human risk of MP intake. Most of the MPs identified, 72%, were microfibers with a range of chemical compositions.

Mussel *Perna perna*, an important food source, is also a biomonitoring species. Utilizing the species as a bioindicator, a group of scientists looked at the existence of microplastic contamination on the Santos estuary, which is the most populated region along the coast of São Paulo state, Brazil. A quick and easy analysis revealed MP ingestion by the mussels (75%) [97]. Other scientists assessed MP accumulation in mussels gathered throughout the Salish Sea and found an average concentration of MPs in wet tissue of 0.75 ± 0.09 MPs/g [98]. A total of 20% of the mussels tested had plastic in their chemical composition, and based on the fraction of chemically identified MPs in the mussels, up to 0.75 MPs/g of the wet mussel tissues were composed of plastics. Compared with oysters, mussels exhibited higher MP concentrations and plastic contamination in the Salish Sea [99]. Polymer particles in mussels were identified as nylon, PP, PS, and PET. In line with earlier research, fibers were consistently the most common morphotype among the various forms of MPs in mussels.

Researchers reported on three approaches for MP detection, identification, and quantification in *M. edulis*: gas chromatography–mass spectrometry (GC/MS), thermal gravimetric analysis, and Fourier transform infrared spectroscopy (FTIR). The techniques utilized to separate and isolate the MPs include density separation, chemical digestion, and filtration [100]. Ding et al. employed μ-FTIR for MP identification [65]. Ta et al. investigated MP accumulation in green mussels (*P. viridis*) and blood cockles (*Tegillarca granosa*), and among 18 types of polymer found, PVC and rayon were commonly identified, with widths ranging from 7 μm to 5,000 μm. [101].

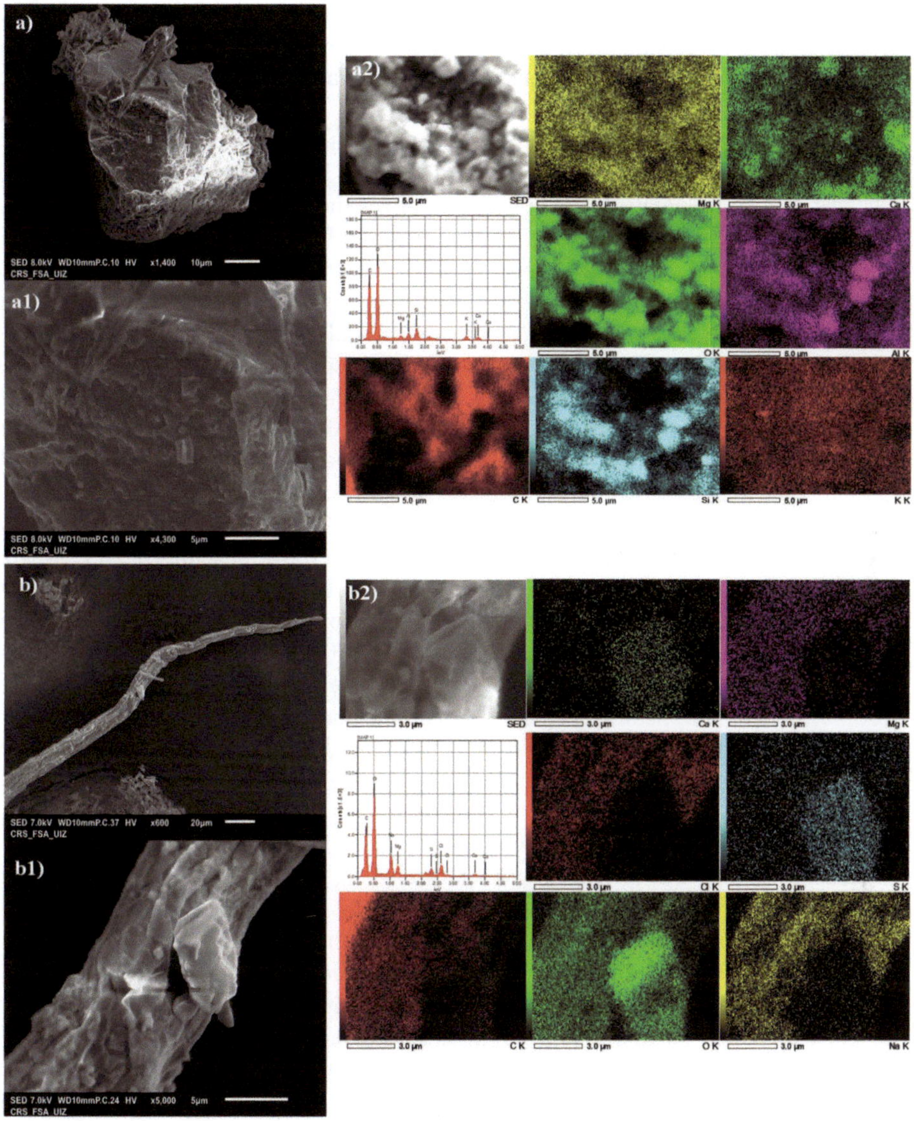

FIGURE 5.5 High-resolution imaging and elemental analyses of the microplastics (both fiber and fragments) extracted from *Mytilus galloprovincialis*: (a) SEM images (a, a1, b, and b1), surface mapping, and EDX elements (a2 and b2) [95].

5.3.2 Asian Clams

Together with oysters and mussels, clams are a popular choice for studying MP pollution in coastal ecosystems due to their ecological relevance and link to human health. Hong Kong was identified as an MP pollution hotspot in recent research in which average MP concentrations were 268 items/kg in mudflat sediments and 16.8 items/

kg in beach sediments [102, 103]. Given the popularity of clam digging as a recreational activity in Hong Kong, worries about possible health risks for recreational harvesters arise when they harvest diseased wild clams from intertidal mudflats for personal use [104].

Researchers have calculated dietary consumption for subnational and national populations in order to address health risks linked with MPs in farmed and market-sold bivalves, but risk evaluations often ignore bivalves that are harvested in the wild [70, 105]. Lam et al. recently examined MP presence and properties in wild edible clams from Starfish Bay and Shui Hau Wan [106]. The reported species were *Anomalocardia flexuosa*, *Anomalocardia squamosa*, *Atactodea striata*, *Gafrarium pectinatum*, *Geloina erosa*, *Marcia hiantina*, *Meretrix lyrate*, and *Venerupis philippinarum*.

MP pollution was discovered in six wild clam species that were collected from two recreational clamming sites in Hong Kong in October 2019 and 2020. All species were found to contain MPs in the concentration of 1.04 n/g, which was in the range of other research findings including for *R. philippinarum* (4.29 n/g) and *Siliqua patula* (0.16 n/g) [107]. *Donax trunculus*, a species of wild clam, was found to be a bioindicator of MP pollution linked to beach urbanization and the growth of the Moroccan tourism destination Taghazout Bay. Across six coastal sites, MP densities in *D. trunculus* ranged from 1.8 to 6 items/gram of wet weight. MP contamination showed significant spatiotemporal variation, correlating with the degree of urbanization. The predominant shapes identified were fragments (9.5%) and fibers (90.5%). The species demonstrated ingestion of various NP sizes, with a notable preference for particles of 0.1 and 0.5 mm [108, 18]. Other clam species recognized as bioindicators include *D. cuneatus* (Location: India; MP abundance: 0.6–1.3 g/wet weight), *Amarilladesma mactroides* (Location: Argentina; MP abundance: 0.3–0.5 g/wet weight), *Siliqua patula* (Location: USA, MP abundance: 0.02–0.16 g/wet weight), and *Ruditapes philippinarum* (Location: South Korea, MP abundance: 0.33 g/wet weight) [64, 109, 107, 110].

Among various shapes that have been examined, clams predominantly exhibited the presence of fibers, which could have resulted from two factors: either the clams' living quarters were heavily contaminated with plastic fibers or fibers are more likely to remain in the digestive tract and form clusters over time [111, 112]. However, little is known about how particle shape affects clams' ability to ingest MPs [53], calling for more research. Particle size has a pivotal influence on microplastic accessibility to organisms [61]. Lam et al. found that MPs < 1 mm constituted the majority, aligning with prior studies [100, 113]. Since MPs and their food sources, such as diatoms and microplankton, have similar size ranges, the prevalence of this size category may be related to clams' selectivity toward certain microplastic sizes [18].

The shoreline zone, acting as a natural biofilter, is acknowledged as a significant sink for MPs [113]. It is not surprising that all of the wild clam species from the studies in Hong Kong were contaminated with MPs, as bivalves' exposure to MPs is heavily dependent on the in situ levels of MP pollution [65], and MPs are clearly present in the tidal flat sediments along Hong Kong's coast [103, 114]. Davidson and Dudas examined MPs in both wild and cultivated *Venerupis philippinarum* (Manila clams) from three operational shellfish farms and three reference beaches (non-shellfish farm locations) in British Columbia (Baynes Sound). They sampled a total of 54 clams and isolated and enumerated the MPs [115].

5.3.3 Oysters

Oysters are crucial to an ecosystem's health and are frequently eaten by humans [49]. Vieira et al. demonstrated their potential as bioindicators of environmental quality [116]. Other researchers have shown that particles as small as 10 μm can enter the digestive gland lining of *Magallana gigas* (Pacific oysters) from the hemolymph-containing circulatory system [70, 83]. Although the eastern oyster (*C. virginica*) is an economically valuable species that consumes MP, oysters are selective suspension feeders, meaning they do not ingest every particle they encounter.

Liao et al. recently measured the MP content in oysters from 22 different locations along Taiwan's coastline and discovered a total of 6,630 MP particles in 660 oysters belonging to *Saccostrea* and *Crassostrea* [117]. MP content ranged from 0.63 to 37.94 items/g, with an average of 3.24 ± 1.02 items/g (wet weight). More than 50% of the MPs were smaller than 100 μm, with fragments accounting for 67% of the total form and fibers for 29%. The MP size range they observed most often in the oysters was 50–290 μm, and other researchers found that oysters from Taiwan's coast and the Asia-Pacific region have similar MP size distributions [105, 118]. PET was the most frequently discovered polymer type. There was more total MP contamination in the wild than in farmed oysters, and there were higher MP concentrations in oysters from the northeastern seas than in other oysters. According to the findings, oysters on Taiwan's coastline had average MP amounts that were comparable with levels in oysters from nearby areas.

Investigators thoroughly characterized *C. gigas* (Pacific oysters) raised in Danang Bay in Vietnam for their ingested MP attributes (quantity, shape, size, and polymer kinds) and MP content ranged from 1.8 to 19 particles/individual calculated in terms of wet weight [119]. The authors found MPs in three primary shapes: fiber (25.84%), fragment (73.71%), and beads (0.45%). Furthermore, the MPs' sizes ranged from 22.4 to 1,318.8 μm, with less than 100 μm accounting for 77.30% of all sizes. There were 15 types of polymers found, with nylon being the most common at 50.56%. The buildup of MPs in this study, along with their polymeric structures and abundance, provides solid grounds for additional research aimed at determining the extent of their possible health risks to humans [119].

Other researchers studied *C. gigas* for its possible role in biomonitoring aged HDPE, PVC, and PP; specifically they investigated the impacts of MPs on developmental deformities focusing on *C. gigas* larvae and spat [130]. Pediveliger larvae (17-day-old) were exposed to a microplastic cocktail in a controlled laboratory setting over the course of a seven-day fixation phase. After being subjected to 10 mg MPs/L, the larvae showed significantly lower settlement success (49%) than did the control group (61%).

There were no visible deformities or anomalies during the transformation. After 11 months of observation, the authors discovered that during the first 28 days, there was a considerable decrease in spit growth compared with the control, but then the oysters that had been exposed before began to develop faster; by the end of the research, they had outpaced the control group. This almost year-long study highlighted the possible effects of aquaculture-derived MPs on the growth of *C. gigas* larvae [130].

An estuary's function as a transitional area between riverine and marine habitats is crucial for understanding the origins, accumulation, and movement of MPs. While estuaries are hotspots for MP contamination, less is known about the relationship between MP pollution and aquatic life. Li et al. investigated *Saccostrea cucullate* (wild oysters) for MP pollution at 11 sampling sites in South China (Pearl River Estuary) and found from 1.5 to 7.2 items/gram of tissue wet weight [72]. They found 69% fibers in the MPs they found in the oysters, and 83.9% of the MPs were smaller than 100 μm, with sizes ranging from 20 to 5,000 μm. Oysters are recommended as microplastic monitoring agents in light of the findings.

Another group investigated *S. cucullate* along the Bornean coastline for biomonitoring MPs [121]. The Brunei Estuarine System and the South China Sea were compared for their respective local pollution levels and their relatively unpolluted open-shore beaches. The authors used three morphologies (fragments, fibers, and pellets) to define 16 different MP color varieties. Sample analyses revealed that fragments predominated (74.9%), of which PP was abundant (31.7% of the total). MP deposition levels varied per site, ranging from 0.43 to 7.20 particles/g of oyster tissue. The Brunei system accumulation peaked in proximity to the anticipated source, Bandar, and declined precipitously in the direction of the ocean, indicating the effects of seawater dilution, environmental sequestration (local sinks), and current flow. Because of this localized variance in estuary MP loading, it is not advisable to draw conclusions from small samples.

For organisms under study to serve as reliable bioindicators of MPs, they must indiscriminately consume the bulk of the plastic particles they encounter. In contrast to the studies we have been discussing, the authors of the following studies could not encourage oysters as MP biomonitoring species. One research group exposed *C. virginica* and *M. edulis* to PS microspheres and nylon microfibers to determine the proportions of these particles rejected in pseudofeces and expelled in fecal matter [111]. The microsphere rejection percentage for both species rose from around 30% for the smallest spheres to a maximum of 98% for the largest. The authors observed constant ingestion of microfibers regardless of length, and more of the largest microspheres were rejected than the longest microfibers. The results imply that the physical properties of the particles affect the quantities and kinds of MPs found in the gut of bivalve organisms. As such, bivalves might not be good bioindicators of MP pollution in the ecosystem; instead, it's best to look into other marine species.

Mladinich et al. recently looked into a relationship between MP in oysters on a recreational oyster bed and the surrounding environmental compartments in Long Island Sound (USA) to establish how much and what kinds of MP were present in sediment, water, marine snow, and oyster samples [122]. The counted a total of 86 MPs among the 885 probable particles they found in the ambient medium. The maximum MP count found in a single oyster was nine, a sign of low MP concentrations in both the oysters and their surroundings. Oysters and the surrounding environmental compartments also shared very few polymers with the exception of PET. Sediments included the most MPs (42 total) of any environmental category. These data assisted in identifying the types of MPs (polymer shape, composition, and size) that oysters are exposed to as well as the MPs that the oysters swallowed. As further indicated by the low quantities of MPs detected and the misalignment of polymers between the oysters and their surroundings, oysters are not an effective bioindicator species for MP pollution.

5.4 MARINE SPONGE

In the context of MP research, sponges are a vital component of filter-feeder benthic communities, but members of the phylum Porifera have received very little attention. Some investigators proposed that sponges are sentinels to keep an eye on the condition of marine ecosystems, particularly with regard to MPs [123]. Table 5.3

TABLE 5.3
Porifera Species as MP and NP Biomonitoring Agents

S. No.	Species	Overview	Location	References
1	*Haliclona implexiformis*	Sessile benthic organisms are good bioindicators for coastal exosystems. *H. implexiformis* absorbed MPs to a moderate degree. Serves as a potential bioindicator owing to its abundance in accessible areas.	–	[264]
2	*Halichondria melanadocia*	The sponge *H. melanadocia* absorbs MPs based on exposure. This and its common occurrence in accessible areas make it a convenient MP bioindicator.	–	[264]
3	*Amorphinopsis atlantica*	*A. atlantica* shows the highest MP contamination of its peers. It expressed bioaccumulation and displayed considerable difference between pristine and disturbed state. Can potentially be an MP bioindicator.	–	[264]
4	Coral reef demosponges	Absorb very fine particles; it was extrapolated that sponges can on an average incorporate 10,000 microparticles in their tissue independent of the material. Along with other factors, this suggests strong MP biomonitor potential.	North Sulawesi, Indonesia	[131]
5	*Petrosia ficiformis*	The filter feeder sponge *P. ficiformis* retains particles form the water column, still showing 30.6% inoculated MP after 72 hours, indicating potential as an MP bioindicator.	–	[135]
6	*Haliclona (Rhizoniera) scotti*	Large amounts of MPs were unexpectedly detected including materials in protective clothing used in Antarctica. Thus, sponges are great MP bioindicators even in remote areas.	Tethys Bay, Antarctica	[133]
7	*Chondrosia reniformis* *Ircinia variabilis* *Petrosia ficiformis*, *Sarcotragus spinosulus*	Sponges are suspension-feeding organisms that are highly susceptible to MPs; 100% of samples were contaminated, with no meaningful differences among the MPs. This implies that uptake of environmental pollutants in sponges is influenced solely by the environment and not the sponge species and suggests sponges as valuable MP bioindicators in the future.	Moroccan Mediterranean coast	[132]

offers an overview of different Porifera species used as MP and NP biomonitoring agents.

According to Hammel and Nickel [124], particles are absorbed in one of two ways: (1) through endocytosis through the exopinacoderm, which are external cells that make up the outermost layer of the sponge body, or (2) through phagocytosis by choanocytes, which are cells that beat their flagella to create water flow within the sponge body. Shallow-water sponges, which are generally located near rocky shorelines throughout a large geographic range, most likely have greater quantities of MPs in them.

According to Corcoran et al., the degradation of bigger plastic items caused by a mix of mechanical forces and UV light intensifies this exposure in coastal regions [125]. Because of the high rate at which they filter out particles, sponges have the capacity to accumulate large amounts of foreign matter and display high levels of contaminants [126]. Their susceptibility to water pollution is well documented [127–129]. Additionally, sponges employ a non-selective feeding strategy [130], potentially heightening their vulnerability to MP infestation. Although the consumption of MPs by sponges has been the subject of few investigations, what is known indicates that sponges accumulate these foreign particles in their fiber skeletons and other internal tissues. Among other researchers, Girard et al. recently reported on sponge consumption of anthropogenic debris, including PS particles [131]. Interestingly, the scientists observed an no apparent relationship between the particle quantities the sponges swallowed and the chemical makeup of those particles. Girard et al. proposed that the primary variables driving the divergences in particle accumulation seen in marine sponges were changes in the composition and concentration of particles in the surrounding water.

Krikech et al. studied four sponge species to determine the existence and quantity of MPs: *Ircinia variabilis*, *Chondrosia reniformis*, *Petrosia ficiformis*, and *Sarcotragus spinosulus* [132]. These sponges were gathered from four locations on Morocco's Mediterranean coast, and their geographic distribution was investigated. The authors employed SEM with energy-dispersive X-ray spectroscopy in conjunction with a novel extraction technology developed in Italy, and they found that MPs were present in every examined sponge specimen. With MP abundance ranging from 3.9×10^5 to 1.0×10^6 particles/gram of dry weight of sponge tissue, the four species of sponges displayed different patterns. There were notable differences between the sampling sites, but Krikech et al. found no species-specific variations. These findings suggested that sponge MP absorption in aquatic environments is impacted by environmental pollution. Respectively, the largest and smallest lowest MPs were discovered in *P. ficiformis* and *C. reniformis*, with mean diameters of 2.57 and 1.84 μm (Figure 5.6). The study provided evidence in favor of using marine sponges as bioindicators of MPs. Apart from these, other Cnidarian species have also been reported for their MP biomonitoring properties, particularly jellyfish (Table 5.4).

Researchers studied a variety of marine sponge species from Antarctica (Tethys Bay) to determine the extent of PES and PA microplastic contamination [133]. The authors concentrated on three distinct analytes: 1–6–4-hexanediamine (HMDA), 6-aminohexanoic acid, and terephthalic acid (TPA), which are produced by the hydrolytic depolymerization of polyesters and polyamides. Only HMDA, of the

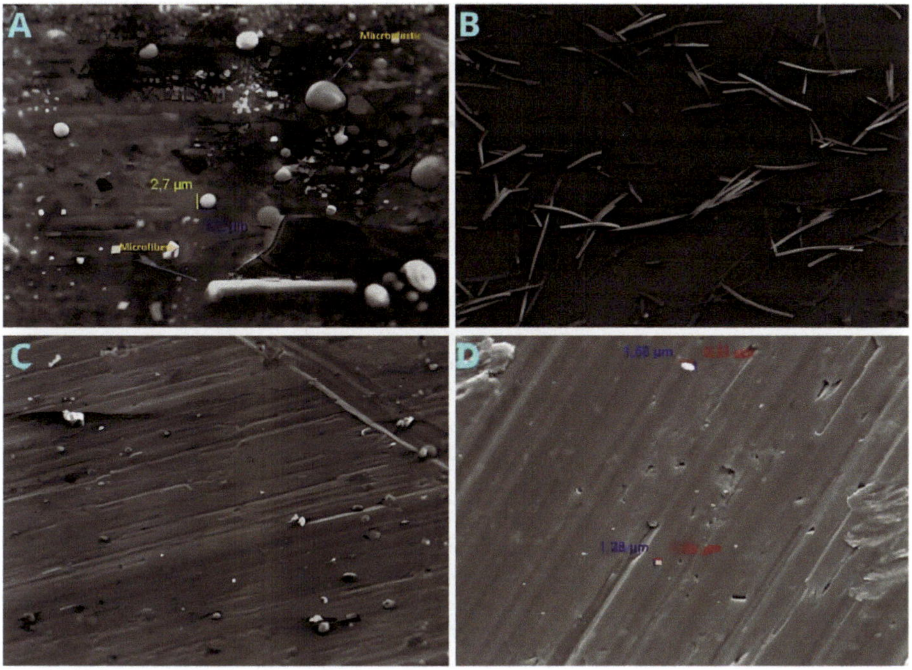

FIGURE 5.6 Scanning electron microscopy images of plastic particles identified in *Petrosia ficiformis* at different magnifications: (A) 1500×, particle size larger than >10 µm; (B) 100×, siliceous spicules; (C) 200×, and (D) 1500× [132].

TABLE 5.4
Cnidarian Species as MP and NP Biomonitoring Agents

S. No.	Species	Overview	Location	References
1	Beadlet anemone	MP accumulation is different between surface seawater and sediment. Sea anemone can be a crucial bioindicator for understanding MP pathways and MP accumulation in the ecosystem.	–	[265]
2	*Anemonia viridis*	*A. viridis* readily uptakes MPs with limited preference for particle shape and size and no influence of water temperature or presence of food. The MPs were trapped in the secreted mucous. Authors highlighted *A. viridis*'s potential as an MP bioindicator.	–	[266]

(Continued)

TABLE 5.4 *(Continued)*
Cnidarian Species as MP and NP Biomonitoring Agents

S. No.	Species	Overview	Location	References
3	*Cassiopea xamachana*	In 115 specimens over three estuaries, 77% of species contained MPs that varied greatly in characterization over the locations. Authors hypothesize that benthic jellyfish could be MP bioindicators near offshore systems.	Florida estuaries	[267]
4	*Aurelia aurita*	Jellyfish blooms indicate the degradation of the ecosystem including MP accumulation in the pelagic food web.	–	[268]
5	*Chrysaora pentastoma* medusae	In *C. pentostoma* medusae, 83% absorbed wastewater MPs, but the composition of the ingested MPs differed greatly from that of the MPs present in the wastewater, which indicated *C. pentostoma* as a poor MP bioindicator.	Wastewater Treatment	[269]

two amine monomers, was present in specimens of *Haliclona scotti* over the quantification limit; the authors measured 27 μg/kg of nylon 6,6. The authors calculated concentrations of either PBAT or PPTA equivalent to 2.5–4.1 mg/kg based on the TPA concentrations found in all sponge species. They found TPA in the acid hydrolysate, which is surprising because PET is the most popular textile fiber, and PET can only be efficiently depolymerized under alkaline circumstances by PISA. The findings showed that sponges can be useful markers of MP pollution kind and extent since they can collect and concentrate MPs from vast amounts of filtered marine water. Even in isolated locations, they offer early alerts when pollution levels are rising.

The composition of the foreign particle content in 15 demosponges from coral reefs near Bangka Island (North Sulawesi, Indonesia) was examined using advanced light microscopy, Raman spectroscopy, and traditional histology techniques [131]. Two-photon microscopy confirmed that the analyzed sponges gathered and absorbed the very fine sediment (<200 μm) absent from the nearby sand into their spongin fibers (skeletal elements) and ectosome (outer epithelia) (Figure 5.7). The investigators found a total of 34 different types of particles, including particulate cotton, titanium dioxide, and degraded manmade products like PS. Eight specimens were reported as having ingested MPs, with concentrations varying from 91 to 612 particles/gram of dry sponge tissue. The authors estimated that on average, sponges can incorporate 10,000 microparticulate contaminants into their tissue given their potential for large weight.

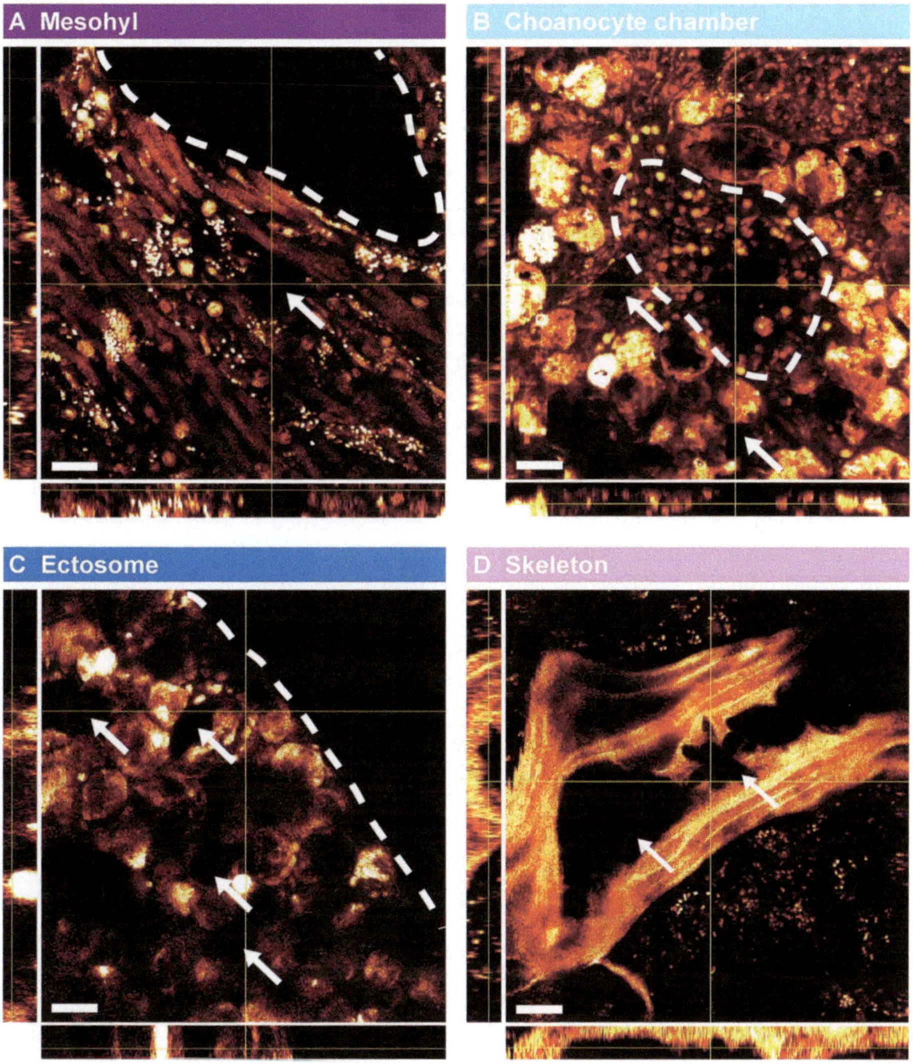

FIGURE 5.7 Two-photon images showing the embedded microplastic particles in different tissue sections of the sponge tissue [131].

Saliu et al. examined the feasibility of employing marine sponges to sample plastic particles and ultimately devised a comprehensive analytical strategy that leveraged the advantages of Py-GC-MS and microspectroscopy, in that order, to determine the masses of the smaller range found in the sponges as well as to identify and tally particles larger than 25 μm [134]. The research team determined the polymer that made up the particle using infrared microspectroscopy for particles bigger than 25 μm and mass spectrometry combined with pyrolysis for particles smaller than 25 μm. They found plastic contamination in 70% of the sponges, with an average of 1.2 particles/g tissue in the 25–150 μm size range and a maximum of 4.8 μg/g of samples containing plastic.

In 2021, Celis-Hernández and colleagues evaluated the amounts of MPs in three distinct environments and three species of marine sponges (*Amorphinopsis atlantica*, *Halichondria melanadocia*, and *Halclona implexiformis*) to determine if these organisms accumulated MPs and could serve as MP sources. MPs were present in each of the three compartments, and the mean concentrations varied as follows: in marine sponges, from 1,861 to 3,456 items/kilogram dry weight; in water, from 130–287 items/L; and in, sediment from 6 to 11 things/kilogram. Particularly in the disturbed research region, *A. atlantica* had the highest MP concentrations, measuring 5,000 items/kilogram of dry weight, 670 items/L in water, and 28 items/kilogram in sediment. MP buildup was present in all three species of sponges, with notable distinctions between pristine and disturbed areas. This suggests their potential use as bioindicators of MP pollution. However, the impact of MP ingestion on biological processes remains a research question yet to be thoroughly explored.

Marchi et al. aimed to evaluate the ingestion of PE MPs by *Petrosia ficiformis*, a Mediterranean sponge, through laboratory tests, specifically to determine how MPs influence key biological processes after varying exposure times (24 hours and 72 hours) and whether the sponges can subsequently eliminate the particles [135]. The findings showed that MP absorption increased with time, with sponge samples containing 30.6% of the supplied particles after 72 hours. Filtration and respiration rates were impaired by MPs, and these effects continued for 72 hours after the sponges were moved to uncontaminated water. The investigation showed that a significant determinant of MP toxicity in sponges is the length of exposure; additionally, marine sponges can absorb foreign particles, which makes them promising bioindicators for evaluating microplastic pollution.

5.5 MARINE CRUSTACEANS

5.5.1 Barnacles

Barnacles are members of class Hexanauplia, subclass Cirripedia, which includes the three infraclasses Thoracica, Acrothoracica, and Rhizocephala. Investigators have found 67 species of barnacles in Hong Kong out of the more than 1,000 globally known species globally. They inhabit a range of marine environments, including rough, exposed, and protected shorelines; rocks; mangrove tree roots; piers; and even floating marine debris. Because they are so resilient to the environment, barnacles are frequently utilized in pollution monitoring, especially for heavy metals [136]. Barnacles have been used in Hong Kong to monitor trace metals and polybrominated diphenyl ethers in coastal waters [157]. In a recent study, Xu et al. conducted studies on barnacle species collected from different sites in Hong Kong for active biomonitoring of MPs [114].

From July to September 2018, the study team searched 30 coasts in Hong Kong for goose barnacles (*Capitulum mitella*) and acorn barnacles (*Fistulobalanus albicostatus*, *Tetraclita japonica japonica*, *Amphibalanus amphitrite*). Hong Kong is a notable hotspot for MP pollution [102]. The riverine flow from the Pearl River, China's third-largest river and the main source of MP immigration, is situated west of Hong Kong. Furthermore, because Hong Kong's eastern and southern beaches border the South China Sea, the city collects marine debris and MPs from the open ocean [102, 138].

Xu et al. [114] chose *A. amphitrite*, *T. japonica japonica*, *F. albicostatus*, and *Capitulum mitella*—the four species of barnacles most frequently seen in Hong Kong—from 30 shorelines for this study, but *F. albicostatus* and *A. amphitrite* are usually found on mudflats or piers. The goal of choosing barnacle species from various habitat types is to give a more thorough picture of the MP pollution in Hong Kong waterways as seen through the eyes of the creatures themselves. After the MPs were removed from the tissues, the authors measured their amounts and compositions. MPs that ranged in number from 0 to 8.63 particles/g of wet weight were the most common type, which were fibers. There were eight different kinds of polymers found, with cellophane being the most common (58%) among them [114].

The results obtained by Davies et al. contradicted the applicability of acorn barnacle as an agent for biomonitoring MPs (microfibers in particular). *B. glandula* was gathered by the research team from four sites close to Bamfield, British Columbia, Canada [139]. For a whole day in unfiltered seawater, half of the barnacles were subjected to high dosage of PES microfibers. The GI tracts of the plastic treatment group had an average of 1.2 ± 1.9 fibers per barnacle before depuration, compared with 0.3 ± 0.6 fibers in the control group, according to measurements taken of feeding rates before and after exposure. Compared with 15% in the control condition, 50% of the barnacles in the plastic treatment devoured at least one microfiber before depuration. Following consumption of microfiber, there was no discernible short-term impact on feeding rate. When subjects were examined right after exposure and when they were undergoing depuration during the 48-hour post-plastic exposure depuration period, there was no difference in their egestion. These findings indicate that even in high concentrations, acorn barnacles only ingest a small number of microfibers, suggesting a low danger of MP contamination and possibly making them unsuitable as indicator species.

As sessile organisms that take up the particles from surrounding waterways, barnacles have the potential to function as bioindicators, providing information about the origin of MPs. Adult barnacles have been shown in lab studies to be able to consume MPs, but *Amphibalanus amphitrite nauplii* have been found to consume NPs [140]. At three field study sites in Thailand, Thushari et al. found 0.23–0.43 particles per gram of tissue in *Balanus amphitrite* [141]. Furthermore, 385 gooseneck barnacles (*Lepas spp.*) obtained from the North Pacific Subtropical Gyre showed evidence of plastics in their gastrointestinal systems, with PE making up the majority of the materials identified [142].

Scotti et al. [143] collected pelagic gooseneck barnacles from fixed moorings in the Capo Milazzo Marine Protected Area (MPA) as part of a recent study, and of the 120 specimens collected, 30% had fibers and fragments in their digestive tracts. Approximately 85.9% of the debris that was consumed was fibers, most of which were natural (11.7%) or synthetic (30.6%). A significant portion of the debris had lengths between 1 and 2 mm (33.3%), and transparent fibers were more common. The largest fiber concentrations were seen in barnacles from the western part of the MPA, which is more impacted by the input of urban effluent. These creatures' opportunistic diet was highlighted by the variety of fiber types, sizes, forms, and colors that they displayed, as well as the fact that several bivalve and ostracod shells were found inside their digestive systems [143]. Given their widespread distribution, abundance, and ease of sampling in macrofouling, we advise that this barnacle species be included in monitoring programs as a straightforward and reasonably priced way to evaluate the presence of MPs and microfibers in coastal marine waters and thereby monitor the ecological status of pelagic and coastal ecosystems.

5.5.2 Lobsters

Crustaceans from family Palinuridae are renowned as some of the most valuable seafood globally, often serving as a pivotal component of the export economy in certain regions [145]. Recent scientific studies have underscored the heightened MP vulnerability of certain decapod crustaceans with a strong connection to the seabed. Among these are *Palinurus elephas* (European spiny lobster) and *Nephrops norvegicus* (Norwegian langoustine). The latter has just recently been found to be exposed to MPs and NPs in the Aegean Sea, despite the former being well-known as a flagship species for MP pollution in EU waters. This highlights the urgent need for more comprehensive geographical data [146].

As a benthic decapod, *N. norvegicus* is found in temperate and cold waters throughout Europe. It is a member of family Nephropidae. Like the European spiny lobster, langoustine is a valuable commodity with a retail price that is on par with other crustaceans like lobster, spiny lobster, and deep-sea shrimp. With a broad bathymetric distribution and a maximum depth of 800 meters, langoustine is a major component of muddy bottom trophic webs. It is found mostly in the deep waters of the Mediterranean (>200 meters). Langoustines' constant scavenging on the seafloor exposes them to sediment-water fluxes and resuspended sediment, which may result in the inadvertent ingestion of MPs; it also allows them to interact with other benthic species. Consequently, langoustine has been identified as a reliable bioindicator for assessing MP contamination in the deep seabed [17, 147]. Previous findings have suggested using *N. norvegicus* as a species to assess MP contamination in the benthic environment [147, 148]. This decapod crustacean inhabits the Mediterranean regions and Northeast Atlantic, extending from Norway and Iceland in the northern reaches to Greece and Morocco in the southern areas [149].

The frequency of MP occurrence in benthic sediments and in *N. norvegicus* throughout six functional units in the Northeast Atlantic was assessed by researchers who looked at the relationship between organism's microplastic levels, biological characteristics, and surroundings [112]. The amounts, sizes, shapes, and polymer types found in *N. norvegicus* closely matched those in the environmental samples around it. The polymers were identified to be PP, PS, and PA. In contrast to other regional studies, the authors of this study found significantly lower MP abundance in the northeast Atlantic, suggesting that the MP contamination in *N. norvegicus* may have its origins locally. Larger organisms also showed reduced microplastic abundances, suggesting that *N. norvegicus* did not accumulate any particles.

Due to inherent limitations, *N. norvegicus* alone currently falls short of providing comprehensive quantitative data on microplastic contamination in marine environments [163]. Nevertheless, it can supplement well-known species like *Caretta caretta* and *Fulmarus gracilis* by acting as the invertebrate counterpart for detecting MP contamination. In addition, *N. norvegicus* occupies a crucial place in the ecological trophic chain at a mid-trophic level. Its consumption of MPs provides information about the makeup of litter in the benthic environment and can be used to predict the possible negative impacts of these newly developing pollutants on other benthic species. Thus, *N. norvegicus* can do more than find MPs, but it can also reveal details about the ecological effects MPs may have on other organisms and offer important information about possible consequences for human use.

Furthermore, unlike *N. norvegicus*, *P. elephas* is described as an omnivorous and scavenging species that inhabits shallow Mediterranean waters down to a depth of 200 meters [150]. MPs are primarily encountered by spiny lobsters and langoustines because of their same trophic habits. All information on particle occurrence in these species, with rare exceptions [112, 151], comes from their separation from stomach contents or from the digestion of the complete digestive system, highlighting their common exposure pathway [112, 152, 153]. The research team led by Kampouris et al. investigated the ingestion and accumulation of MPs and NPs in tissue samples from *P. elephas* [146]. Between the male and female spiny lobsters, the team removed a total of 4,102 plastic particles from their stomach tissues; of the 63 samples examined, only three of them contained no plastic particles. The authors found from 20 to 273 MP particles in the specimens' GI tracts, with 98% of those constituting fragments; they found NPs in 48.1% of gills and 22.6% of stomachs and found 43 different types of polymers in both tissues. The findings on the accumulation of MPs and NPs suggested the potential of *P. elephas* as a biomonitoring agent for these pollutants.

Other investigators conducted a comparative analysis of MP contamination in *N. norvegicus* and *P. elephas* and found 100% prevalence in both species' specimens, although *P. elephas* was significantly more contaminated and consumed larger MPs with a unique polymeric composition [154]. Both species' scavenging habits, primarily single-use plastic, could account for this exposure to MPs. The combined findings show that both species are impacted by plastic pollution.

5.5.3 Shrimp

Researchers have extensively documented MP and NP occurrence in shrimp [155]. In the Mediterranean Sea, Bordbar et al. presented the first evidence of nylon filament presence in the stomach of the economically significant *Plesionika narval* [156]. In the Persian Gulf, *Metapenaeus affinis* (white shrimp) had an average abundance of 1.02 items/g of MPs in digestive tissues [157]. Fibers and films were the most common MP morphologies that Keshavarzifard et al. found; the most common polymers were PET, PP, and PS. Hariharan et al. then investigated the consumption of weathered PE MPs by whiteleg shrimp (*Penaeus vannamei*) [158]. Crustacean feeding behaviors, which include scavenging, deposit feeding, suspension feeding, and predation, make them especially vulnerable to consuming MPs [159]. Shrimp, notably the brown shrimp *Crangon crangon* [160], may ingest MPs under the misperception they are food rather than accidentally.

The midgut gland of shrimp, which comprises fibrillary cells (F-cells), resorption cells (R-cells), and blister-like cells (B-cells), is part of the digestive system, which is separated into the foregut, midgut, and hindgut. Figure 5.8 is a schematic representation of the effects of MPs and NPs on shrimp and their route of entry via endocytosis. MPs, especially those smaller than 0.1 μm, can interfere with nutrient absorption, potentially causing oxidative stress, cytotoxic reactions, decreased enzyme activity,

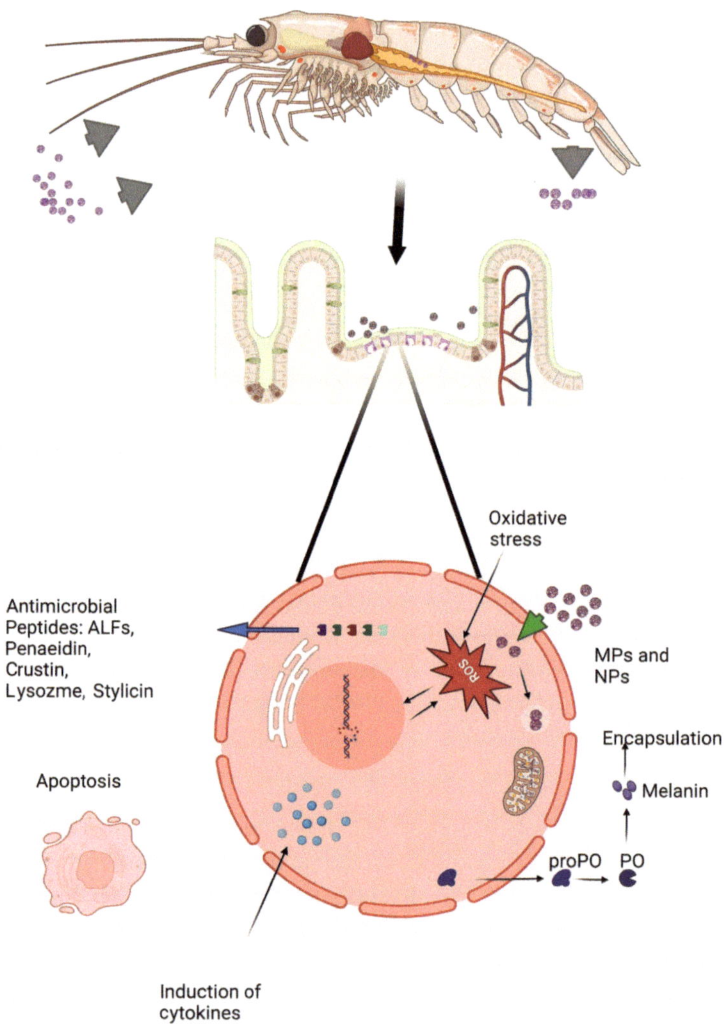

FIGURE 5.8 Schematic representation of the impacts of microplastics and nanoplastics on the cellular level in the gut of shrimp [155].

Note: ALFs = Antilipopolysaccharide Factors

and increased immunotoxicity. Shrimp survival rates, a key biomonitoring parameter for MPs, can be significantly affected [161–163].

The gills of shrimp are essential for respiration and water pollutant filtering, but they can be harmed by MPs and NPs. Toxic substances can be absorbed by particles from water, leading to decreased oxygen intake, physical blockages, and decreased filtration effectiveness. Stress, stunted growth, problems with reproduction, and even death can result [164]. MP accumulation in shrimp gills can result in altered oxygen consumption rates, ammonia excretion, and gene expression. Additionally, the ingestion of the particles has been reported to cause abnormalities in the gut and hepatopancreas, among other organs in shrimp [164–166].

The first line of defense against free radicals in shrimp involves antioxidant enzymes such as glutathione peroxidase, catalase, superoxide dismutase, and glutathione S-transferase. Exposure to NPs or MPs may cause oxidative damage, inhibiting the antioxidant defense system and leading to elevated H_2O_2, lipid peroxidation, and MDA levels. Chronic exposure to MPs results in reduced antioxidant enzyme activity [158, 159, 162].

Examining gene expression in shrimp offers a credible approach for biomonitoring MPs and NPs. Timilsina et al. described modulating various genes in diverse species of shrimp exposed to MPs and NPs [155]. Shrimp immune system modulation can result from either up- or down-regulation of certain genes, and plastic particle accumulation can trigger immunological response involving cellular processes such as encapsulation, phagocytosis, nodule formation, and apoptosis. Furthermore, different signaling pathways, such as the clotting cascade system, the prophenoloxidase system, and antimicrobial peptides, activate different humoral components. These pathways involve multiple genes that cooperate and operate separately to oppose outside stimuli.

Prophenoloxidase is a unique enzyme that is essential to *P. vannamei* pathogen clearance and wound healing. Specific genes like clap and cstb exhibit induction in the presence of NPs, leading to increased molting and the initiation of apoptotic pathways [167]. In fairy shrimp [168] and artemia [169], the stress imposed by PS NPs produces heat shock proteins, which are generated as a cellular response to stress. Gene expression, swimming behavior, and molting are all influenced by these proteins. Consequently, a thorough analysis and evaluation of alterations in diverse parameters could substantiate the viability of shrimp as effective biomonitoring organisms for MPs and NPs. Table 5.5 gives a detailed overview of species from phylum Arthropoda proposed for biomonitoring micro- and nanoplastics.

5.6 MARINE ZOOPLANKTON

In aquatic ecosystems and worldwide biogeochemical cycles, zooplankton—which includes a variety of filter-feeding creatures like cladocerans, jellyfish, copepods, tunicates, and krill—plays a critical role. When these organisms graze, they unintentionally swallow plastic particles, which could be harmful to their health and behavior. Zooplankton feed fish that are important to the economy, graze on primary and secondary producers, and help aquatic ecosystems recycle nutrients, transmit energy, and move materials [170]. According to studies conducted in laboratories,

TABLE 5.5
Arthropoda Species as MP and NP Biomonitoring Agents

S. No.	Species	Overview	Location	References
1	*Amphibalanus amphitrite*	*A. amphitrite* correlates positively with MP content in sediments. Thus, this species has high potential as an MP bioindicator.	Hong Kong	[114]
2	*Nephrops norvegicus*	This species is a bioindicator of environmental contaminants because of its availability and depth distribution among other factors. More MPs are found in this species than in any other Mediterranean species, and the MPs found in *N. norvegicus* mimic the surrounding environment, arguing for it as an MP bioindicator.	North East Atlantic west Sardinia, Italy	[112, 147, 154]
3	*Palinurus elephas*	The spiny lobster *P. elephas* shows even more MP contamination than the Norwegian lobster. The species is so sensitive to pollution, it was 100% contaminated, making *P. elephas* an excellent MP bioindicator.	West Sardinia, Italy	[154]
4	*Norwegian langoustine*	*N. langoustine* partially digests MPs, with digestion bottlenecked by its stomach size; larger particles are fragmented into smaller particles in the stomach. Due to their ability to absorb microplastics around them, they are considered good bioindicators.	–	[151]
5	*Aristeus antennatus*	In the stomachs of shrimp *A. antennatus*, 67% of the samples showed MP contamination, suggesting this economically important species as an MP bioindicator in the deep sea.	Sardinia Island	[147]
6	*Metapenaeus affinis*	*M. affinis* lives in an area under high anthropogenic pressure, and the shrimp showed 1.02 g MPs per gram of digestive tissue. Thus, the species could be a potential MP bioindicator in the marine environment.	Persian Gulf	[157]

zooplankton MP consumption can have negative consequences on their growth, fertility, and eating habits [16, 171]. Table 5.6 provides an overview of the various zooplankton species that could serve as biomonitors of MPs and NPs. The small size of MPs makes them accessible to a wide range of animals, including zooplankton, which may lead to bioaccumulation and biomagnification across food webs.

TABLE 5.6
Zooplankton Species as MP and NP Biomonitoring Agents

S. No.	Species	Overview	Location	References
1	39 zooplankton species from 28 taxonomic orders	Zooplankton are one of the most susceptible species to MP pollution, influenced largely by MP presence in the seawater. No major difference was noted between various zooplankton over various seasons.	Bohai Sea China	[183]
2	9 Zooplankton species from 28 taxonomic orders	Intentional or accidental zooplankton MP ingestion can introduce toxicity into the food chain. Crustacean zooplankton are at greater risk of NP contamination than their counterparts, suggesting zooplankton as appropriate bioindicators of MP contamination.	–	[183]
3	10 Zooplankton groups	Zooplankton are one of the most sensitive groups to MPs as their prey are the same size as microplastics. Bioaccumulation correlated negatively with number of zooplankton, which exhibited significant biological dilution.	East China Sea	[186]
4	Siphonophorae, Copepoda, Euphausiacea, Amphipoda	With increased MP distribution closer to coastal cities, MP distribution has become uneven. MP retention varies by taxon, but zooplankton are repositories of MP contamination in coastal water settings. MP retention in zooplankton can be used as an index for future ecological risk assessments.	Yellow Sea	[270]
5	28 taxonomic orders of zooplankton	Zooplankton represent an entry point for MPs to enter the food web because they ingest the MPs themselves and are in turn the food source for multiple other species. Some researchers found negative impacts of MPs on zooplankton, but others found no impact. Their indiscriminate feeding modes contribute to their MP ingestion, but not all zooplankton are noted to ingest MPs.	–	[184]

(Continued)

TABLE 5.6 *(Continued)*
Zooplankton Species as MP and NP Biomonitoring Agents

S. No.	Species	Overview	Location	References
6	*Themisto libellula, Themisto abyssorrum, Apherusa glacialis Calanus hyperboreus, Calanus glacialis/ finmarchicus*	Some of the highest levels of MP contamination have been reported in the Arctic straits. Here, the diverse ecosystem is supported by zooplankton at the base. The zooplankton preferred the smaller MPs to those present in the water samples. Increased global plastic consumption and global warming releasing ice-bound MPs are putting strain on this ecosystem.	Fram Strait in the Arctic	[176]
7	Zooplankton	MPs were identified in 61% of 152 samples, and concentrations were higher in densely populated areas. Plankton surveys worldwide can help monitor plastic particle presence in the ocean and are important tools for addressing marine litter under the Marine Strategy Framework Directive (2008/56/EC).	Portuguese coastal waters	[174]

Even though there are large global data sets on marine zooplankton biomass and abundance, more is needed to better utilize available zooplankton data at the policy level to preserve and manage marine ecosystem services [172, 173]. Developing MP contamination indicators based on zooplankton research can be essential in monitoring progress against plastic pollution, in line with recent initiatives by the UN Environment Assembly to form a global convention to combat plastic pollution. This novel strategy presents fresh angles and difficulties for the study of MPs. For the first time, Frias et al. suggested using plankton surveys from regular global monitoring campaigns to track plastic particles both at the water's surface and deep within the water column. The authors determine that this tactic is a vital and economical means of addressing marine litter [174]. Larger species' ingestion of zooplankton, a fundamental component of the food chain, enables MPs to move up the trophic scale [175].

The impacts of polymer kinds, forms, and sizes on zooplankton growth, reproduction, and survival—as well as possible consequences—have been studied in recent years, and utilizing zooplankton as a bioindicator for MPs and smaller NPs seems to be a promising approach, especially since common monitoring tools like nets and stereo microscopes are available. Indeed, global zooplankton sampling provides a wealth of data for databases and models including abundance, biomass, and DNA. Fish that eat particles larger than 300 μm have a wider variety of polymer kinds, forms, and colors than other creatures, which makes them useful MP bioindicators.

However, despite focusing on smaller MPs, crustacean zooplankton pose a greater risk of ingesting and exposure, including to NPs. A preponderance of particles between 6 and 50 μm was seen in recent studies evaluating MP consumption in different Arctic zooplankton species, which mirrored concurrent water samples. There was also a minor preference for particles smaller than 20 μm [176]. Figure 5.9 presents an overview of suggested qualities and benefits for zooplankton as a viable bioindicator, taking into account the presence and abundance of smaller MPs in water samples [173].

The epipelagic zone is a preferred feeding location for zooplanktonic organisms and is recognized for having a subsurface chlorophyll maxima at depths of 50–150 meters [177]. Prior analyses of MPs' vertical distribution have repeatedly found that the epipelagic zone has larger concentrations [178, 179], whereas concentrations in deeper waters decrease or show new peaks in the mesopelagic zone. These variations are influenced by physical, geographical, and seasonal factors [162, 180, 181], highlighting the heightened MP exposure of zooplankton, as indicated in prior studies and underscored in plastic research guidelines. We can increase the efficacy of zooplankton monitoring globally by combining cutting-edge methods like analyzing plastic particle ingestion in natural settings and MP zooplankton retention and excretion rates and combining ecotoxicological model studies with existing water quality assessments. Currently, the main advantages of using zooplankton as a tool for MP research include the abilities to construct toxicity models, assess ingestion in natural habitats, compile long-term statistics on abundance and biomass, and conduct ecotoxicology studies.

Refining the empirical relationship between biological conditions (such as zooplankton ingestion) and habitat quality (particularly, plastic particle pollution) within

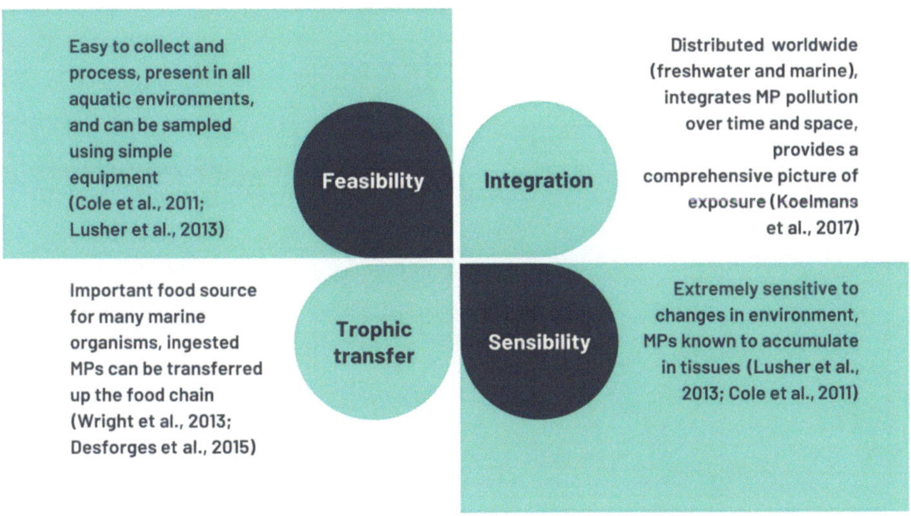

FIGURE 5.9 Key characteristics and benefits of using zooplankton creatures as a microplastics bioindicator [173].

a chosen region is necessary to improve the long-term study framework. Measuring MP ingestion by zooplankton in natural settings presents challenges due to the minute particle sizes and the intricate marine environment. Small plastic particle analysis techniques are still being debated, and there are a number of drawbacks to consider, such as sample contamination and the possibility of particle loss or damage during processing. Richon et al. discovered two primary characteristics that impact zooplankton consumption of MPs: high particle pollution in water and high grazing activity. According to recent research, plastic microparticles are concentrated at a depth of 50 meters, which corresponds with the zooplankton's preferred feeding zone [177, 182]. Figure 5.10 is an illustrative representation of a mathematical model describing how zooplankton species consume MPs in the wild. MPs sink via the species' ingestion and egestion. Although previous researchers have employed diverse techniques to verify that zooplanktonic organisms have consumed MPs [183, 184, 176, 185], but the scarcity of these studies emphasizes the necessity for additional thorough inquiry.

In a recent study on zooplankton consumption in the Arctic that included information on MPs in seawater, the researchers found some interesting trends [176]. Ingestion rates of sea ice-associated species were higher (>1 particle/individual) than those of copepods (0.2–0.01 particles/individual). Furthermore, the consumed MPs were mostly fragments of different polymer kinds and less than 50 μm in size. The researchers noted that the particles that zooplankton consumed were marginally smaller than those discovered in water samples, indicating that zooplankton exhibited some degree of selectivity. There is a wide range of recorded MP ingestion levels in natural environments: 0.13–0.35 particles/individual [186]; 0.056–0.117 particles/individual [183]; 0.01–1.8 particles/individual [62, 176]. Moreover, a scarcity of research has demonstrated a relationship between MPs absorbed by zooplankton

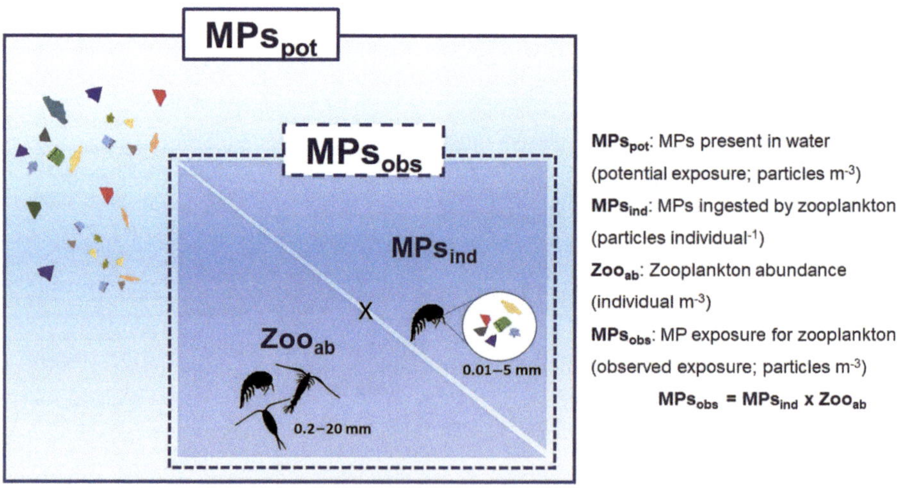

FIGURE 5.10 A mathematical model to describe how zooplankton consume microplastics in the wild [173].

communities and those found in saltwater, leading to knowledge that is geographically constrained and difficulties in drawing comparisons because of methodological differences.

The Arctic's Fram Strait is home to some of the highest MP concentrations seen in marine ecosystems. Botterell et al. investigated MP ingestion in various species of Arctic zooplankton, gathering samples during research cruises in the boreal summers of 2018 and 2019 using Bongo and MOCNESS nets [176]. The five Arctic zooplankton species they investigated had all consumed MPs; most of the MPs that were consumed were less than 50 µm, were made up only of pieces, and had different kinds of polymers. The MPs that the zooplankton swallowed were smaller than those that were discovered in water samples, suggesting that the zooplankton was using a selective mechanism. The authors linked adverse impacts on zooplankton growth, development, and fertility to high MP ingestion.

Zheng et al. used a unique methodology to investigate the features of MPs in copepods over four different seasons in Jiaozhou Bay [183]. The authors investigated shape, abundance, size, and chemical composition and identified significant seasonal differences in MP/copepod values. Ingested MP sizes in copepods ranged from 90 to 2485 µm, with fibers constituting the majority among all particle shapes ingested. The primary components were PES and cellophane. Results obtained from the study provide crucial insights into the parameters and properties of MPs in copepod species, thereby forming a critical foundation for future ecological risk assessments.

Sun et al. looked at the traits and bioaccumulation of MPs in East China Sea zooplankton and found three different forms of MPs: fibers, pellets, and pieces, with fibers the most common [186]. The fragments, fibers and pellets exhibited average lengths of 20.3 µm, 295.2 µm, and 82.4 µm, respectively. The bioaccumulated concentration of MPs in the zooplankton varied among the species, with the highest concentration in Copepoda being 0.13 pieces/zooplankton and that in Pteropoda being 0.35 pieces/zooplankton. Sun et al. found 19 polymers in the zooplankton, but the accumulated MP concentration showed a negative correlation with zooplankton abundance, indicating a significant biological dilution effect.

5.7 BIRDS, AMPHIBIANS, REPTILES, AND MAMMALS

There has been little research discussion on MP presence and frequency in marine top predators like cetaceans [187, 148]. Table 5.7 gives an overview of different large predator species that have been studied as microplastic/nanoplastic biomonitoring agents.

Cetaceans are often regarded as ideal sentinel species in oceanic ecosystems due to their high trophic level, long longevity, and charming demeanor that tends to evoke public attention and community action. Given the abundance of plastics in the world's seas and the heightened risk of accumulation that these apex predators confront, researchers have focused attention on the interactions between plastics and cetaceans [188–191]. The impact of macro-debris on cetaceans, including fishing gear, plastic bags, and sheets, has been extensively documented and discussed cetaceans [192–194].

The relationship between cetaceans and MPs is, however, poorly understood. Occasionally found in coastal and estuary areas, *Sousa chinensis*, often known as

TABLE 5.7
Amphibian, Aves, Reptile, and Mammal Species as MP and NP Biomonitoring Agents

S. No.	Species	Overview	Location	References
1	*Ootarids* *Otaria byronia* *Arctocephalus phillippii* *Arctocephalus australis* *A. australis un-named*	*Pinniped* are sentinel species for ocean health. Scat samples were processed from six rookeries that displayed considerable variation in MP contamination influenced by the movement of currents.	Southeastern Pacific Ocean Peru and Chile coastline	[203, 272]
2	*Caretta caretta*	Sea turtles such as *Caretta caretta* are very sensitive to the negative effects of MP contamination. Dead specimens were necropsied, and over 60% had plastic contamination in their digestive tracts. Plastic ingestion by the species indicates it as a bioindicator for MP contamination.	Balearic Islands	[222]
3	*Sousa chinensis*	Low-trophic-level organisms such as cetaceans also experience high MP contamination. A sample of stranded Indo-Pacific humpback dolphins' stomach were observed sampled from both urban and rural areas. Heavy MP contamination was observed, but there was less in *S. chinensis* from rural areas. This species can be used as an endpoint biomonitoring species for MP contamination.	Pearl River Estuary, China	[195]
4	Mediterranean monk seals	The feces were collected and analyzed, and even top-level predators such as Monk seals are not left untouched by MP pollution. Seal feces can be used to identify plastic additives and MPs.	Western coast of Zakynthos Island, Ionian Sea, Greece	[213]
5	*Stenella coeruleoalba*	Western Mediterranean striped dolphins' intestinal contents were observed and analysed over 3 decades, and more than 90% showed MP contamination; overall MP content was low. Striped dolphins' use as a MP bioindicator is limited due to the difficulties in sampling and processing time.	Western Mediterranean Sea	[201]
6	*Sterna hirundo* and *Sterna dougallii*	MP load in common terns was similar across all life stages, and there were no significant regional differences. In contrast, roseate terns showed significantly more MP ingestion that varied interspecifically as well as geographically. Roseate terns also showed little difference by life stage, which suggests dietary differences and also suggests terns as a potential MP contamination bioindicator.	Northwestern Atlantic	[219]

7	*Fulmarus glacialis*	The norther fulmar ingests more MPs than other sea birds due to its gut morphology and feeding habits. Analysis of the gizzards and the proventricular contents revealed that more than 90% of the birds had ingested at least one piece of plastic. Since the fulmar has faced MP contamination fordecades, it is a suitable candidate for MP biomonitoring in the Arctic.	North Sea region	[197]
8	*Spheniscus magellanicus*	This is the first case where stranded juvenile Magellan penguins were studied for debris pollution. Microparticles were observed in 100% of the studied species and more than 60% were microplastics. Stranded juvenile Magellan penguins are promising bioindicators for microplastic pollution.	West south Atlantic coasts	[220]

the Indo-Pacific humpback dolphin, is a species of cetacean that spans the Indian and Pacific Oceans. Researchers examined the stomachs of 12 dolphins that had become stranded close to China's Pearl River Estuary for MP pollution, and all samples contained MPs, in quantities that ranged greatly, from 11 to 145 items/individual [25, 195]. PP and PE fibers were the most common MPs; their sizes typically ranged from 1 to 5 mm, and their prevalent hues were clear or white. The study authors suggested cetacean species as potential contributors to microplastic food chains, making them potential biomonitoring species in marine and estuarine ecosystems.

In a different study, between 2010 and 2013, seabirds that were abandoned along the Brazilian shore had plastic garbage lodged in their digestive systems, and there was more debris in birds feeding in intermediate (3–6 m) and deep (20–100 m) seas than in birds feeding at the surface (less than 2 m) [196], suggesting that evaluating the debris presence in organisms that mostly feed at the ocean's surface provides a limited perspective of the risks that this kind of pollution presents to marine life. There have been reports of northern fulmars (*Fulmarus glacialis*), consuming plastic recently [197]. With an average of ten pieces, 91% of the birds had consumed at least one piece of plastic, most often hard fragments and PE. The most common species of dolphin in the seas of the western Mediterranean is *Stenella coeruleoalba* (white-striped dolphin), which can be found in all oceans' temperate and subtropical regions.

A discernible shift toward more neritic species, such as *M. merluccius* (juvenile hake) and *Illex coindetii* (Southern shortfin squid), has been observed in recent decades, despite Mediterranean *S. coeruleoalba*'s traditional preference for oceanic cephalopod species in their diet [198, 199]. Like other marine animals, these dolphins consume MPs by inhaling at the air–water interface, by trophic transfer, or by direct intake from the water column [148, 200]. Striped dolphins are frequently exposed to MPs, primarily in the form of fibers. Nonetheless, in comparison with their body size, the amount of MPs in their bodies seems to be minimal. As a result, the main issue is not the sheer amount of MPs physically present in dolphins' digestive tracts but rather the chemicals that may hurt the health and population of these animals.

Novillo et al. examined the existence of MPs in the digestive tracts of striped dolphins from the western Mediterranean during a three-decade period and found MPs in 90.5% of the dolphins [201]. Of these MPs, primary pellets made up fragmented MPs (23.87%) and fibers (73.6%). The number of MPs per dolphin was quite low and showed significant variability despite the high occurrence. Due to sample challenges and lengthy processing times, the use of striped dolphins as markers of microplastic abundance in the pelagic biota at sea in the western Mediterranean is limited. However, keeping an eye on them is essential to determining the risks to their community.

Plastic waste in particular has become a major anthropogenic hazard to marine creatures in recent years. By addressing factors including the effect of marine litter on biota, the Marine Strategy Framework Directive seeks to improve the environmental quality of European seas. Only lately have studies been carried out in European waters to determine the frequency of marine mammal absorption of litter, mostly by analyzing the digestive tracts of dead and stranded marine mammals [193, 202, 203].

Whales, among mammals, are often considered as biomonitoring species. The North Pacific Gyre (NPG), which is close to East Asia, is home to some baleen whale species, such as fin whales, that use it as a nursery, foraging ground, and route for migration. Several research investigations have documented cases of baleen whale stranding and bycatch in this area [204, 205], and concerns have been expressed for marine species that use the NPG as a core habitat or for migration due to the number of plastics that drift to the surface [206].

Because of plastic pollution, baleen whale species in the NPG are thought to be potentially endangered [207–209]. Few researchers have documented fin whales in the area consuming plastic debris, despite the sharp rise in plastic pollution [210], and in this regard, a recent study on fin whale (*Balaenoptera physalus*) MP consumption is notable; the investigation included a necropsy on the carcass of a young fin whale that revealed 45 human artifacts in the digestive tract, one of which was a fishing line (118 cm long and 1.15 mm thick) that had been discovered in the stomach [211]. During the necropsy, more fishing lines, fragments of Styrofoam, and green filaments were extracted from the small intestine.

Using a three-pronged approach, researchers promoted the use of fecal matter as a comprehensive method for evaluating the occurrence and effects of plastic intake [148], and Hernandez-Milian et al. monitored fecal matter to examine the frequency of MPs consumed by *M. monachus* (Mediterranean monk seal), a top predator in Greek waters [213]. Fecal matter examination offers valuable insights into trends in MP presence over time, and Hernandez-Milian et al. chose this noninvasive technique because the species is in conservation status. As one of the top predators in the Mediterranean Sea and its environs, including western Greece, the Mediterranean monk seal plays a vital role. However, studies using the digestive systems of dead animals are complicated by the small population of seals in the eastern Mediterranean.

Since MPs are not as frequently retained along the GI tract of monk seals as they are in other marine vertebrates, investigators assumed that the MPs found in the monk seal feces came from their prey [214, 215]. The researchers examined the possibility of plastic additives such phthalates and biomarkers of pollutant exposure like porphyrins in monk seal excrement for the first time, and in a different study, researchers studied MP prevalence in pinniped feces in the northern Adriatic and Chilean seas [216]. However, they found much more MP content in the monk seal feces from the Zakynthos caves: 100% of the fecal samples from monk seals contained MPs and plastic additives (phthalates), demonstrating a link between the two.

The prevalence of microlitter in the digestive systems of different marine creatures has been the subject of much research conducted over the past ten years; most findings imply that microlitter is expelled after going through the digestive system in its entirety, which could be the case for larger microliter that cannot transport through cell membranes [217, 218]. For very small litter (like nanolitter) or additives from ingested polymers that may be released during digestion, this might not be the case. For example, 87.5% of fecal samples from *Sterna hirundo* (common terns) and *S. dougallii* (roseate terns) nesting in the northwestern Atlantic had MP fiber loads higher than the average load observed in controls, according to an examination of their fecal matter [219]. Analyses of plastic debris found in the GI tracts of young Magellan penguins left stranded on the coastlines of the west and south Atlantic revealed that 100% of the birds under study had anthropogenic macro-, meso-, and microparticles, with an average debris abundance of 33–200 items/bird; 97% of the microparticles, or 9% of the total waste, were fibers, the most common of which were PP (27.8%) and PE (21.6% [220]. The study highlighted Magellan penguins as potentially useful bioindicators of South Atlantic plastic contamination.

The loggerhead sea turtle, *C. caretta*, stands out as the most common turtle in the Mediterranean Sea. Reptiles have been focal points in research as sentinel species for MP biomonitoring [221]. Given its wide range of habitat preferences, feeding habits, migratory patterns, and susceptibility to plastic ingestion, *C. caretta* is ideally positioned to serve as a sentinel species for the global assessment of plastic pollution in the environment. Its extensive distribution across subtropical and temperate regions further contribute to this advantage.

Solomando et al. investigated plastic ingestion by conducting necropsies on over 45 stranded *C. caretta* loggerhead sea turtles in the Balearic Islands and examining the contents of their digestive tracts for plastic debris [222]. The authors found that 61% of the turtles had plastic things in them, with an average of 12 items per turtle. The scientists also observed litter in the feces pellets of 67 living people and found plastic elements in 46 of them (an average of 9.7 plastic elements per person). The primary elements ingested included plastic sheets, prevalent in 65.3% of the analyzed turtles. High-density PE (42.3%) and PP (33.8%) constituted the major types of plastic. The significance of *C. caretta* as a bioindicator organism for marine pollution was highlighted in this study, which was the first evidence of substantial plastic debris in loggerhead turtles in the Balearic Islands [222].

In addition to sea turtles, toads and tadpoles have been identified as potential sentinel species for MP and NP biomonitoring. In one study, *Xenopus tropicalis* tadpoles absorbed, accumulated, and eliminated PS microspheres [223]. Within one hour of exposure, microspheres were found in the tadpoles' gills and digestive tract; six hours later, they were found in their feces. Tadpoles accumulated microspheres in a concentration-dependent manner, but after a day following clearance, the number of microspheres significantly dropped because of their filter-feeding behavior. According to the research, aquatic vertebrate species consume more MPs if the number of MPs in the environment is rising while food supplies are getting smaller.

Separately, Cruz-Santiago et al. conducted a study that documented the effectiveness of using nondestructive biomarkers in *Rhinella marina* (giant toads) in assessing the presence of cell lipid peroxidation, enzyme activity, and death (apoptosis) [224]. Glutathione S-transferase activity and the proportion of erythrocyte apoptosis were higher, although delta-aminolevulinic dehydratase activity was decreased. Toad biomarkers showed variations in industrial and rural zones, and correlations with responses changed, indicating that this species of toad might be used as a biomonitor for pollution from MP and NP particles.

5.8 MARINE BIOINDICATORS FOR NP RISK ASSESSMENTS

The presence of NPs in the environment, including particles below 1 μm, remains unknown, posing challenges for extraction and quantification from complex biological matrices [225–227]. Their low concentrations and limited research on real environmental samples impede detection [228–230]. Utilizing marine model species aligned with regulatory standards is common, but lab-based models may not reflect the natural world complexities, particularly concerning NPs [231, 232]. A comprehensive four-tiered method has been proposed to select bioindicators, considering specific natural resources, ecosystem health, and human well-being ("One Health"). Target species, key species, those most at risk, and economically important species play distinct roles in understanding NPs 'impact on marine habitats [231, 232] (Figure 5.11).

Employing ascidians (sea squirts) as a bioindicator offers a practical approach to monitoring NP pollution. Ascideans, with their stationary nature and efficient particle retention, provide a more accurate representation of NP interactions. Researchers have often focused on phyto- and zooplankton, with limited research on benthic flora and endobenthic fauna [233]. Toxicity tests on embryonic and larval stages of ascidians, microcrustaceans, bivalves, and sea urchins have shown sensitivity to NPs, impacting recruitment and survival predictions. Effects range from altered behavior and growth delay to deformities in sea urchins and ascidian larvae [234–241]. NPs also affect immune responses in bivalves and sea urchins [242–244].

Seagrass and marine vascular plants are crucial bioindicators, yet research on NPs impacts on terrestrial plants is lacking [245–247]. Coral bleaching studies emphasize reduced photosynthetic capacity in symbiotic dinoflagellates due to NPs exposure [248, 249]. In polar marine ecosystems, sea ice interfaces with NPs affect phytoplankton populations [250–252].

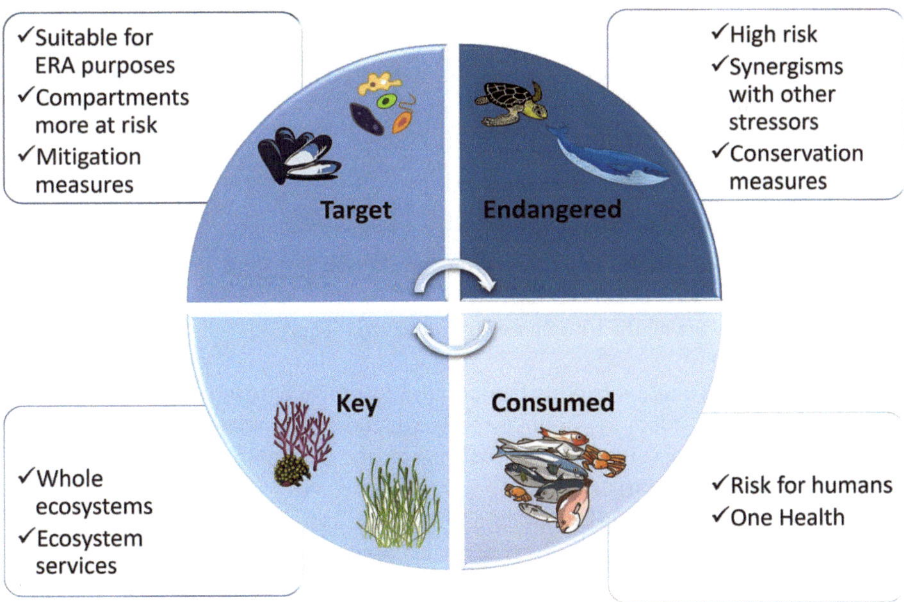

FIGURE 5.11 A suggested framework for choosing nanoplastic bioindicators in marine environments: an analysis of their relationships and significance [232].

Antarctic krill exposed to NPs exhibit physiological changes, altered swimming ability, and modified exuviae discharge. Fecal pellets containing NPs impact krill settling, influencing particulate organic carbon transport in the Southern Ocean [253, 254]. Pelagic ecosystems in the Arctic and Antarctic, including copepods, amphipods, and *Salpa thompsoni*, may be crucial in climate change scenarios [255]. Algal species, diatoms, krill, and lower-trophic-level organisms can serve as essential biomonitoring agents for NPs, emphasizing the need for further research [255].

5.9 CONCLUSION AND PERSPECTIVES

The escalating plastic pollution crisis has heightened the concentration of micro- and nanosized plastic particles, significantly impacting ecosystems. To address environmental health concerns, bioindicators, species sensitive to pollutants and used for biome monitoring, have gained prominence. These indicators help identify contamination levels when traditional testing is impractical. In marine environments, various organisms, including bivalves (oysters, mussels, scallops), fish, arthropods, and sponges, serve as bioindicators.

Bivalves, particularly filter feeders, are crucial for reflecting MPs in their surroundings. Recent studies, however, raise questions about their reliability due to differential MP ingestion and egestion. Fish, extensively researched bioindicators, show microplastic presence in gills and the GI tract, with benthic species exhibiting higher contamination. Economic fish species consumed by humans also display significant MP contamination. Arthropods and marine sponges, known for nonselective particle uptake, are considered excellent bioindicators. Zooplankton as biomarkers introduce MP toxicity into the food chain. Challenges include standardizing microplastic extraction processes, adopting a multispecies approach, and addressing contamination in remote areas like Antarctica, necessitating regulatory measures against plastic pollution and sustainable practices.

5.9.1 Perspectives

As research on bioindicators in marine environment on plastic pollution continues, we must direct our attention to assessing the knowledge gaps and lacunae that are present in the current scenario. The perspective aims to highlight these gaps and propose potential areas for future research and investigation.

- The existing studies lack a standardized response across various bioindicators, which hinders the comparability of studies. Standardized bioindicator responses concerning different MP types and concentrations will facilitate cross-study comparison.
- Limited validation of bioindicators for MP exposure is another issue. Various species are proposed as potential bioindicators, but only a few studies have been conducted to confirm if the species are viable. Validating bioindicators will ensure their applicability for environmental pollution testing.
- Bioindicator studies are usually species specific. Examining the collective multispecies impacts of MPs in an ecosystem will help elucidate the effects of MPs in that ecosystem.

- It is imperative to conduct longitudinal studies to understand the impact of plastic pollution concerning bioaccumulation and transfer between trophic level because of insufficient knowledge of long-term cumulative MP effects on marine organisms.
- Studies on species such as dolphins are not easily replicable due to the nature of the bioindicator organism. This inherent inaccessibility and difficulty replicating studies limits the potential of further studying certain fauna as bioindicators for environmental pollution assessment and highlights the critical need of proposed bioindicators that are accessible.

5.10 CREDIT AUTHORSHIP CONTRIBUTION STATEMENT

Devlina Das Pramanik: conceptualization, supervision, collection of resources, formal analyses, writing—original draft, preparation, writing—review and editing. Roudrashwa Nag: collection of resources, formal analyses, manuscript editing.

5.11 DECLARATION OF COMPETING INTEREST

The authors declare that they have no known competing financial interests or personal relationships that could have appeared to influence the work reported in this paper.

ACKNOWLEDGMENTS

The authors are grateful to the guest editors for providing the opportunity to contribute to the book chapter. The authors would like to acknowledge Amity Institute of Biotechnology at Amity University for providing the necessary resources toward our successful completion of the book chapter.

REFERENCES

1. Gall, S. and Thompson, R.C., 2015. The impact of debris on marine life. *Marine Pollution Bulletin*, 92 (1–2), 170–179.
2. Lu, Y., Zhang, Y., Deng, Y., Jiang, W., Zhao, Y., Geng, J., Ding, L., and Ren, H., 2016. Uptake and accumulation of polystyrene microplastics in zebrafish (Danio rerio) and toxic effects in liver. *Environmental Science & Technology*, 50 (7), 4054–4060.
3. Dawson, A.L., Huston, W.M., Kawaguchi, S., King, C.K., Cropp, R.A., Wild, S.J., Eisenmann, P., Townsend, K.A., and Nash, S.B., 2018. Uptake and depuration kinetics influence microplastic bioaccumulation and toxicity in Antarctic krill (Euphausia superba). *Environmental Science & Technology*, 52 (5), 3195–3201.
4. Carpenter, E. and Smith, K.L., 1972. Plastics on the Sargasso Sea surface. *Science*, 175 (4027), 1240–1241.
5. Eriksen, M., Lebreton, L., Carson, H.S., Thiel, M., Moore, C.J., Borerro, J.C., Galgani, F., Ryan, P.G., and Reisser, J., 2014. Plastic pollution in the world's Oceans: More than 5 trillion plastic pieces weighing over 250,000 tons afloat at sea. *PLOS One*, 9 (12), e111913.
6. Barletta, M., Lima, A.R.A., and Costa, M.F., 2019. Distribution, sources and consequences of nutrients, persistent organic pollutants, metals and microplastics in South American estuaries. *Science of the Total Environment*, 651, 1199–1218.

7. Brown, E., MacDonald, A., Allen, S., and Allen, D., 2023. The potential for a plastic recycling facility to release microplastic pollution and possible filtration remediation effectiveness. *Journal of Hazardous Materials Advances*, 10, 100309.
8. Arthur, C., Baker, J.E., and Bamford, H.A., 2009. *Proceedings of the International Research Workshop on the Occurrence, Effects, and Fate of Microplastic Marine Debris, September 9–11, 2008, University of Washington Tacoma, Tacoma, WA, USA*. National Oceanic and Atmospheric Administration (NOS-OR&R-30).
9. Van Sebille, E., Wilcox, C., Lebreton, L., Maximenko, N., Hardesty, B.D., Van Franeker, J.A., Eriksen, M., Siegel, D.A., Galgani, F., and Law, K.L., 2015. A global inventory of small floating plastic debris. *Environmental Research Letters*, 10 (12), 124006.
10. Yang, D., Shi, H., Lan, L., Li, J., Jabeen, K., and Kolandhasamy, P., 2015. Microplastic pollution in table salts from China. *Environmental Science & Technology*, 49 (22), 13622–13627.
11. Wesch, C., Bredimus, K., Paulus, M.P., and Klein, R., 2016. Towards the suitable monitoring of ingestion of microplastics by marine biota: A review. *Environmental Pollution*, 218, 1200–1208.
12. Bucci, K., Túlio, M., and Rochman, C.M., 2020. What is known and unknown about the effects of plastic pollution: A meta-analysis and systematic review. *Ecological Applications*, 30 (2).
13. Ehlers, S.M., Maxein, J., and Koop, J.H.E., 2020. Low-cost microplastic visualization in feeding experiments using an ultraviolet light-emitting flashlight. *Ecological Research*, 35 (1), 265–273.
14. Zhou, Q., Zhang, J., Fu, J., Shi, J., and Jiang, G., 2008. Biomonitoring: An appealing tool for assessment of metal pollution in the aquatic ecosystem. *Analytica Chimica Acta*, 606 (2), 135–150.
15. GESAMP, 2019. *Guidelines for the Monitoring and Assessment of Plastic Litter and Microplastics in the Ocean*. GESAMP Reports and Studies 99. GESAMP Joint Group of Experts on the Scientific Aspects of Marine Environmental Protection.
16. Cole, M., Lindeque, P.K., Fileman, E.S., Halsband, C., Goodhead, R.M., Moger, J., and Galloway, T.S., 2013. Microplastic ingestion by zooplankton. *Environmental Science & Technology*, 47 (12), 6646–6655.
17. Carreras-Colom, E., Constenla, M., Soler-Membrives, A., Cartes, J.E., Baeza, M., Padrós, F., and Carrassón, M., 2018. Spatial occurrence and effects of microplastic ingestion on the deep-water shrimp Aristeus antennatus. *Marine Pollution Bulletin*, 133, 44–52.
18. Li, H.-X., Shi, M., Tian, F., Lin, L., Liu, S., Hou, R., Peng, J., and Xu, X., 2022. Microplastics contamination in bivalves from the Daya Bay: Species variability and spatio-temporal distribution and human health risks. *Science of the Total Environment*, 841, 156749.
19. Neves, D., Sobral, P., Ferreira, J.L., and Pereira, T.R., 2015. Ingestion of microplastics by commercial fish off the Portuguese coast. *Marine Pollution Bulletin*, 101 (1), 119–126.
20. Limonta, G., Mancia, A., Benkhalqui, A., Bertolucci, C., Abelli, L., Fossi, M.C., and Panti, C., 2019. Microplastics induce transcriptional changes, immune response and behavioral alterations in adult zebrafish. *Scientific Reports*, 9 (1).
21. Wang, W., Ge, J., and Yu, X., 2020. Bioavailability and toxicity of microplastics to fish species: A review. *Ecotoxicology and Environmental Safety*, 189, 109913.
22. Tsangaris, C., Digka, N., Valente, T., Aguilar, À., Borrell, A., De Lucia, G.A., Gambaiani, D., Garcia-Garin, O., Kaberi, H., Martin, J., Mauriño, E., Miaud, C., Palazzo, L., Pérez-del-Olmo, A., Raga, J.A., Sbrana, A., Silvestri, C., Skylaki, E., Vighi, M., Wongdontree, P., and Matiddi, M., 2020. Using Boops boops (osteichthyes) to assess microplastic ingestion in the Mediterranean Sea. *Marine Pollution Bulletin*, 158, 111397.

23. Miller, E., Sedlak, M., Lin, D., Box, C., Holleman, C., Rochman, C.M., and Sutton, R., January 2021. Recommended best practices for collecting, analyzing, and reporting microplastics in environmental media: Lessons learned from comprehensive monitoring of San Francisco Bay. *Journal of Hazardous Materials*, 409, 124770. https://doi.org/10.1016/j.jhazmat.2020.124770.
24. Justino, A.K.S., Lenoble, V., Pelage, L., Ferreira, G.V.B., Passarone, R., Frédou, T., and Lucena-Frédou, F., 2021. Microplastic contamination in tropical fishes: An assessment of different feeding habits. *Regional Studies in Marine Science*, 45, 101857.
25. Zhang, F., Wang, X., Xu, J., Zhu, L., Peng, G., Xu, P., and Li, D., 2019. Food-web transfer of microplastics between wild caught fish and crustaceans in East China Sea. *Marine Pollution Bulletin*, 146, 173–182.
26. Barboza, L.G.A., Lopes, C., Oliveira, P., Bessa, F., Otero, V., Henriques, B., Raimundo, J., Caetano, M., Vale, C., and Guilhermino, L., 2020. Microplastics in wild fish from North East Atlantic Ocean and its potential for causing neurotoxic effects, lipid oxidative damage, and human health risks associated with ingestion exposure. *Science of the Total Environment*, 717, 134625.
27. Kılıç, E. and Yücel, N., 2022. Microplastic occurrence in the gastrointestinal tract and gill of bioindicator fish species in the northeastern Mediterranean. *Marine Pollution Bulletin*, 177, 113556.
28. Alomar, C., Sureda, A., Capó, X., Guijarro, B., Tejada, S., and Deudero, S., 2017. Microplastic ingestion by Mullus surmuletus Linnaeus, 1758 fish and its potential for causing oxidative stress. *Environmental Research*, 159, 135–142.
29. Güven, O., Gökdağ, K., Jovanović, B., and Kıdeyş, A.E., 2017. Microplastic litter composition of the Turkish territorial waters of the Mediterranean Sea, and its occurrence in the gastrointestinal tract of fish. *Environmental Pollution*, 223, 286–294.
30. Gündoğdu, S., Çevik, C., and Ataş, N.T., 2020. Occurrence of microplastics in the gastrointestinal tracts of some edible fish species along the Turkish coast. *Turkish Journal of Zoology*, 44 (4), 312–323.
31. Cheung, L.T.O., Lui, C.Y., and Fok, L., 2018. Microplastic contamination of wild and captive flathead grey mullet (Mugil cephalus). *International Journal of Environmental Research and Public Health*, 15 (4), 597.
32. Borges-Ramírez, M.M., Mendoza-Franco, E.F., Escalona-Segura, G., and Osten, J.R.-V., 2020. Plastic density as a key factor in the presence of microplastic in the gastrointestinal tract of commercial fishes from Campeche Bay, Mexico. *Environmental Pollution*, 267, 115659.
33. Halstead, J.E., Smith, J.A., Carter, E.A., Lay, P.A., and Johnston, E.L., 2018. Assessment tools for microplastics and natural fibres ingested by fish in an urbanised estuary. *Environmental Pollution*, 234, 552–561.
34. Naidoo, T., Smit, A.J., and Glassom, D., 2016. Plastic ingestion by estuarine mullet-Mugil cephalus(Mugilidae) in an urban harbour, KwaZulu-Natal, South Africa. *African Journal of Marine Science*, 38 (1), 145–149.
35. Jabeen, K., Su, L., Li, J., Yang, D., Tong, C., Mu, J., and Shi, H., 2017. Microplastics and mesoplastics in fish from coastal and fresh waters of China. *Environmental Pollution*, 221, 141–149.
36. Roch, S., Friedrich, C., and Brinker, A., 2020. Uptake routes of microplastics in fishes: practical and theoretical approaches to test existing theories. *Scientific Reports*, 10 (1).
37. Costa, E., Piazza, V., Lavorano, S., Faimali, M., Garaventa, F., and Gambardella, C., 2020. Trophic transfer of microplastics from copepods to jellyfish in the marine environment. *Frontiers in Environmental Science*, 8.
38. Da Costa Araújo, A.P. and Malafaia, G., 2021. Microplastic ingestion induces behavioral disorders in mice: A preliminary study on the trophic transfer effects via tadpoles and fish. *Journal of Hazardous Materials*, 401, 123263.

39. Abbasi, S., Soltani, N., Keshavarzi, B., Moore, F., Turner, A., and Hassanaghaei, M., 2018. Microplastics in different tissues of fish and prawn from the Musa Estuary, Persian Gulf. *Chemosphere*, 205, 80–87.
40. Capillo, G., Savoca, S., Panarello, G., Mancuso, M., Branca, C., Romano, V., D'Angelo, G., Bottari, T., and Spanò, N., 2020. Quali-quantitative analysis of plastics and synthetic microfibers found in demersal species from Southern Tyrrhenian Sea (Central Mediterranean). *Marine Pollution Bulletin*, 150, 110596.
41. Koongolla, J.B., Lin, L., Pan, Y.-F., Yang, C., Sun, D., Liu, S., Xu, X., Maharana, D., Huang, J., and Li, H., 2020. Occurrence of microplastics in gastrointestinal tracts and gills of fish from Beibu Gulf, South China Sea. *Environmental Pollution*, 258, 113734.
42. Atamanalp, M., Köktürk, M., Uçar, A., Duyar, H.A., Özdemir, S., Parlak, V., Esenbuğa, N., and Alak, G., 2021. Microplastics in tissues (brain, gill, muscle and gastrointestinal) of Mullus barbatus and Alosa immaculata. *Archives of Environmental Contamination and Toxicology*, 81 (3), 460–469.
43. Valente, T., Pelamatti, T., Avio, C.G., Camedda, A., Costantini, M.L., De Lucia, G.A., Jacomini, C., Piermarini, R., Regoli, F., Sbrana, A., Ventura, D., Silvestri, C., and Matiddi, M., 2022. One is not enough: Monitoring microplastic ingestion by fish needs a multispecies approach. *Marine Pollution Bulletin*, 184, 114133.
44. Giani, D., Baini, M., Galli, M., Casini, S., and Fossi, M.C., 2019. Microplastics occurrence in edible fish species (Mullus barbatus and Merluccius merluccius) collected in three different geographical sub-areas of the Mediterranean Sea. *Marine Pollution Bulletin*, 140, 129–137.
45. Avio, C.G., Pittura, L., D'Errico, G., Abel, S.M., Amorello, S., Marino, G., Gorbi, S., and Regoli, F., 2020. Distribution and characterization of microplastic particles and textile microfibers in Adriatic food webs: General insights for biomonitoring strategies. *Environmental Pollution*, 258, 113766.
46. Bianchi, J., Valente, T., Scacco, U., Cimmaruta, R., Sbrana, A., Silvestri, C., and Matiddi, M., 2020. Food preference determines the best suitable digestion protocol for analysing microplastic ingestion by fish. *Marine Pollution Bulletin*, 154, 111050.
47. Labropoulou, M. and Eleftheriou, A., 1997. The foraging ecology of two pairs of congeneric demersal fish species: Importance of morphological characteristics in prey selection. *Journal of Fish Biology*, 50 (2), 324–340.
48. Digka, N., Tsangaris, C., Torre, M., Anastasopoulou, A., and Zeri, C., 2018. Microplastics in mussels and fish from the Northern Ionian Sea. *Marine Pollution Bulletin*, 135, 30–40.
49. Zhang, D., Liu, X., Huang, W., Li, J., Wang, C., Zhang, D., and Zhang, C., 2020. Microplastic pollution in deep-sea sediments and organisms of the Western Pacific Ocean. *Environmental Pollution*, 259, 113948.
50. Anastasopoulou, A., Viršek, M.K., Varezić, D.B., Digka, N., Fortibuoni, T., Koren, Š., Mandić, M., Mytilineou, C., Pešić, A., Ronchi, F., Šiljić, J., Torre, M., Tsangaris, C., and Tutman, P., 2018. Assessment on marine litter ingested by fish in the Adriatic and NE Ionian Sea macro-region (Mediterranean). *Marine Pollution Bulletin*, 133, 841–851.
51. Aytan, Ü., Esensoy, F.B., Şentürk, Y., Arifoğlu, E., Karaoğlu, K., Ceylan, Y., and Valente, A., 2021. Plastic occurrence in commercial fish species of the Black Sea. *Turkish Journal of Fisheries and Aquatic Sciences*, 22 (Special Issue).
52. Sbrana, A., Valente, T., Scacco, U., Bianchi, J., Silvestri, C., Palazzo, L., De Lucia, G.A., Valerani, C., Ardizzone, G., and Matiddi, M., 2020. Spatial variability and influence of biological parameters on microplastic ingestion by Boops boops (L.) along the Italian coasts (Western Mediterranean Sea). *Environmental Pollution*, 263, 114429.
53. Zhang, C., Wang, S., Sun, D., Pan, Z.-K., Zhou, A., Xie, S., and Wang, J., 2020. Microplastic pollution in surface water from East Coastal areas of Guangdong, South China and preliminary study on microplastics biomonitoring using two marine fish. *Chemosphere*, 256, 127202.

54. Whitfield, A.K., Panfili, J., and Durand, J.-D., March 27, 2012. A global review of the cosmopolitan flathead mullet mugil cephalus linnaeus 1758 (Teleostei: Mugilidae), with emphasis on the biology, genetics, ecology and fisheries aspects of this apparent species complex. *Reviews in Fish Biology and Fisheries*, 22 (3), 641–681. https://doi.org/10.1007/s11160-012-9263-9.
55. Park, S.Y. and Ha, S., 2015. Reduction of Escherichia coli and Vibrio parahaemolyticus counts on freshly sliced shad (Konosirus punctatus) by combined treatment of slightly acidic electrolyzed water and ultrasound using response surface methodology. *Food and Bioprocess Technology*, 8 (8), 1762–1770.
56. Bellas, J., Martínez-Armental, J., Martínez-Cámara, A., Besada, V., and Martínez-Gómez, C., 2016. Ingestion of microplastics by demersal fish from the Spanish Atlantic and Mediterranean coasts. *Marine Pollution Bulletin*, 109 (1), 55–60.
57. Miccoli, A., Mancini, E., Saraceni, P.R., Della Ventura, G., Scapigliati, G., and Picchietti, S., 2022. First evidence of in vitro cytotoxic effects of marine microlitter on Merluccius merluccius and Mullus barbatus, two Mediterranean commercial fish species. *Science of the Total Environment*, 813, 152618.
58. Rios, M.F., Hernández-Moresino, R.D., and Galván, D.E., 2020. Assessing urban microplastic pollution in a benthic habitat of Patagonia Argentina. *Marine Pollution Bulletin*, 159, 111491.
59. Ferreira, G.V.B., Barletta, M., and Lima, A.R.A., 2019. Use of estuarine resources by top predator fishes. How do ecological patterns affect rates of contamination by microplastics? *Science of the Total Environment*, 655, 292–304.
60. Lehtiniemi, M., Hartikainen, S., Näkki, P., Engström-Öst, J., Koistinen, A., and Setälä, O., 2018. Size matters more than shape: Ingestion of primary and secondary microplastics by small predators. *Food Webs*, 17, e00097.
61. Wright, S., Thompson, R.C., and Galloway, T.S., 2013. The physical impacts of microplastics on marine organisms: A review. *Environmental Pollution*, 178, 483–492.
62. Desforges, J., Galbraith, M., and Ross, P.S., 2015. Ingestion of microplastics by zooplankton in the Northeast Pacific Ocean. *Archives of Environmental Contamination and Toxicology*, 69 (3), 320–330.
63. Lusher, A., Welden, N., Sobral, P., and Cole, M., 2017. Sampling, isolating and identifying microplastics ingested by fish and invertebrates. *Analytical Methods*, 9 (9), 1346–1360.
64. Cho, Y., Shim, W.J., Jang, M., Han, G.M., and Hong, S.H., 2021. Nationwide monitoring of microplastics in bivalves from the coastal environment of Korea. *Environmental Pollution*, 270, 116175.
65. Ding, J., Sun, C., He, C., Li, J., Ju, P., and Li, F., 2021. Microplastics in four bivalve species and basis for using bivalves as bioindicators of microplastic pollution. *Science of the Total Environment*, 782, 146830.
66. Said, R.M., Nassar, S.E., and Mohamed, A.A., 2022. Assessment of microplastic and trace element pollution in the Southeastern Mediterranean coasts, Egypt, using shellfish Arca noae as a bioindicator. *Marine Pollution Bulletin*, 177, 113493.
67. Farrington, J.W., Tripp, B.W., Tanabe, S., Subramanian, A., Sericano, J.L., Wade, T.L., and Knap, A.H., 2016. Edward D. Goldberg's proposal of "the mussel watch": Reflections after 40 years. *Marine Pollution Bulletin*, 110 (1), 501–510.
68. Beyer, J., Green, N.W., Brooks, S., Allan, I., Ruus, A., Gomes, T., Bråte, I.L.N., and Schøyen, M., 2017. Blue mussels (Mytilus edulis spp.) as sentinel organisms in coastal pollution monitoring: A review. *Marine Environmental Research*, 130, 338–365.
69. Goldberg, E.D., 1975. The mussel watch—a first step in global marine monitoring. *Marine Pollution Bulletin*, 6 (7), 111.
70. Van Cauwenberghe, L. and Janssen, C.R., 2014. Microplastics in bivalves cultured for human consumption. *Environmental Pollution*, 193, 65–70.

71. Leslie, H.A., Brandsma, S.H., Van Velzen, M.J.M., and Vethaak, A.D., 2017. Microplastics en route: Field measurements in the Dutch river delta and Amsterdam canals, wastewater treatment plants, North Sea sediments and biota. *Environment International*, 101, 133–142.
72. Li, J., Liu, H., and Chen, J., 2018. Microplastics in freshwater systems: A review on occurrence, environmental effects, and methods for microplastics detection. *Water Research*, 137, 362–374.
73. Qu, X., Su, L., Li, H., Liang, M., and Shi, H., 2018. Assessing the relationship between the abundance and properties of microplastics in water and in mussels. *Science of the Total Environment*, 621, 679–686.
74. Railo, S., Talvitie, J., Setälä, O., Koistinen, A., and Lehtiniemi, M., 2018. Application of an enzyme digestion method reveals microlitter in Mytilus trossulus at a wastewater discharge area. *Marine Pollution Bulletin*, 130, 206–214.
75. Li, J., Qu, X., Su, L., Zhang, W., Yang, D., Kolandhasamy, P., Li, D., and Shi, H., 2016. Microplastics in mussels along the coastal waters of China. *Environmental Pollution*, 214, 177–184.
76. Karlsson, T., Vethaak, A.D., Almroth, B.C., Ariese, F., Van Velzen, M., Hassellöv, M., and Leslie, H.A., 2017. Screening for microplastics in sediment, water, marine invertebrates and fish: Method development and microplastic accumulation. *Marine Pollution Bulletin*, 122 (1–2), 403–408.
77. Li, L., Su, L., Cai, H., Rochman, C.M., Li, Q., Kolandhasamy, P., Peng, J., and Shi, H., 2019. The uptake of microfibers by freshwater Asian clams (Corbicula fluminea) varies based upon physicochemical properties. *Chemosphere*, 221, 107–114.
78. Jaouani, R., Roman, C., Decaix, J., Lagarde, F., and Châtel, A., 2023. Effect of aging of microplastics on gene expression levels of the marine mussel Mytilus edulis: Comparison in vitro/in vivo exposures. *Marine Pollution Bulletin*, 189, 114767.
79. Von Moos, N.R., Burkhardt-Holm, P., and Köhler, A., 2012. Uptake and effects of microplastics on cells and tissue of the blue mussel Mytilus edulis L. after an experimental exposure. *Environmental Science & Technology*, 46 (20), 11327–11335.
80. Pittura, L., Avio, C.G., Giuliani, M., D'Errico, G., Keiter, S., Cormier, B., Gorbi, S., and Regoli, F., 2018. Microplastics as vehicles of environmental PAHs to marine organisms: Combined chemical and physical hazards to the mediterranean mussels, Mytilus galloprovincialis. *Frontiers in Marine Science*, 5.
81. Santana, M., Moreira, F.T., and Turra, A., 2017. Trophic transference of microplastics under a low exposure scenario: Insights on the likelihood of particle cascading along marine food-webs. *Marine Pollution Bulletin*, 121 (1–2), 154–159.
82. Porter, A.H., Lyons, B.P., Galloway, T.S., and Lewis, C., 2018. Role of marine snows in microplastic fate and bioavailability. *Environmental Science & Technology*, 52 (12), 7111–7119.
83. Browne, M.A.O., Dissanayake, A., Galloway, T.S., Lowe, D.M., and Thompson, R.C., 2008. Ingested microscopic plastic translocates to the circulatory system of the mussel, Mytilus edulis (L.). *Environmental Science & Technology*, 42 (13), 5026–5031.
84. Ward, J.E. and Kach, D.J., 2009. Marine aggregates facilitate ingestion of nanoparticles by suspension-feeding bivalves. *Marine Environmental Research*, 68 (3), 137–142.
85. Wegner, A., Besseling, E., Foekema, E.M., Kamermans, P., and Koelmans, A.A., 2012. Effects of nanopolystyrene on the feeding behavior of the blue mussel (Mytilus edulis L.). *Environmental Toxicology and Chemistry*, 31 (11), 2490–2497.
86. Van Cauwenberghe, L., Devriese, L., Galgani, F., Robbens, J., and Janssen, C.R., 2015. Microplastics in sediments: A review of techniques, occurrence and effects. *Marine Environmental Research*, 111, 5–17.
87. Kolandhasamy, P., Su, L., Li, J., Qu, X., Jabeen, K., and Shi, H., 2018. Adherence of microplastics to soft tissue of mussels: A novel way to uptake microplastics beyond ingestion. *Science of the Total Environment*, 610–611, 635–640.

88. Avio, C.G., Gorbi, S., and Regoli, F., 2015. Experimental development of a new protocol for extraction and characterization of microplastics in fish tissues: First observations in commercial species from Adriatic Sea. *Marine Environmental Research*, 111, 18–26.
89. Capolupo, M., Franzellitti, S., Valbonesi, P., Lanzas, C.S., and Fabbri, E., 2018. Uptake and transcriptional effects of polystyrene microplastics in larval stages of the Mediterranean mussel Mytilus galloprovincialis. *Environmental Pollution*, 241, 1038–1047.
90. Paul-Pont, I., Lacroix, C., González-Fernández, C., Hégaret, H., Lambert, C., Goïc, N.L., Frère, L., Cassone, A.-L., Sussarellu, R., Fabioux, C., Guyomarch, J., Albentosa, M., Huvet, A., and Soudant, P., 2016. Exposure of marine mussels Mytilus spp. to polystyrene microplastics: Toxicity and influence on fluoranthene bioaccumulation. *Environmental Pollution*, 216, 724–737.
91. Magni, S., Gagné, F., André, C., Della Torre, C., Auclair, J.L., Hanana, H., Parenti, C.C., Bonasoro, F., and Binelli, A., 2018. Evaluation of uptake and chronic toxicity of virgin polystyrene microbeads in freshwater zebra mussel Dreissena polymorpha (Mollusca: Bivalvia). *Science of the Total Environment*, 631–632, 778–788.
92. Khan, M.B. and Prezant, R.S., 2018. Microplastic abundances in a mussel bed and ingestion by the ribbed marsh mussel Geukensia demissa. *Marine Pollution Bulletin*, 130, 67–75.
93. Mladinich, K., Holohan, B.A., Shumway, S.E., Brown, K., and Ward, J.E., 2022. Determining the properties that govern selective ingestion and egestion of microplastics by the blue mussel (Mytilus edulis) and Eastern oyster (Crassostrea virginica). *Environmental Science & Technology*, 56 (22), 15770–15779.
94. Franco-Martínez, L., Mateo, S.V., Cerón, J.J., Romero, D., Oliveira, M.E.D.E.S., Teles, M., and Martínez-Subiela, S., 2019. Evaluation of C-reactive-like protein in Mytilus galloprovincialis. *Ecological Indicators*, 106, 105537.
95. Abelouah, M.R., Romdhani, I., Ben-Haddad, M., Hajji, S., De-La-Torre, G.E., Gaaied, S., Barra, I., Bannı, M., and Alla, A.A., 2023. Binational survey using Mytilus galloprovincialis as a bioindicator of microplastic pollution: Insights into chemical analysis and potential risk on humans. *Science of the Total Environment*, 870, 161894.
96. Ferreira, O., Barboza, L.G.A., Rudnitskaya, A., Moreirinha, C., Vieira, L.R., Botelho, M.J., Vale, C., Fernandes, J.O., Cunha, S.C., and Guilhermino, L., 2023b. Microplastics in marine mussels, biological effects and human risk of intake: A case study in a multi-stressor environment. *Marine Pollution Bulletin*, 197, 115704.
97. Santana, M., Ascer, L.G., Custódio, M.R., Moreira, F.T., and Turra, A., 2016. Microplastic contamination in natural mussel beds from a Brazilian urbanized coastal region: Rapid evaluation through bioassessment. *Marine Pollution Bulletin*, 106 (1–2), 183–189.
98. Harris, L.S.T., Phan, S., DiMarco, D., Padilla-Gamiño, J.L., Luscombe, C.K., and Carrington, E., 2023. Microparticles in marine mussels at regional and localized scales across the Salish Sea, Washington. *Marine Pollution Bulletin*, 196, 115609.
99. Martinelli, J.C., Phan, S., Luscombe, C.K., and Padilla-Gamiño, J.L., 2020. Low incidence of microplastic contaminants in Pacific oysters (Crassostrea gigas Thunberg) from the Salish Sea, USA. *Science of the Total Environment*, 715, 136826.
100. Liu, Y., Li, R., Yu, J., Ni, F., Sheng, Y., Scircle, A., Cizdziel, J.V., and Zhou, Y., 2021. Separation and identification of microplastics in marine organisms by TGA-FTIR-GC/MS: A case study of mussels from coastal China. *Environmental Pollution*, 272, 115946.
101. Ta, A.T., Pupuang, P., Babel, S., and Wang, L.P., 2022. Investigation of microplastic contamination in blood cockles and green mussels from selected aquaculture farms and markets in Thailand. *Chemosphere*, 303, 134918.
102. Fok, L. and Cheung, P.K., 2015. Hong Kong at the Pearl River Estuary: A hotspot of microplastic pollution. *Marine Pollution Bulletin*, 99 (1–2), 112–118.
103. Lo, H.-S., Xu, X., Wong, C., and Cheung, S.G., 2018. Comparisons of microplastic pollution between mudflats and sandy beaches in Hong Kong. *Environmental Pollution*, 236, 208–217.

104. So, K.J., Cheang, C.C., Hui, T.Y., and Chan, J.K.Y., September 20, 2021. Understanding the behavioural gap between perceived and actual environmental behaviour: Investigating the clam-harvesting pattern in Hong Kong SAR, China. *Journal of Cleaner Production*, 316, 128259. https://doi.org/10.1016/j.jclepro.2021.128259.
105. Cho, Y., Shim, W.J., Jang, M., Han, G.M., and Hong, S.H., February 2019. Abundance and characteristics of microplastics in market bivalves from South Korea. *Environmental Pollution*, 245, 1107–1116. https://doi.org/10.1016/j.envpol.2018.11.091.
106. Lam, T.W.L., Tsui, Y.C.J., Cheng, Y.-S., Hin, A.T., and Fok, L., 2023. Microplastic contamination in edible clams from popular recreational clam-digging sites in Hong Kong and implications for human health. *Science of the Total Environment*, 875, 162576.
107. Baechler, B.R., Stienbarger, C.D., Horn, D.A., Joseph, J., Taylor, A.R., Granek, E.F., and Brander, S.M., 2019. Microplastic occurrence and effects in commercially harvested North American finfish and shellfish: Current knowledge and future directions. *Limnology and Oceanography Letters*, 5 (1), 113–136.
108. Ben-Haddad, M., Abelouah, M.R., Hajji, S., De-La-Torre, G.E., Oualid, H.A., Rangel-Buitrago, N., and Alla, A.A., 2022. The wedge clam Donax trunculus L., 1758 as a bioindicator of microplastic pollution. *Marine Pollution Bulletin*, 178, 113607.
109. Sathish, N., Jeyasanta, K.I., and Patterson, J., 2020. Monitoring of microplastics in the clam Donax cuneatus and its habitat in Tuticorin coast of Gulf of Mannar (GoM), India. *Environmental Pollution*, 266, 115219.
110. Truchet, D.M., López, A.D.F., Ardusso, M.G., Rimondino, G.N., Buzzi, N.S., Malanca, F.E., Spetter, C.V., and Severini, M.D.F., 2021. Microplastics in bivalves, water and sediments from a touristic sandy beach of Argentina. *Marine Pollution Bulletin*, 173, 113023.
111. Ward, J.E., Zhao, S., Holohan, B.A., Mladinich, K., Griffin, T.W., Wozniak, J., and Shumway, S.E., 2019. Selective ingestion and egestion of plastic particles by the blue mussel (Mytilus edulis) and Eastern Oyster (Crassostrea virginica): Implications for using bivalves as bioindicators of microplastic pollution. *Environmental Science & Technology*, 53 (15), 8776–8784.
112. Joyce, H., Frias, J., Kavanagh, F., Lynch, R., Pagter, E., White, J., and Nash, R., 2022. Plastics, prawns, and patterns: Microplastic loadings in Nephrops norvegicus and surrounding habitat in the North East Atlantic. *Science of the Total Environment*, 826, 154036.
113. Ogbuagu, C.C., Kassem, H., Udiba, U.U., Stead, J.L., and Cundy, A., 2022. Role of saltmarsh systems in estuarine trapping of microplastics. *Scientific Reports*, 12 (1).
114. Xu, X., Wong, C., Tam, N.F.-Y., Liu, H.M., and Cheung, S.G., 2020. Barnacles as potential bioindicator of microplastic pollution in Hong Kong. *Marine Pollution Bulletin*, 154, 111081.
115. Davidson, K. and Dudas, S.E., 2016. Microplastic ingestion by wild and cultured Manila Clams (Venerupis philippinarum) from Baynes Sound, British Columbia. *Archives of Environmental Contamination and Toxicology*, 71 (2), 147–156.
116. Vieira, K.S., Neto, J.A.B., Crapez, M.A.C., Gaylarde, C.C., Da Silva Pierri, B., Saldaña-Serrano, M., Bainy, A.C.D., Nogueira, D.J., and Da Fonseca, E.M., 2021. Occurrence of microplastics and heavy metals accumulation in native oysters Crassostrea Gasar in the Paranaguá estuarine system, Brazil. *Marine Pollution Bulletin*, 166, 112225.
117. Liao, C.-P., Chiu, C., and Huang, H.-H., 2021. Assessment of microplastics in oysters in coastal areas of Taiwan. *Environmental Pollution*, 286, 117437.
118. Jang, M., Shim, W.J., Cho, Y., Han, G.M., Song, Y.K., and Hong, S.H., 2020. A close relationship between microplastic contamination and coastal area use pattern. *Water Research*, 171, 115400.

119. Manh, V., Dang, T.T., Le, X.T., Nguyen, D.T., Phung, T.V., Vu, D.N., and Pham, H.V., 2022. Abundance of microplastics in cultured oysters (Crassostrea gigas) from Danang Bay of Vietnam. *Marine Pollution Bulletin*, 180, 113800.
120. Bringer, A., Cachot, J., Dubillot, E., Lalot, B., and Thomas, H., 2021. Evidence of deleterious effects of microplastics from aquaculture materials on pediveliger larva settlement and oyster spat growth of Pacific oyster, Crassostrea gigas. *Science of the Total Environment*, 794, 148708.
121. Lee, H.C., Khan, M.M., Yusli, A. 'Aqilah, Jaya, N.A., and Marshall, D.J., 2022. Microplastic accumulation in oysters along a Bornean coastline (Brunei, South China Sea): Insights into local sources and sinks. *Marine Pollution Bulletin*, 177, 113478.
122. Mladinich, K., Holohan, B.A., Shumway, S.E., and Ward, J.E., 2023. The relationship between microplastics in eastern oysters (Crassostrea virginica) and surrounding environmental compartments in Long Island Sound. *Marine Environmental Research*, 189, 106040.
123. De Mestre, C., Maher, W.A., Roberts, D., Broad, A., Krikowa, F., and Davis, A.R., 2012. Sponges as sentinels: Patterns of spatial and intra-individual variation in trace metal concentration. *Marine Pollution Bulletin*, 64 (1), 80–89.
124. Hammel, J.U. and Nickel, M., November 19, 2014. A new flow-regulating cell type in the demosponge Tethya Wilhelma—functional cellular anatomy of a leuconoid canal system. *PLOS One*, 9 (11). https://doi.org/10.1371/journal.pone.0113153.
125. Corcoran, P.L., Biesinger, M.C., and Grifi, M.D., 2009. Plastics and beaches: A degrading relationship. *Marine Pollution Bulletin*, 58 (1), 80–84.
126. Bennett, H., Bell, J.J., Davy, S.K., Webster, N.S., and Francis, D.S., 2018. Elucidating the sponge stress response; lipids and fatty acids can facilitate survival under future climate scenarios. *Global Change Biology*, 24 (7), 3130–3144.
127. Gentric, C., Réhel, K., Dufour, A., and Sauleau, P., 2015. Bioaccumulation of metallic trace elements and organic pollutants in marine sponges from the South Brittany Coast, France. *Journal of Environmental Science and Health, Part A*, 51 (3), 213–219.
128. Rizzi, C., Seveso, D., Galli, P., and Villa, S., 2020. First record of emerging contaminants in sponges of an inhabited island in the Maldives. *Marine Pollution Bulletin*, 156, 111273.
129. Roveta, C., Annibaldi, A., Afghan, A., Calcinai, B., Di Camillo, C.G., Gregorin, C., Illuminati, S., Mantas, T.P., Truzzi, C., and Puce, S., 2021. Biomonitoring of heavy metals: The unexplored role of Marine Sessile Taxa. *Applied Sciences*, 11 (2), 580.
130. Dolenec, T., Lojen, S., Kniewald, G., Dolenec, M., and Rogan, N., 2007. Nitrogen stable isotope composition as a tracer of fish farming in invertebrates Aplysina aerophoba, Balanus perforatus and Anemonia sulcata in central Adriatic. *Aquaculture*, 262 (2–4), 237–249.
131. Girard, E.B., Fuchs, A.C.D., Kaliwoda, M., Lasut, M.T., Ploetz, E., Schmahl, W.W., and Wörheide, G., 2021. Sponges as bioindicators for microparticulate pollutants? *Environmental Pollution*, 268, 115851.
132. Krikech, I., Conti, G.O., Pulvirenti, E., Rapisarda, P., Castrogiovanni, M., Maisano, M., Pennec, G.L., Leermakers, M., Ferrante, M., Cappello, T., and Ezziyyani, M., 2023. Microplastics (≤ 10 μm) bioaccumulation in marine sponges along the Moroccan Mediterranean coast: Insights into species-specific distribution and potential bioindication. *Environmental Research*, 235, 116608.
133. Corti, A., Pagano, G., Lo Giudice, A., Papale, M., Rizzo, C., Azzaro, M., Vinciguerra, V., Castelvetro, V., and Giannarelli, S., 2023. Marine sponges as bioindicators of pollution by synthetic microfibers in Antarctica. *Science of the Total Environment*, 902, 166043.
134. Dolenec, T., Lojen, S., Kniewald, G., Dolenec, M., and Rogan, N., 2007. Nitrogen stable isotope composition as a tracer of fish farming in invertebrates Aplysina aerophoba, Balanus perforatus and Anemonia sulcata in central Adriatic. *Aquaculture*, 262 (2–4), 237–249.

135. De Marchi, L., Renzi, M., Anselmi, S., Pretti, C., Guazzelli, E., Martinelli, E., Cuccaro, A., Oliva, M., Magri, M., and Bulleri, F., 2022. Polyethylene microplastics reduce filtration and respiration rates in the Mediterranean sponge Petrosia ficiformis. *Environmental Research*, 211, 113094.
136. Powell, M.I. and White, K., 1990. Heavy metal accumulation by barnacles and its implications for their use as biological monitors. *Marine Environmental Research*, 30 (2), 91–118.
137. Chen, L., Lam, J.C.W., Zhang, X., Pan, K., Guo, C., Lam, P.K.S., Wang, W., Liu, H., and Qian, P., 2015. Relationship between metal and polybrominated diphenyl ether (PBDE) body burden and health risks in the barnacle Balanus amphitrite. *Marine Pollution Bulletin*, 100 (1), 383–392.
138. Cheung, P.K., Cheung, L.T.O., and Fok, L., 2016. Seasonal variation in the abundance of marine plastic debris in the estuary of a subtropical macro-scale drainage basin in South China. *Science of the Total Environment*, 562, 658–665.
139. Davies, H.L., Robb, H., Cox, K., Covernton, G.A., Eastham, T.M., Alexander, H.J., and Juanes, F., 2021. A preliminary analysis of ingestion and egestion of microplastic fibres in the acorn barnacle Balanus glandula. *Journal of Experimental Marine Biology and Ecology*, 542–543, 151589.
140. Bhargava, S., Lee, S.S.C., Ying, L.S.M., Neo, M.L., Teo, S.L., and Valiyaveettil, S., 2018. Fate of nanoplastics in marine larvae: A case study using Barnacles, Amphibalanus amphitrite. *ACS Sustainable Chemistry & Engineering*, 6 (5), 6932–6940.
141. Thushari, G.G.N., Senevirathna, J.D.M., Yakupitiyage, A., and Chavanich, S., 2017. Effects of microplastics on sessile invertebrates in the Eastern Coast of Thailand: An approach to coastal zone conservation. *Marine Pollution Bulletin*, 124 (1), 349–355.
142. Goldstein, M.C. and Goodwin, D.S., 2013. Gooseneck barnacles (Lepasspp.) ingest microplastic debris in the North Pacific Subtropical Gyre. *PeerJ*, 1, e184.
143. Scotti, G., D'Alessandro, M., Esposito, V., Vivona, P., and Panti, C., 2023. Anthropogenic fibers and microplastics in the pelagic gooseneck barnacle Lepas (Lepas) anatifera in Capo Milazzo Marine Protected Area (Tyrrhenian Sea): A first characterization. *Ecological Indicators*, 152, 110368.
145. Higgs, N.D., Newton, J., and Attrill, M.J., 2016. Caribbean spiny lobster fishery is underpinned by trophic subsidies from chemosynthetic primary production. *Current Biology*, 26 (24), 3393–3398.
146. Kampouris, T.E., Syranidou, E., Seridou, P., Gagoulis, K., Batjakas, I.E., and Kalogerakis, N., January 1, 2023. MPs and NPS intake and heavy metals accumulation in tissues of Palinurus Elephas (J.C. Fabricius, 1787), from NW Aegean Sea, Greece. *Environmental Pollution*, 316, 120725. https://doi.org/10.1016/j.envpol.2022.120725.
147. Cau, A., Avio, C.G., Dessì, C., Follesa, M.C., Moccia, D., Regoli, F., and Pusceddu, A., 2019. Microplastics in the crustaceans Nephrops norvegicus and Aristeus antennatus: Flagship species for deep-sea environments? *Environmental Pollution*, 255, 113107.
148. Fossi, M.C., Pedà, C., Compa, M., Tsangaris, C., Alomar, C., Claro, F., Ioakeimidis, C., Galgani, F., Hema, T., Deudero, S., Romeo, T., Battaglia, P., Andaloro, F., Caliani, I., Casini, S., Panti, C., and Baini, M., 2018. Bioindicators for monitoring marine litter ingestion and its impacts on Mediterranean biodiversity. *Environmental Pollution*, 237, 1023–1040.
149. Johnson, M.P., Lordan, C., and Power, A.M., 2013. Habitat and ecology of Nephrops norvegicus. *Advances in Marine Biology*, 27–63.
150. Groeneveld, J.C., Goñi, R., and Díaz, D., 2013. Palinurus species. In *Lobsters: Biology, Management, Aquaculture and Fisheries*, Second Edition, 326–356. Blackwell Scientific Publications.
151. Cau, A., Avio, C.G., Dessì, C., Moccia, D., Pusceddu, A., Regoli, F., Cannas, R., and Follesa, M.C., 2020. Benthic crustacean digestion can modulate the environmental fate of microplastics in the deep sea. *Environmental Science & Technology*, 54 (8), 4886–4892.

152. Avio, C.G., Pittura, L., D'Errico, G., Abel, S.M., Amorello, S., Marino, G., Gorbi, S., and Regoli, F., 2020. Distribution and characterization of microplastic particles and textile microfibers in Adriatic food webs: General insights for biomonitoring strategies. *Environmental Pollution*, 258, 113766.
153. Hara, J., Frias, J., and Nash, R., 2020. Quantification of microplastic ingestion by the decapod crustacean Nephrops norvegicus from Irish waters. *Marine Pollution Bulletin*, 152, 110905.
154. Cau, A., Gorule, P.A., Bellodi, A., Carreras-Colom, E., Moccia, D., Pittura, L., Regoli, F., and Follesa, M.C., 2023. Comparative microplastic load in two decapod crustaceans Palinurus elephas (Fabricius, 1787) and Nephrops norvegicus (Linnaeus, 1758). *Marine Pollution Bulletin*, 191, 114912.
155. Timilsina, A., Adhikari, K., Yadav, A.K., Joshi, P., Ramena, G., and Bohara, K., 2023. Effects of microplastics and nanoplastics in shrimp: Mechanisms of plastic particle and contaminant distribution and subsequent effects after uptake. *Science of the Total Environment*, 894, 164999.
156. Bordbar, L., Kapiris, K., Kalogirou, S., and Anastasopoulou, A., 2018. First evidence of ingested plastics by a high commercial shrimp species (Plesionika narval) in the eastern Mediterranean. *Marine Pollution Bulletin*, 136, 472–476.
157. Keshavarzifard, M., Vazirzadeh, A., and Sharifinia, M., 2021. Occurrence and characterization of microplastics in white shrimp, Metapenaeus affinis, living in a habitat highly affected by anthropogenic pressures, northwest Persian Gulf. *Marine Pollution Bulletin*, 169, 112581.
158. Hariharan, G., Ramachandran, R., Anandavelu, I., Robin, R.S., and Ramachandran, R., 2022. Ingestion and toxic impacts of weathered polyethylene (wPE) microplastics and stress defensive responses in whiteleg shrimp (Penaeus vannamei). *Chemosphere*, 300, 134487.
159. Saborowski, R., Korez, Š., Riesbeck, S., Weidung, M., Bickmeyer, U., and Gutow, L., 2022. Shrimp and microplastics: A case study with the Atlantic ditch shrimp Palaemon varians. *Ecotoxicology and Environmental Safety*, 234, 113394.
160. Schmidt, N., Korez, Š., Saborowski, R., and Gutow, L., 2021. Sand grains in the stomach of brown shrimp, Crangon crangon: Crunchy garnish, supportive macerator, or simply dirt? *Journal of Sea Research*, 170, 102020.
161. Gray, A.D. and Weinstein, J.E., 2017. Size- and shape-dependent effects of microplastic particles on adult daggerblade grass shrimp (Palaemonetes pugio). *Environmental Toxicology and Chemistry*, 36 (11), 3074–3080.
162. Li, Y., Liu, Z., Li, M., Jiang, Q., Wu, D., Huang, Y., Jiao, Y., Zhang, M., and Zhao, Y., 2020. Effects of nanoplastics on antioxidant and immune enzyme activities and related gene expression in juvenile Macrobrachium nipponense. *Journal of Hazardous Materials*, 398, 122990.
163. Li, Y., Liu, Z., Yang, Y., Jiang, Q., Wu, D., Huang, Y., Jiao, Y., Chen, Q., Huang, Y., and Zhao, Y., 2021. Effects of nanoplastics on energy metabolism in the oriental river prawn (Macrobrachium nipponense). *Environmental Pollution*, 268, 115890.
164. Hsieh, S., Wu, Y., Xu, R.-Q., Chen, Y., Chen, C.-W., Singhania, R.R., and Dong, C.-D., 2021. Effect of polyethylene microplastics on oxidative stress and histopathology damages in Litopenaeus vannamei. *Environmental Pollution*, 288, 117800.
165. Xing, Y., Zhu, X., Duan, Y., Huang, J., Nan, Y., and Zhang, J., 2023. Toxic effects of nitrite and microplastics stress on histology, oxidative stress, and metabolic function in the gills of Pacific white shrimp, Litopenaeus vannamei. *Marine Pollution Bulletin*, 187, 114531.
166. Severini, M.D.F., Buzzi, N.S., López, A.D.F., Colombo, C.V., Sartor, G.L.C., Rimondino, G.N., and Truchet, D.M., 2020. Chemical composition and abundance of microplastics in the muscle of commercial shrimp Pleoticus muelleri at an impacted coastal environment (Southwestern Atlantic). *Marine Pollution Bulletin*, 161, 111700.

167. Bergami, E., Bocci, E.B., Vannuccini, M.L., Monopoli, M.P., Salvati, A., Dawson, K.A., and Corsi, I., 2016. Nano-sized polystyrene affects feeding, behavior and physiology of brine shrimp Artemia Franciscana larvae. *Ecotoxicology and Environmental Safety*, 123, 18–25.
168. Bergami, E., Emerenciano, A.K., Pinto, L.P., Joviano, W.R., Font, A., De Godoy, T.A., Silva, J.R.M.C., González-Aravena, M., and Corsi, I., 2022. Behavioural, physiological and molecular responses of the Antarctic fairy shrimp Branchinecta gaini (Daday, 1910) to polystyrene nanoplastics. *NanoImpact*, 28, 100437.
169. Varó, I., Perini, D.A., Torreblanca, A., García, Y., Bergami, E., Vannuccini, M.L., and Corsi, I., 2019. Time-dependent effects of polystyrene nanoparticles in brine shrimp Artemia Franciscana at physiological, biochemical and molecular levels. *Science of the Total Environment*, 675, 570–580.
170. Lomartire, S., Marques, J.C., and Gonçalves, A.M.M., 2021. The key role of zooplankton in ecosystem services: A perspective of interaction between zooplankton and fish recruitment. *Ecological Indicators*, 129, 107867.
171. Lee, J.-M., Hong, S., Song, Y.K., Hong, S.H., Jang, Y.C., Jang, M., Heo, N.W., Han, G.M., Lee, M., Kang, D., and Shim, W.J., 2013. Relationships among the abundances of plastic debris in different size classes on beaches in South Korea. *Marine Pollution Bulletin*, 77 (1–2), 349–354.
172. Moriarty, R. and O'Brien, T., 2013. Distribution of mesozooplankton biomass in the global ocean. *Earth System Science Data*, 5 (1), 45–55.
173. Alfonso, M.B., Lindsay, D.J., Arias, A.H., Nakano, H., Jandang, S., and Isobe, A., 2023. Zooplankton as a suitable tool for microplastic research. *Science of the Total Environment*, 905, 167329.
174. Frias, J., Otero, V., and Sobral, P., 2014. Evidence of microplastics in samples of zooplankton from Portuguese coastal waters. *Marine Environmental Research*, 95, 89–95.
175. Torres, L.G., Brander, S.M., Parker, J.I., Bloom, E.M., Norman, R., Van Brocklin, J.E., Lasdin, K.S., and Hildebrand, L., 2023. Zoop to poop: Assessment of microparticle loads in gray whale zooplankton prey and fecal matter reveal high daily consumption rates. *Frontiers in Marine Science*, 10.
176. Botterell, Z.L.R., Bergmann, M., Hildebrandt, N., Krumpen, T., Steinke, M., Thompson, R.C., and Lindeque, P.K., 2022. Microplastic ingestion in zooplankton from the Fram Strait in the Arctic. *Science of the Total Environment*, 831, 154886.
177. Ward, P., Atkinson, A., Venables, H.J., Tarling, G.A., Whitehouse, M.J., Fielding, S., Collins, M.A., Korb, R.E., Black, A., Stowasser, G., Schmidt, K., Thorpe, S., and Enderlein, P., 2012. Food web structure and bioregions in the Scotia Sea: A seasonal synthesis. *Deep Sea Research Part II: Topical Studies in Oceanography*, 59–60, 253–266.
178. Choy, C.A., Robison, B.H., Gagne, T., Erwin, B., Firl, E., Halden, R.U., Hamilton, J.A., Katija, K., Lisin, S., Rolsky, C., and Van Houtan, K.S., 2019. The vertical distribution and biological transport of marine microplastics across the epipelagic and mesopelagic water column. *Scientific Reports*, 9 (1).
179. Fischer, R., Lobelle, D., Kooi, M., Koelmans, A.A., Onink, V., Laufkötter, C., Amaral-Zettler, L., Yool, A., and Van Sebille, E., 2022. Modelling submerged biofouled microplastics and their vertical trajectories. *Biogeosciences*, 19 (8), 2211–2234.
180. Song, Y.K., Hong, S.H., Eo, S., Jang, M., Han, G.M., Isobe, A., and Shim, W.J., 2018. Horizontal and vertical distribution of microplastics in Korean coastal waters. *Environmental Science & Technology*, 52 (21), 12188–12197.
181. Zobkov, M., Esiukova, E., Zyubin, A., and Самусев, И.Г., 2019. Microplastic content variation in water column: The observations employing a novel sampling tool in stratified Baltic Sea. *Marine Pollution Bulletin*, 138, 193–205.
182. Rowlands, E., Galloway, T.S., Cole, M., Peck, V.L., Posacka, A.M., Thorpe, S., and Manno, C., 2023. Vertical flux of microplastic, a case study in the Southern Ocean, South Georgia. *Marine Pollution Bulletin*, 193, 115117.

183. Zheng, S., Zhao, Y., Liangwei, W., Liang, J., Liu, T., Zhu, M., Li, Q., and Sun, X., 2020. Characteristics of microplastics ingested by zooplankton from the Bohai Sea, China. *Science of the Total Environment*, 713, 136357.
184. Botterell, Z.L.R., Beaumont, N., Dorrington, T., Steinke, M., Thompson, R.C., and Lindeque, P.K., 2019. Bioavailability and effects of microplastics on marine zooplankton: A review. *Environmental Pollution*, 245, 98–110.
185. Botterell, Z.L.R., Beaumont, N., Cole, M., Hopkins, F.E., Steinke, M., Thompson, R.C., and Lindeque, P.K., 2020. Bioavailability of microplastics to marine zooplankton: Effect of shape and infochemicals. *Environmental Science & Technology*, 54 (19), 12024–12033.
186. Sun, X., Liu, T., Zhu, M., Liang, J., Zhao, Y., and Zhang, B., 2018. Retention and characteristics of microplastics in natural zooplankton taxa from the East China Sea. *Science of the Total Environment*, 640–641, 232–242.
187. Bossart, G.D., 2010. Marine mammals as sentinel species for oceans and human health. *American College of Veterinary Pathologists*, 48 (3).
188. Hernández-González, A., Saavedra, C., Gago, J., Covelo, P., Santos, M.B., and Pierce, G.J., 2018. Microplastics in the stomach contents of common dolphin (Delphinus delphis) stranded on the Galician coasts (NW Spain, 2005–2010). *Marine Pollution Bulletin*, 137, 526–532.
189. Moore, R.C., Loseto, L.L., Noël, M., Etemadifar, A., Brewster, J.D., MacPhee, S.A., Bendell, L.I., and Ross, P.S., 2020. Microplastics in beluga whales (Delphinapterus leucas) from the Eastern Beaufort Sea. *Marine Pollution Bulletin*, 150, 110723.
190. Nelms, S.E., Barnett, J., Brownlow, A., Davison, N.J., Deaville, R., Galloway, T.S., Lindeque, P.K., Santillo, D., and Godley, B.J., 2019. Microplastics in marine mammals stranded around the British coast: Ubiquitous but transitory? *Scientific Reports*, 9 (1).
191. Panti, C., Baini, M., Lusher, A., Hernandez-Milan, G., Rebolledo, E., Unger, B., Syberg, K., Simmonds, M., and Fossi, M.C., 2019. Marine litter: One of the major threats for marine mammals. Outcomes from the European Cetacean society workshop. *Environmental Pollution*, 247, 72–79.
192. Baulch, S. and Perry, C.P., 2014. Evaluating the impacts of marine debris on cetaceans. *Marine Pollution Bulletin*, 80 (1–2), 210–221.
193. Lusher, A., Hernández-Milian, G., Berrow, S., Rogan, E., and O'Connor, I., 2018. Incidence of marine debris in cetaceans stranded and bycaught in Ireland: Recent findings and a review of historical knowledge. *Environmental Pollution*, 232, 467–476.
194. Unger, B., Rebolledo, E., Deaville, R., Gröne, A., IJsseldijk, L.L., Leopold, M.F., Siebert, U., Spitz, J., Wohlsein, P., and Herr, H., 2016. Large amounts of marine debris found in sperm whales stranded along the North Sea coast in early 2016. *Marine Pollution Bulletin*, 112 (1–2), 134–141.
195. Zhang, X., Luo, D., Yu, R.-Q., Xie, Z., He, L., and Wu, Y., 2021. Microplastics in the endangered Indo-Pacific humpback dolphins (Sousa chinensis) from the Pearl River Estuary, China. *Environmental Pollution*, 270, 116057.
196. Tavares, D.C., De Moura, J.F., Merico, A., and Siciliano, S., 2017. Incidence of marine debris in seabirds feeding at different water depths. *Marine Pollution Bulletin*, 119 (2), 68–73.
197. Collard, F., Bangjord, G., Herzke, D., and Gabrielsen, G.W., 2022. Plastic burdens in Northern fulmars from Svalbard: Looking back 25 years. *Marine Pollution Bulletin*, 185, 114333.
198. Meissner, A.M., MacLeod, C.D., Richard, P., Ridoux, V., and Pierce, G.J., 2011. Feeding ecology of striped dolphins, Stenella coeruleoalba, in the north-western Mediterranean Sea based on stable isotope analyses. *Journal of the Marine Biological Association of the United Kingdom*, 92 (8), 1677–1687.
199. Aznar, F.J., Míguez-Lozano, R., Ruiz, B.L., De Castro, A., Raga, J.A., and Blanco, C., 2017. Long-term changes (1990–2012) in the diet of striped dolphins Stenella coeruleoalba from the western Mediterranean. *Marine Ecology Progress Series*, 568, 231–247.

200. Lusher, A., Hernández-Milian, G., O'Brien, J., Berrow, S., O'Connor, I., and Officer, R., 2015. Microplastic and macroplastic ingestion by a deep diving, oceanic cetacean: The True's beaked whale Mesoplodon mirus. *Environmental Pollution*, 199, 185–191.
201. Novillo, O., Raga, J.A., and Tòmàs, J., 2020. Evaluating the presence of microplastics in striped dolphins (Stenella coeruleoalba) stranded in the Western Mediterranean Sea. *Marine Pollution Bulletin*, 160, 111557.
202. Hernández-Milian, G., Lusher, A., MacGabban, S., and Rogan, E., 2019. Microplastics in grey seal (Halichoerus grypus) intestines: Are they associated with parasite aggregations? *Marine Pollution Bulletin*, 146, 349–354.
203. Van Franeker, J.A., Rebolledo, E., Heße, E., IJsseldijk, L.L., Kühn, S., Leopold, M.F., and Mielke, L., 2018. Plastic ingestion by harbour porpoises Phocoena phocoena in the Netherlands: Establishing a standardised method. *AMBIO: A Journal of the Human Environment*, 47, 387–397.
204. Aguilar, À. and García-Vernet, R., 2018. *Fin Whale*, 368–371. Elsevier eBooks.
205. Mizroch, S.A., Rice, D.W., Zwiefelhofer, D., Waite, J.M., and Perryman, W.L., 2009. Distribution and movements of fin whales in the North Pacific Ocean. *Mammal Review*, 39 (3), 193–227.
206. Gibbs, S.E., Kent, C.S., Slat, B., Morales, D., Fouda, L., and Reisser, J., 2019. Cetacean sightings within the Great Pacific garbage patch. *Marine Biodiversity*, 49 (4), 2021–2027.
207. Besseling, E., Foekema, E.M., Van Franeker, J.A., Leopold, M.F., Kühn, S., Rebolledo, E., Heße, E., Mielke, L., IJzer, J., Kamminga, P., and Koelmans, A.A., 2015. Microplastic in a macro filter feeder: Humpback whale Megaptera novaeangliae. *Marine Pollution Bulletin*, 95 (1), 248–252.
208. Di Beneditto, A.P.M. and Da Silva Oliveira, A., 2019. Debris ingestion by carnivorous consumers: Does the position in the water column truly matter? *Marine Pollution Bulletin*, 144, 134–139.
209. Fossi, M.C., Coppola, D., Baini, M., Giannetti, M., Guerranti, C., Marsili, L., Panti, C., De Sabata, E., and Clò, S., 2014. Large filter feeding marine organisms as indicators of microplastic in the pelagic environment: The case studies of the Mediterranean basking shark (Cetorhinus maximus) and fin whale (Balaenoptera physalus). *Marine Environmental Research*, 100, 17–24.
210. Lebreton, L., Slat, B., Ferrari, F., Sainte-Rose, B., Aitken, J., Marthouse, R., Hajbane, S., Cunsolo, S., Schwarz, A., Levivier, A., Noble, K., Debeljak, P., Maral, H., Schoeneich-Argent, R., Brambini, R., and Reisser, J., 2018. Evidence that the Great Pacific Garbage Patch is rapidly accumulating plastic. *Scientific Reports*, 8 (1).
212. Im, J., Joo, S., Lee, Y.-R., Kim, B.-Y., and Kim, T., 2020. First record of plastic debris ingestion by a fin whale (Balaenoptera physalus) in the sea off East Asia. *Marine Pollution Bulletin*, 159, 111514.
213. Hernández-Milian, G., Tsangaris, C., Anestis, A., Fossi, M.C., Baini, M., Caliani, I., Panti, C., Bundone, L., and Panou, A., 2023. Monk seal faeces as a non-invasive technique to monitor the incidence of ingested microplastics and potential presence of plastic additives. *Marine Pollution Bulletin*, 193, 115227.
214. Grigorakis, S., Mason, S.A., and Drouillard, K.G., 2017. Determination of the gut retention of plastic microbeads and microfibers in goldfish (Carassius auratus). *Chemosphere*, 169, 233–238.
215. Sun, X., Li, Q., Shi, Y., Zhao, Y., Zheng, S., Liang, J., Liu, T., and Tian, Z., 2019. Characteristics and retention of microplastics in the digestive tracts of fish from the Yellow Sea. *Environmental Pollution*, 249, 878–885.
216. Pérez-Venegas, D., Toro-Valdivieso, C., Ayala, F., Brito, B., Iturra, L., Arriagada, M., Seguel, M., Barrios, C., Sepúlveda, M., Oliva, D., Cárdenas-Alayza, S., Urbina, M.A., Jorquera, A., Castro-Nallar, E., and Galbán-Malagón, C., 2020. Monitoring the occurrence of microplastic ingestion in Otariids along the Peruvian and Chilean coasts. *Marine Pollution Bulletin*, 153, 110966.

217. Lusher, A. and Covernton, G.A., 2022. *Microplastics in Fish and Seafood Species*, 367–388. Wiley Online Library.
218. Zantis, L.J., Carroll, E.L., Nelms, S.E., and Bosker, T., 2021. Marine mammals and microplastics: A systematic review and call for standardisation. *Environmental Pollution*, 269, 116142.
219. Caldwell, A., Brander, S.M., Wiedenmann, J., Clucas, G.V., and Craig, E., 2022. Incidence of microplastic fiber ingestion by common terns (Sterna hirundo) and Roseate Terns (S. dougallii) breeding in the Northwestern Atlantic. *Marine Pollution Bulletin*, 177, 113560.
220. Recabarren-Villalón, T., Ronda, A.C., La Sala, L., Sanhueza, C., Díaz, L., Pirani, L.S.R., Picone, A., Romano, R.M., Petracci, P.F., and Arias, A., 2023. First assessment of debris pollution in the gastrointestinal content of juvenile Magellanic penguins (Spheniscus magellanicus) stranded on the west south Atlantic coasts. *Marine Pollution Bulletin*, 188, 114628.
221. Casale, P., Broderick, A.C., Camiñas, J.A., Cardona, L., Carreras, C., Demetropoulos, A., Fuller, W.J., Godley, B.J., Hochscheid, S., Kaska, Y., Lazar, B., Margaritoulis, D., Panagopoulou, A., Rees, A.F., Tòmàs, J., and Türkozan, O., 2018. Mediterranean sea turtles: Current knowledge and priorities for conservation and research. *Endangered Species Research*, 36, 229–267.
222. Solomando, A., Pujol, F., Sureda, A., and Pinya, S., 2022. Ingestion and characterization of plastic debris by loggerhead sea turtle, Caretta caretta, in the Balearic Islands. *Science of the Total Environment*, 826, 154159.
223. Hu, L., Su, L., Xue, Y., Mu, J., Zhu, J., Xu, J., and Shi, H., 2016. Uptake, accumulation and elimination of polystyrene microspheres in tadpoles of Xenopus tropicalis. *Chemosphere*, 164, 611–617.
224. Cruz-Santiago, O., Pérez-Maldonado, I.N., González-Mille, D.J., Espinosa-Reyes, G., Martínez-Toledo, Á., and Ilizaliturri-Hernández, C.A., 2021. Nondestructive biomarkers in giant toad (Rhinella marina) to assess the effect of complex mixture of pollutants in Coatzacoalcos River, Mexico. *Environmental Toxicology and Pharmacology*, 82, 103558.
225. Miller, M.E., Hamann, M., and Kroon, F.J., 2020. Bioaccumulation and biomagnification of microplastics in marine organisms: A review and meta-analysis of current data. *PLOS One*, 15 (10), e0240792.
226. Valsesia, A., Parot, J., Ponti, J., Méhn, D., Marino, R., Melillo, D., Muramoto, S., Verkouteren, M., Hackley, V.A., and Colpo, P., 2021. Detection, counting and characterization of nanoplastics in marine bioindicators: A proof of principle study. *Microplastics and Nanoplastics*, 1 (1).
227. Hartmann, N.B., Hüffer, T., Thompson, R.C., Hassellöv, M., Verschoor, A., Daugaard, A.E., Rist, S., Karlsson, T., Brennholt, N., Cole, M., Herrling, M.P., Heß, M., Ivleva, N.P., Lusher, A., and Wagner, M., 2019. Are we speaking the same language? Recommendations for a definition and categorization framework for plastic debris. *Environmental Science & Technology*, 53 (3), 1039–1047.
228. Ter Halle, A., Jeanneau, L., Martignac, M., Jardé, É., Pedrono, B., Brach, L., and Gigault, J., 2017. Nanoplastic in the North Atlantic subtropical gyre. *Environmental Science & Technology*, 51 (23), 13689–13697.
229. Materić, D., Kasper-Giebl, A., Kau, D., Anten, M., Greilinger, M., Ludewig, E., Van Sebille, E., Röckmann, T., and Holzinger, R., 2020. Micro- and nanoplastics in Alpine snow: A new method for chemical identification and (semi)quantification in the Nanogram range. *Environmental Science & Technology*, 54 (4), 2353–2359.
230. Enfrin, M., Lee, J., Gibert, Y., Basheer, F., Kong, L., and Dumée, L.F., 2020. Release of hazardous nanoplastic contaminants due to microplastics fragmentation under shear stress forces. *Journal of Hazardous Materials*, 384, 121393.
231. Corsi, I., Bellingeri, A., Eliso, M.C., Grassi, G., Liberatori, G., Murano, C., Sturba, L., Vannuccini, M.L., and Bergami, E., 2021. Eco-interactions of engineered nanomaterials in the marine environment: Towards an eco-design framework. *Nanomaterials*, 11 (8), 1903.

232. Corsi, I., Bellingeri, A., and Bergami, E., 2023. Progress in selecting marine bioindicators for nanoplastics ecological risk assessment. *Ecological Indicators*, 154, 110836.
233. Chae, Y. and An, Y., 2017. Effects of micro- and nanoplastics on aquatic ecosystems: Current research trends and perspectives. *Marine Pollution Bulletin*, 124 (2), 624–632.
234. Bergami, E., Pugnalini, S., Vannuccini, M.L., Manfra, L., Faleri, C., Savorelli, F., Dawson, K.A., and Corsi, I., 2017. Long-term toxicity of surface-charged polystyrene nanoplastics to marine planktonic species Dunaliella tertiolecta and Artemia franciscana. *Aquatic Toxicology*, 189, 159–169.
235. Jeong, C.B., Kang, H.M., Lee, M.C., Kim, D.-H., Han, J., Hwang, D.S., Souissi, S., Lee, S.J., Shin, K., Park, H.G., and Lee, J.S., 2017. Adverse effects of microplastics and oxidative stress-induced MAPK/Nrf2 pathway-mediated defense mechanisms in the marine copepod Paracyclopina nana. *Scientific Reports*, 7 (1).
236. Balbi, T., Camisassi, G., Montagna, M.K., Fabbri, R., Franzellitti, S., Carbone, C., Dawson, K.A., and Canesi, L., 2017. Impact of cationic polystyrene nanoparticles (PS-NH_2) on early embryo development of Mytilus galloprovincialis: Effects on shell formation. *Chemosphere*, 186, 1–9.
237. Tallec, K., Huvet, A., Di Poi, C., González-Fernández, C., Lambert, C., Petton, B., Goïc, N.L., Berchel, M., Soudant, P., and Paul-Pont, I., 2018. Nanoplastics impaired oyster free living stages, gametes and embryos. *Environmental Pollution*, 242, 1226–1235.
238. Della Torre, C., Bergami, E., Salvati, A., Faleri, C., Cirino, P., Dawson, K.A., and Corsi, I., 2014. Accumulation and embryotoxicity of polystyrene nanoparticles at early stage of development of sea urchin embryos paracentrotus lividus. *Environmental Science & Technology*, 48 (20), 12302–12311.
239. Pinsino, A., Bergami, E., Della Torre, C., Vannuccini, M.L., Addis, P., Secci, M., Dawson, K.A., Matranga, V., and Corsi, I., 2017. Amino-modified polystyrene nanoparticles affect signalling pathways of the sea urchin (Paracentrotus lividus) embryos. *Nanotoxicology*, 11 (2), 201–209.
240. Eliso, M.C., Bergami, E., Manfra, L., Spagnuolo, A., and Corsi, I., 2020. Toxicity of nanoplastics during the embryogenesis of the ascidian Ciona robusta (Phylum Chordata). *Nanotoxicology*, 14 (10), 1415–1431.
241. Eliso, M.C., Bergami, E., Bonciani, L., Riccio, R., Belli, G., Belli, M., Corsi, I., and Spagnuolo, A., 2023. Application of transcriptome profiling to inquire into the mechanism of nanoplastics toxicity during Ciona robusta embryogenesis. *Environmental Pollution*, 318, 120892.
242. Marques-Santos, L.F., Grassi, G., Bergami, E., Faleri, C., Balbi, T., Salis, A., Damonte, G., Canesi, L., and Corsi, I., 2018. Cationic polystyrene nanoparticle and the sea urchin immune system: Biocorona formation, cell toxicity, and multixenobiotic resistance phenotype. *Nanotoxicology*, 12 (8), 847–867.
243. Murano, C., Agnisola, C., Caramiello, D., Castellano, I., Casotti, R., Corsi, I., and Palumbo, A., 2020. How sea urchins face microplastics: Uptake, tissue distribution and immune system response. *Environmental Pollution*, 264, 114685.
244. Murano, C., Donnarumma, V., Corsi, I., Casotti, R., and Palumbo, A., 2021. Impact of microbial colonization of polystyrene microbeads on the toxicological responses in the sea urchin paracentrotus lividus. *Environmental Science & Technology*, 55 (12), 7990–8000.
245. Cullen-Unsworth, L.C. and Unsworth, R.K.F., 2013. Seagrass meadows, ecosystem services, and sustainability. *Environment: Science and Policy for Sustainable Development*, 55 (3), 14–28.
246. Reynolds, L.K., Waycott, M., McGlathery, K.J., and Orth, R.J., 2016. Ecosystem services returned through seagrass restoration. *Restoration Ecology*, 24 (5), 583–588.

247. Nordlund, L.M., Jackson, E.L., Nakaoka, M., Samper-Villarreal, J., Beca-Carretero, P., and Creed, J.C., 2018. Seagrass ecosystem services—what's next? *Marine Pollution Bulletin*, 134, 145–151.
248. Ripken, C., Khalturin, K., and Shoguchi, E., 2020. Response of coral reef dinoflagellates to nanoplastics under experimental conditions suggests downregulation of cellular metabolism. *Microorganisms*, 8 (11), 1759.
249. Marangoni, L.F.B., Béraud, É., and Ferrier-Pagès, C., 2022. Polystyrene nanoplastics impair the photosynthetic capacities of Symbiodiniaceae and promote coral bleaching. *Science of the Total Environment*, 815, 152136.
250. Hoffmann, L., Eggers, S.L., Allhusen, E., Katlein, C., and Peeken, I., 2020. Interactions between the ice algae Fragillariopsis cylindrus and microplastics in sea ice. *Environment International*, 139, 105697.
251. Bellingeri, A., Casabianca, S., Capellacci, S., Faleri, C., Paccagnini, E., Lupetti, P., Koelmans, A.A., Penna, A., and Corsi, I., 2020. Impact of polystyrene nanoparticles on marine diatom Skeletonema marinoi chain assemblages and consequences on their ecological role in marine ecosystems. *Environmental Pollution*, 262, 114268.
252. González-Fernández, C., Grand, F.L., Bideau, A., Huvet, A., Paul-Pont, I., and Soudant, P., 2020. Nanoplastics exposure modulate lipid and pigment compositions in diatoms. *Environmental Pollution*, 262, 114274.
253. Bergami, E., Manno, C., Cappello, S., Vannuccini, M.L., and Corsi, I., 2020. Nanoplastics affect moulting and faecal pellet sinking in Antarctic krill (Euphausia superba) juveniles. *Environment International*, 143, 105999.
254. Manno, C., Fielding, S., Stowasser, G., Murphy, E.J., Thorpe, S., and Tarling, G.A., 2020. Continuous moulting by Antarctic krill drives major pulses of carbon export in the north Scotia Sea, Southern Ocean. *Nature Communications*, 11 (1).
255. Rowlands, E., Galloway, T.S., and Manno, C., 2021. A Polar outlook: Potential interactions of micro- and nano-plastic with other anthropogenic stressors. *Science of the Total Environment*, 754, 142379.
256. Palazzo, L., Coppa, S., Camedda, A., Cocca, M., De Falco, F., Vianello, A., Massaro, G., and De Lucia, G.A., 2021. A novel approach based on multiple fish species and water column compartments in assessing vertical microlitter distribution and composition. *Environmental Pollution*, 272, 116419.
257. Filgueiras, A.V., Preciado, I., Cartón, A., and Gago, J., 2020. Microplastic ingestion by pelagic and benthic fish and diet composition: A case study in the NW Iberian shelf. *Marine Pollution Bulletin*, 160, 111623.
258. Rios-Fuster, B., Alomar, C., Compa, M., Guijarro, B., and Deudero, S., 2019. Anthropogenic particles ingestion in fish species from two areas of the western Mediterranean Sea. *Marine Pollution Bulletin*, 144, 325–333.
259. Solomando, A., Cohen-Sánchez, A., Box, A., Montero, I., Pinya, S., and Sureda, A., 2022. Microplastic presence in the pelagic fish, Seriola dumerili, from Balearic Islands (Western Mediterranean), and assessment of oxidative stress and detoxification biomarkers in liver. *Environmental Research*, 212, 113369.
260. Morgan, E., Hutchinson, D., and Gaion, A., 2021. Plastic ingestion by the small-spotted catshark (Scyliorhinus canicula) from the South West Coast of the United Kingdom. *Bulletin of Environmental Contamination and Toxicology*, 106 (6), 910–915.
261. Su, L., Cai, H., Kolandhasamy, P., Wu, C., Rochman, C.M., and Shi, H., 2018. Using the Asian clam as an indicator of microplastic pollution in freshwater ecosystems. *Environmental Pollution*, 234, 347–355.
262. Atıcı, A.A., 2021. The first evidence of microplastic uptake in natural freshwater mussel, Unio stevenianus from Karasu River, Turkey. *Biomarkers*, 27 (2), 118–126.

263. Ferreira, G.V.B., Barletta, M., and Lima, A.R.A., 2019. Use of estuarine resources by top predator fishes: How do ecological patterns affect rates of contamination by microplastics? *Science of the Total Environment*, 655, 292–304.
264. Celis-Hernández, O., Ávila, E., Ward, R.D., Rodríguez-Santiago, M.A., and Aguirre-Téllez, J.A., 2021. Microplastic distribution in urban vs pristine mangroves: Using marine sponges as bioindicators of environmental pollution. *Environmental Pollution*, 284, 117391.
265. Tümerkan, E.T.A., Köse, E., Aksu, S., Mol, O., Kantamaneni, K., Başkurt, S., Çınar, E., and Emiroğlu, Ö., 2024. Beadlet anemone: A novel bio-indicator of microplastic pollution in the marine environment. *Journal of Environmental Management*, 349, 119538.
266. Savage, G., Porter, A.H., and Simpson, S.D., 2022. Uptake of microplastics by the snakelocks anemone (Anemonia viridis) is commonplace across environmental conditions. *Science of the Total Environment*, 836, 155144.
267. Iliff, S.M., Wilczek, E.R., Harris, R.J., Bouldin, R.M., and Stoner, E.W., 2020. Evidence of microplastics from benthic jellyfish (Cassiopea xamachana) in Florida estuaries. *Marine Pollution Bulletin*, 159, 111521.
268. Goldstein, J. and Javidpour, J., 2023. Selective feeding protects moon jellyfish (Aurelia aurita s.l.) from overloading with microplastics. *Marine Pollution Bulletin*, 188, 114702.
269. Sucharitakul, P., Pitt, K.A., and Welsh, D.T., 2021. Assessment of microplastics in discharged treated wastewater and the utility of Chrysaora pentastoma medusae as bioindicators of microplastics. *Science of the Total Environment*, 790, 148076.
270. Sun, X., Liang, J., Zhu, M., Zhao, Y., and Zhang, B., 2018. Microplastics in seawater and zooplankton from the Yellow Sea. *Environmental Pollution*, 242, 585–595.
271. Dowarah, K., Arunkumar, P., Thirunavukkarasu, C., Jayakumar, S., and Devipriya, S.P., 2020. Quantification of microplastics using Nile Red in two bivalve species Perna viridis and Meretrix meretrix from three estuaries in Pondicherry, India and microplastic uptake by local communities through bivalve diet. *Marine Pollution Bulletin*, 153, 110982.
272. Pérez-Venegas, D., Toro-Valdivieso, C., Ayala, F., Brito, B., Iturra, L., Arriagada, M., Seguel, M., Barrios, C., Sepúlveda, M., Oliva, D., Cárdenas-Alayza, S., Urbina, M.A., Jorquera, A., Castro-Nallar, E., and Galbán-Malagón, C., 2020. Monitoring the occurrence of microplastic ingestion in Otariids along the Peruvian and Chilean coasts. *Marine Pollution Bulletin*, 153, 110966.

6 Microplastics in Soils
Sources, Fate, Toxicity, and Remediation Technologies

Thies Thiemann

6.1 INTRODUCTION: BACKGROUND AND DRIVING FORCES

Microplastics (MPs) are plastic particles that are less than 5 mm in size. MPs can be categorized as primary or secondary. Primary MPs are produced for purposes such as exfoliating agents in personal care products [1], constituents in industrial abrasives [2] or drug delivery systems [3]. Secondary MPs derive from larger plastic pieces through mechanical [4] and chemical degradation [5], including photo-oxidative [6] and biodegradation [7]. MPs have become ubiquitous in our environment. Over the last 20 years, most reports of MP contamination have come from studies of the aquatic environment, specifically of the marine environment. Oceans have become a major MP sink and depository, with significant input of plastic in general, and of MPs in particular, from land into the oceans [8]. Oceans for the most part are no longer viewed by the public as the pristine environment they once were.

In addition, oceans are seen as a source of sea food for human consumption [9] but also as a source of animal feed in the form of fishmeal [10]. MPs have been found in hundreds of species of marine organisms [11–13]. Jambeck et al. calculated that in 2010, 275 MT (million tons) of plastic were produced in the 192 coastal countries of the world, of which 4.8–12.7 MT entered the oceans [14]. This leaves over 90% of the total, namely over 250 MT plastic on land, some of which is in landfills but much of which can be found in the soil. This is in accordance with a recent study that showed possibly 4 to 23 times more MPs on land than in aquatic environments [15]. Nizetto et al. estimated that up to 700,000 tons of MPs are introduced annually in Europe and North America to agricultural soil via manure application alone [16].

The use of plastic mulch, which has been practiced since the mid-1950s [17] to improve soil climate and water use efficiency [18], continues to increase [19] and is a significant source of MPs in the soil. MPs are also transported from outside the boundaries of agricultural lands such as from bordering roads [20, 21], commercial/industrial land, or residential areas. Furthermore, there is aeolian deposition of MPs on soil [22, 23] and delivery of MPs through irrigation [24, 25], especially when grey water is used. Overall, soil, especially agricultural land, increasingly represents a major sink for MPs [16, 26–29], with an indication that MPs in the soil over longer periods of time form nanoplastics (NPs) [30, 31], plastic particles less than 1 μm in

DOI: 10.1201/9781003468141-6

size. This grave situation is reflected in the increasing number of papers appearing in the last five years studying MPs in the terrestrial environment, including reviews [15, 32–36].

MPs have been reported to alter physical properties of the soil [37, 38] and to change the soil structure, affecting the soil functioning [28]. Also, it has become clear that MPs are present in produce grown on land [39, 40], which represents the largest food source for humans and animals, and therefore, it has been noted that the presence of MPs in the soil affects not only soil biodiversity [41] but also food safety and human health [42]. In this chapter, I look at MP abundance, distribution in soil, sources, toxicity, fates, and interactions with soil biota as well as possibilities for MP mitigation and remediation.

6.2 MP SOURCES, ABUNDANCES, AND SOIL DISTRIBUTION

Some of the main sources of MP pollution in agricultural lands are solid biological substances and composts [43–45], including sewage sludge in some regions [46], irrigating sewage waters [24, 42, 47], mulch films [18, 41, 48, 49], polymer-coated fertilizers [43, 50], and pesticides [43, 51]. Sewage sludge is a significant by-product of wastewater treatment and can undergo thermal processing, composting, agricultural use, and landfilling. Due to legal regulations, sewage sludge is increasingly reused rather than landfilled. This is to minimize the generated waste in line with the zero-waste strategy of many countries.

Sewage sludge ends up in direct agricultural application in China (48%), USA (52%), Ireland (79%), Bulgaria (49%), France (37%), and Czechia (46%). In addition, sewage sludge can find its way into composting [52]. Sludge can hold a significant amount of MPs because most of the MPs in wastewater are retained at the treatment plant [53] and become entrained in the sludge [54–57]. Compost fertilizer in agriculture has also been identified as one of the primary sources of MP contamination in soil.

One significant source of MPs is plastic film degraded from plastic mulching. Plastic mulching is used in agriculture not only to protect plants against inclement weather, weeds, birds and insects but also to regulate the soil temperature and humidity [58], as well as to control the amount of solar radiation hitting the soil and the growing plants. This increases cash crop yields by 20%–60% [59, 60]. Plastic mulch films, which are between 0.0152 and 0.0508 mm in thickness [58, 61], have the advantage over natural mulching materials such as straw, bark, or wood chips in that they do not introduce weeds to the soil and are abundantly available.

In 2006, it was estimated that 700,000 tons/year of low-density polyethylene (LDPE) were used for mulching films, with the amount increasing year by year [62], although guidelines in certain countries ask for a certain percentage of recyclable mulch material. Altogether, by 2001, it was estimated that 1 million tons of plastic mulching film were used worldwide on an annual basis [63]. China in particular has seen a strong increase in plastic mulching, from 0.32 million tons in 1991 to 1.25 million tons in 2011 [59, 64]. This goes hand in hand with a higher absolute intensity of use, from 7.0 kg/ha in 1991 to 34.8 kg/ha in 2011 in different regions in China such as in Xianjiang [59, 60].

Common mulch films are made of polyethylene (PE), including linear LDPE and high-density polyethylene (HDPE) [65]. Mulches containing polyvinyl chloride (PVC) have been banned in many countries including the United States [66, 67] because of toxicity concerns, but many of the mulch films used in China are still made of PVC [59]. In addition, numerous items used in agriculture are made of plastic such as water pipes, silage and bale nets, shade nets and wind barriers, and plastic greenhouses [68], and all of these are potential sources of MPs. Agricultural activities utilized 3.4% of the 50.7 million tons of plastic used in Europe in 2019 [68, 69].

From roads adjoining agricultural land, abraded tire material enters the soil [21, 70, 71]. Here, Baensch-Baltruschat et al. report that 1,600–2,800 tons of road and tire wear material are moved annually per acre of road-adjoining land [72]. Apart from micro-tires and road wear material, other plastics are entering agricultural lands from urban areas through general littering, waste dumping, and inappropriate management of landfill sites [68, 73]. Small plastic particles can easily be transported from landfills, waste dumps, and urban areas onto agricultural lands through air movement [74, 75]. The use of recycled water for irrigation is another potential source of MPs [76–78], with recycled irrigation water analyzed in Fuerteventura, Spain, showing as many as 40.0 ± 19.0 items/L [76]. In the United Arab Emirates, where a significant fraction of the water used to irrigate the greeneries alongside roads and in parks is grey water, there always has been the worry of soil contamination with MPs.

MP movement is not unidirectional, however; driven by water and air, MPs can redistribute within the soil but also leave the soil altogether in the direction of aquatic environments such as groundwater, lakes, rivers, and the ocean. These movements are most pronounced during water run-offs [79] and in cases of soil erosion [80] through wind and water, where often the final receiving body is surface water [81]. The vertical and horizontal redistribution of MPs within the soil is more complex, as it can also be accomplished by MP transport through soil organisms, both invertebrate and vertebrate. In this regard, it has been hypothesized that aged MPs are more likely to be transported as they are more palatable to soil invertebrates due to biofilm formation on their surface as they become colonized by microorganisms [82].

Apart from the bioturbation, human disturbance of the soil such as through tillage can contribute to the vertical movement of MPs in the soil [67]. Direct MP leaching through the soil has been found to be more difficult and depends on the nature of the soil; soils with a large proportion of macropores such as loamy soils with bio-pores, leptosols, coarse sandy soils, and organic soils will allow the vertical movement of larger MPs [67]. On average, the particles that leach through soil pores are no larger than 1.91 μm [67, 83, 84]. However, investigators have found leaching of MPs 2.6–20 μm in different soils, including in sand [85, 86]. Although NPs (<100 nm) are smaller than many soil meso- and macropores [87], it is difficult for these particles to leach through soils, and soils retain many of these particles [67].

Büks and Kaupenjohann [88] gave a comprehensive overview of studies conducted on MP concentrations in agricultural soils from all over the world, concentrating on sites that had been exposed to sewage sludge and/or mulching films. The studied sites showed an average concentration of <13,000 items/kg dry soil (4.5 mg/kg dry soil).

MP concentrations in soils of urban residential areas were on average 10 times higher than those in soils from rural sites. Most studies on soil MP concentrations come from Europe (Spain, Germany, Austria, and Scandinavia) and from China. Studies from China showed a median MP concentration of 1,076 items/kg dry soil, while studies from Europe revealed a median concentration of 2,914 items/ kg dry soil [88].

Once entrained in the soil, plastic materials can have a long lifetime. It is known that plastics can degrade through photo-oxidative [2] and hydrolytic [89] processes, and wind and weather can also abrade MPs [90]. Even animals such as insects, including insect larvae, can accelerate the degradation of MPs in the soil upon ingestion through the action of intestinal bacteria [91–93]. In addition, there is microbially driven degradation by soil microorganisms. Nevertheless, all these processes are slow: Albertson et al. found a weight reduction of 0.4 w% for PE after 800 days [94], and Arkatbar et al. [95] reported weight loss of 0.4 w% for polypropylene after one year. Little to no degradation was found for PVC, even after 35 years in soil [96–98]. An additional worry is that with time, plastic particles vertically embed themselves further in the soil strata, where certain mechanisms of degradation such as photo-oxidative breakdown of plastics cease to function.

6.3 EFFECTS OF MPs ON SOIL PROPERTIES

MPs can decrease soil bulk density [99, 100], which leads to better soil aeration, helping root penetration and growth [101], and MPs can affect soil aggregation; soil aggregation is often aided by microbial exudates such as polysaccharide metabolites, and MPs, especially their leachable additives, can have a pronounced effect on certain types of microbial communities in the soil such as on actinobacteria, which are found reduced in numbers and variety in the presence of plastic film fragments in the soil [102, 103]. MPs in the soil can also affect soil pH: PE decreased soil pH [104], while polylactic acid MPs increased it [105]. Soil pH may partially stem from a changed pattern of exuded metabolites from an altered composition of the microbial community in the soil. Conversely, a change of soil pH has a significant impact on the make-up of the microbial community [106, 107].

Water retention capacity is another property of the soil that can be influenced by MPs. However, different types of MPs can influence this property in different ways; for instance, polyester fibers increase potential water-holding capacity [37], whereas PE films increase soil water loss through enhanced evaporation [108]. Often, larger MPs decrease soil water-holding capacity [60]. MPs can also block pores in the soil, decreasing the soil infiltration capability. Especially when MPs are present in large concentrations, this will influence the soil nutrient cycle [109].

6.4 EFFECTS OF MPs ON SOIL BIOTA

Studies have shown that MPs can have a profound effect on soil organisms [110]. Amending soil with PE (2,000 MPs/kg soil) significantly increased urease and catalase activity in the soil after 15 days, while keeping the invertase activity unaffected; 6S rRNA analyses revealed that the alpha diversities (richness, evenness, and diversity) of the microbiota in soil were not obviously changed by the PE amendment [111].

However, the authors observed different taxonomic compositions in the amended soils after 90 days of incubation; PE MPs had been colonized by bacterial assemblages with distinct community structures including plastic-degrading bacteria [111]. MPs in the soil microbiome also influence nutrient transport, bacteria transport, and the decomposition of organic matter in the soil [112–116].

Phthalates are a common additive in plastics, especially in plastic mulch films and used as plasticizers, but phthalates can be released into the soil, where they can be taken up by crops [64, 117, 118] and thus pose a human health risk via the food chain. Phthalates have become so ubiquitous in the environment that they have been isolated from many plants, where for the most part they have entered as contaminants [119]. In humans, phthalates can act as endocrine disruptors [66] and are known to affect Sertoli cells and Leydig cells [120]; some researchers noted decreased sperm motility and concentration upon human exposure to phthalates [120, 121].

Phthalates can also induce the proliferation of estrogen receptor positive MCF-7 human breast cancer cells [122]. In soil, phthalate-containing MPs have been seen to affect microbial activity and diversity [59, 123]. At concentrations >1mg/g, diethyl phthalate decreased bacteria by 47% and pseudomonad activity by 62% [124]. Other additives in plastics serve as flame retardants (boric acid. decabromodiphenyl ether [BDE209], tetrabromobisphenol-A [TBBPA], tris(2-chloroethyl)phosphate), antioxidants (bisphenol-A, octyl- and nonylphenols, 2,6-bis(*tert*-butyl)-4-methylphenol), light and heat stabilizers (barium and calcium nonylphenolate), lubricants (oleamide, erucamide, zinc stearate), and pigments (cobalt(II) diacetate, diazo pigments, anthraquinones and phthalocyanines) [125], where many of these additives have a significant effect on the soil microbial population. For instance, Zhang et al. found that both BDE209 and TBBPA affect soil microorganisms, with fungi more affected than bacteria and actinomycetes the least affected [126]. Upon exposure to BDE209 and TBBPA, urease activity of microorganisms is much reduced, while catalase and saccharase activities are less affected. Rothenbacher and Pecquet noted that TTBPA affects not only soil microorganisms but also earthworms and seedling emergence [127].

Plastic surfaces are known to adsorb metals [128] as well as small organic molecules that include pesticides [129], pharmaceuticals [130], and their metabolites. These adsorbed species can desorb under the right conditions and leach into the neighboring soil, affecting soil organisms. As MPs can also transport microorganisms, including pathogens [131], they can act as vectors for these and can thereby alter the microbial soil environment.

6.5 EFFECTS OF MPs ON SOIL FAUNA

Earthworms have been seen to transport MPs within the soil. Rillig et al. found that PE microbeads placed on top of the soil were transported by the earthworm *Lumbricus terrestris L.* down the soil profile [132], with smaller particles being transported the farthest down. MPs were also found in burrows of *L. terrestris* [133]. Zhu et al. looked at the dispersal of PVC MPs, sized 80–250 μm, by a combined micro-arthropod population of predatory mites *Hypoaspis aculeifer*, oribatid mites *Damaeus exspinosus*, and springtails *Folsomia candida* [134].

Earthworms are also known to ingest MPs. Some scientists have a found a preference of earthworms for certain types of MPs such as for polylactate and polyethylene terephthalate containing MPs [135], while other scientists reported more indiscriminate MP ingestion [136–138]. While earthworms were not affected when ingesting small amounts of MPs (<0.5% w/w) [139], the ingestion of larger amounts of MPs (1%–2% w/w) reflected in higher mortality and slower growth [139, 140]. Significant ingestion of MPs also led to altered burrowing behavior, altered feeding behavior [141], and reduced reproduction [142]. At a molecular level, earthworm exposure to MPs altered catalase (CAT) activity [143–146]. CAT is a common and highly active enzyme in organisms that directly catalyzes the decomposition of H_2O_2 to nontoxic water and O_2. In most of the studies mentioned, MPs often increased CAT activity [143–145], although instances of reduced activity have also been observed [146].

Furthermore, researchers have investigated the interaction between MPs and springtails (collembolans). Zhu et al. studied the effect of MP ingestion on the gut microbiota of common parthenogenetic collembolan *F. candida* [147] and found enhanced bacterial diversity in gut, along with 16.8% reduction in growth and a 28.8% reduction in reproduction. According to Kim and An [148], springtails *F. candida* consume PE MPs of 66 μm or less, and this results in a 74% ± 38% decrease in the velocity and distance of the springtail motions.

6.6 EFFECTS OF MPs ON PLANTS

Changes in soil properties due to MPs can lead to negative effects in plant growth, especially where MPs reduce the water holding capacity of the soil or where it leads to channels or cracks in the soil that release moisture too quickly. Some of the effects of MPs on plants result from actual physical blockage of pores in the plant body during the plant's life cycle, diminishing water uptake [149]. Alternatively, reduced seed germination in the presence of MPs was explained by the cytotoxicity of the MPs themselves [150] or by the cytotoxicity of plasticizers found in MPs [151].

MPs can be taken up through the roots of plants, or they can be adsorbed by plant shoot system, where MPs can be deposited through irrigation or through aeolian transport [152]. There is some evidence that MPs and especially the smaller NPs, absorbed through the roots, can be transported to edible parts of plants [153], again posing the risk of MP transfer to the human food chain. Studies have been published that seem to indicate that certain biodegradable MPs pose greater risks to plants than petroleum-based MPs. This needs to be taken into close consideration when looking for substitute materials for agricultural use.

6.7 EFFECTS OF MPs ON MAMMALS

There are fewer reports on the ingestion of plastic material by land mammals [154] than by marine mammals [155], although researchers have looked at plastic ingestion, including of MPs, by animals in agricultural communities [156–159]. Plastic ingestion contributes to redistribution, especially of MPs, in soils, where they are redeposited with the feces of, for instance, sheep [160], deer [161], and birds such as vultures [162]. MPs have also been found in livestock and poultry manure [163, 164].

Ingestion of MPs by farm animals is especially worrying [165] as in this way MPs may transfer to the human food chain. Already milk products [166], eggs [167], and meat [168] have been shown to contain MPs at various concentrations.

6.8 MITIGATION AND REMEDIATION OF MICROPLASTIC CONTAMINATION

Tao et al. [169] found that accumulated LDPE (up to 0.28%) improved the soil micro-environment for growing horsebeans, and Zhang et al. [170] reported that LDPE remnants (with a maximum content of 0.35%) improved soil fertility when growing corn. Nevertheless, apart from finding substitute materials for silage, bale, and shade nets as well as wind barriers, a number of papers have advocated replacing PE-based mulch film with biodegradable plastic mulch film [58, 171–175].

Interestingly, substituting synthetic hydrocarbon-based, polymeric material in mulching such as LDPE for biodegradable polymer-based material does not always warrant success. Qi et al. [176] showed that residues from one of the more widely used [177] biodegradable films composed of pullulan (37.1%), polyethylene terephthalate (44.6%), and polybutylene terephthalate (18.3%) showed more detrimental effects on wheat growth than residues from a LDPE film. Often, it is also the extent and the rate of biodegradability of the substitute material that are in question. The major reason why the breakdown of biodegradable products in soil cannot be predicted is that many of them are certified to meet the standards of biodegradation only under the conditions of industrial composting, the conditions they have been tested for, but these are not necessarily the conditions the materials are exposed to in agricultural soils. Therefore, Brodhagen et al. [19] advocated for a new standard of testing and certifying such biodegradable products.

Much is made of the ability of microorganisms (e.g., of the phylum Bacillota) belonging to the digestive tracts of certain soil invertebrates [178] to metabolize plastic polymers. While without doubt this will lead to new methods of enzymatic polymer biodegradation, the volume of contaminated soil would only allow in situ remediation effort. Any addition of soil invertebrates with such a remediation in mind should be carried out with great care and with the utmost regard for the soil ecosystem.

REFERENCES

1. Habib, R.Z., Abdoon, M., Al Meqbaali, R., Ghebremedhin, F., Elkashlan, M., Kittaneh, W.F., Cherupurakal, N., Mourad, A.-H.I., Thiemann, T., and Al Kindi, R. 2020 Analysis of microbeads in cosmetic products in the United Arab Emirates. *Environmental Pollution*, 258, p. 113831.
2. Duis, K. and Coors, A. 2016 Microplastics in the aquatic and terrestrial environment: Sources (with a specific focus on personal care products), fate and effects. *Environmental Sciences Europe*, 28, p. 2.
3. Santos, L.H.M.L.M., Rodríguez-Mozaz, S., and Barceló, D. 2021 Microplastics as vectors of pharmaceuticals in aquatic organisms—an overview of their environmental implications. *Environmental Engineering*, 3, p. 100079.
4. Manzoor, S., Naqash, N., Rashid, G., and Singh, R. 2022 Plastic material degradation and formation of microplastic in the environment: A review. *Materials Today: Proceedings*, 56, p. 3254.

5. Webb, H.K., Arnott, J., Crawford, R.J., and Ivanova, E.P. 2013 Plastic degradation and its environmental implications with special reference to poly(ethylene terephthalate). *Polymers*, 5, p. 1.
6. Yousif, E. and Haddad, R. 2013 Photodegradation and photostabilization of polymers, especially polystyrene: Review. *Springerplus*, 2, p. 398.
7. Shah, A.A., Hasan, F., Hameed, A., and Ahmed, S. 2008 Biological degradation of plastics: A comprehensive review. *Biotechnology Advances*, 26, p. 246.
8. Andrady, A.L. 2011. Microplastics in the marine environment. *Marine Pollution Bulletin*, 62, p. 1596.
9. Smith, M., Love, D.C., Rochman, C.M., and Neff, R.A. 2018 Microplastics in seafood and the implications for human health. *Current Environmental Health Reports*, 5, p. 375.
10. Thiele, C.J., Hudson, M.D., Russell, A.E., Saluveer, M., and Sidaoui-Haddad, G. 2021 Microplastics in fish and fishmeal: An emerging environmental challenge? *Scientific Reports*, 11, p. 2045.
11. Miller, M.E., Hamann, M., and Kroon, F.J. 2020 Bioaccumulation and biomagnification of microplastics in marine organisms: A review and meta-analysis of current data. *PLoS One*, 15, e0240792.
12. Ugwu, K., Herrera, A., and Gómez, M. 2021 Microplastics in marine biota: A review. *Marine Pollution Bulletin*, 169, p. 112540.
13. Wootton, N., Reis-Santos, P., and Gillanders, B.M. 2021 Microplastic in fish—a global synthesis. *Reviews in Fish Biology and Fisheries*, 31, p. 753.
14. Jambeck, J.R., Geyer, R., Wilcox, C., Siegler, T.R., Perryman, M., Andrady, A., Narayan, R., and Law, K.L. 2015 Plastic waste inputs from land into the ocean. *Science*, 347, p. 768.
15. De Souza Machado, A.A., Kloas, W., Zarfl, C., Hempel, S., and Rillig, M.C. 2018 Microplastics as an emerging threat to terrestrial ecosystems. *Global Change Biology*, 24, p. 1405.
16. Nizzetto, L., Langaas, S., and Futter, M. 2016 Pollution: Do microplastics spill on to farm soils? *Nature*, 537, p. 488.
17. Emmert, E.M. 1957 Black polyethylene for mulching vegetables. *Proceedings of the American Society for Horticultural Science*, 69, p. 464.
18. Qi., Y., Yang, X., Pelaez, A.M., Lwanga, E.H., Beriot, N., Gertsen, H., Garbeva, P., and Geissen, V. 2018 Macro- and microplastic in soil-plant system: Effect of plastic mulch film residues on wheat (*Triticium aestivum*) growth. *Science of the Total Environment*, 645, p. 1048.
19. Brodhagen, M., Goldberger, J.R., Hayes, D.G., Inglis, D.A., Marsh, T.L., and Miles, C. 2017. Policy considerations for limiting unintended residual plastic in agricultural soils. *Environmental Science & Policy*, 60, p. 81.
20. Habib, R.Z., Ramachandran, T., Hamed, F., Al Kindi, R., Mourad, A.I., and Thiemann, T. 2022 Microplastic in an arid region: Identification, quantification and characterization on and along-side roads in Al Ain, Abu Dhabi, United Arab Emirates. *Journal of Environmental Protection (JEP)*, 13, p. 671.
21. Habib, R.Z., Al Kendi, R., Ghebremedhin, F., Elkashlan, M., Iftikhar, S.H., Poulose, V., Ramachandran, T., Mourad, A.H.I., Hamed, F., and Thiemann, T. 2023 Tire and rubber particles in the environment—a case study from a hot arid region. *Frontiers in Environmental Science—Toxicology, Pollution and the Environment*, 10, p. 1009802.
22. Zhang, Y., Gao, T., Kang, S., and Sillanpää, M. 2019 Importance of atmospheric transport for microplastics deposited in remote areas. *Environmental Pollution*, 254A, p. 112953.
23. Klein, M. and Fischer, E.K. 2019 Microplastic abundance in atmospheric deposition within the Metropolitan area of Hamburg, Germany. *Science of the Total Environment*, 685, p. 96.

24. Jiang, J.J., Hanun, J.N., Chen, K.Y., Hassan, F., Liu, K.T., Hung, Y.H., and Chang, T.W. 2023 Current levels and composition profiles of microplastics in irrigation water. *Environmental Pollution*, 318, p. 120858.
25. Pérez-Reverón, R., González-Sálamo, J., Hernández-Sánchez, C., González-Pleiter, M., Hernández-Borges, J., and Díaz-Peña, F.J. 2022 Recycled wastewater as a potential source of microplastics in irrigated soils from an arid-insular territory (Fuerteventura, Spain). *Science of the Total Environment*, 817, p. 152830.
26. Mahon, A.M., O'Connell, B., Healy, M.G., O'Connor, I., Officer, R., and Nash, R. 2017 Microplastics in sewage sludge: Effects of treatment. *Environmental Science & Technology*, 51, p. 810.
27. Nizzetto, L., Futter, M., and Langaas, S. 2016 Are agricultural soils dumps for microplastics of urban origin? *Environmental Science & Technology*, 50, p. 10777.
28. Rillig M.C. 2012 Microplastic in terrestrial ecosystems and the soil? *Environmental Science & Technology*, 46, p. 6453.
29. Rillig, M.C., Ingraffia, R., and Machado, A.A.D. 2017 Microplastic incorporation into soil in agroecosystems. *Frontiers in Plant Science, Section Agroecology*, 8, p. 01805.
30. Lambert, S. and Wagner, M. 2016 Characterisation of nanoplastics during the degradation of polystyrene. *Chemosphere*, 145, p. 265.
31. Lwanga, E.H., Thapa, B., Yang, X., Gertsen, H., Salánki, T. Geissen, V., and Grabeva, P. 2018 Decay of low-density polyethylene by bacteria extracted from earthworm's guts: a potential for soil restoration. *Science of the Total Environment*, 624, p. 753.
32. Yang, L., Zhang, Y., Kang, S., Wang, Z., and Wu, C. 2021 Microplastics in soil: A review on methods, occurrence, sources, and potential risk. *Science of the Total Environment*, 780, p. 146546.
33. Zhou, Y., Wang, J., Zou, M., Jia, Z., Zhou, S., and Li, Y. 2020. Microplastics in soils: A review of methods, occurrence, fate, transport, ecological and environmental risks. *Science of the Total Environment*, 748, p. 141368.
34. Wang, J., Liu, X., Li, Y., Powell, T., Wang, X., Wang, G., and Zhang, P. 2019 Microplastics as contaminants in the soil environment: A mini-review. *Science of the Total Environment*, 691, p. 848.
35. Xu, B., Liu, F., Cryder, Z., Huang, D., Lu, Z., and He, Y. 2019 Microplastics in the soil environment: Occurrence, risks, interactions and fate—a review. *Critical Reviews in Environmental Science and Technology*, 50(21), p. 2175.
36. Ng, E.-L., Lwanga, E.H., Eldridge, S.M., Johnston, P., Hu, H.-W., Geissen, V., and Chen, D. 2018 An overview of microplastic and nanoplastic pollution in agroecosystems. *Science of the Total Environment*, 627, p. 1377.
37. De Souza Machado, A.A., Lau, C.W., Kloas, W., Bermann, J., Bachelier, J.B., Faltin, E., Becker, R.A.S., and Rillig, M.C. 2019 Microplastics can change soil properties and affect plant performance. *Environmental Science & Technology*, 53, p. 6044.
38. De Souza Machado, A.A., Lau, C.W., Till, J., Kloas, W., Lehmann, A., Becker, R., and Rillig, M.C. 2018 Impacts of microplastics on the soil biophysical environment. *Environmental Science & Technology*, 52(17), p. 9656.
39. Conti, G.O., Ferrante, M., Banni, M., Favara, C., Nicolosi, I., Cristaldi, A., Fiore, M., and Zuccarello, P. 2020 Micro- and nano-plastics in edible fruit and vegetables. The first diet risks assessment for the general population. *Environmental Research*, 187, p. 109677.
40. Tympa, L.-E., Katsara, K., Moschou, P.N., Kenanakis, G., and Papadakis, V.M. 2021 Do microplastics enter our food chain via root vegetables? A Raman based spectroscopic study on *Raphanus sativus*. *Materials*, 14, p. 2329.
41. Tibbett, M., Fraser, T.D., and Duddigan, S. 2020 Identifying potential threats to soil biodiversity. *PeerJ*, 8, p. e9271.

42. Briassoulis, D. 2023 Agricultural plastics as a potential threat to food security, health, and environment through soil pollution by microplastics: Problem definition. *Science of the Total Environment*, 892, p. 164533.
43. Kumar, M., Xiong, X., He, M., Tsang, D.C.W., Gupta, J., Khan, E., Harrad, S., Hou, D., Ok, Y.S., and Bolan, N.S. 2020 Microplastics as pollutants in agricultural soils. *Environmental Pollution*, 265, p. 114980.
44. Vithanage, M., Ramanayaka, S., Hasinthara, S., and Navaratne, A. 2021 Compost as a carrier for microplastics and plastic-bound toxic metals into agroecosystems. *Current Opinion in Environmental Science & Health*, 24, p. 100297.
45. Porterfield, K.K., Hobson, S.A., Neher, D.A., Niles, M.T., and Roy, E.D. 2023 Microplastics in composts, digestates, and food wastes: A review. *Journal of Environmental Quality*, 52, p. 225.
46. Koyuncuoğlu, P. and Erden, G. 2021 Sampling, pre-treatment, and identification methods of microplastics in sewage sludge and their effects in agricultural soils: A review. *Environmental Monitoring Assessment*, 193, p. 175.
47. Quilliam, R.S., Pow, C.J., Shilla, D.J., Mwesiga, J.J., Shilla, D.A., and Woodford, L. 2023 Microplastics in agriculture—a potential novel mechanism for the delivery of human pathogens onto crops. *Frontiers in Plant Science*, 14, p. 1152419.
48. Steinmetz, Z., Wollmann, C., Schaefer, M., Buchmann, C., David, J., Tröger, J., Muñoz, K., Frör, O., and Schaumann, G.E. 2016 Plastic mulching in agriculture. Trading short-term agronomic benefits for long-term soil degradation? *Science of the Total Environment*, 550, p. 690.
49. Huang, Y., Liu, Q., Jia, W., Yan, C., and Wang, J. 2020 Agricultural plastic mulching as a source of microplastics in the terrestrial environment. *Environmental Pollution*, 260, p. 114096.
50. Lawrencia, D., Wong, S.K., Low, D.Y.S., Goh, B.H., Goh, J.K., Ruktanonchai, U.R., Soottitantawat, A., Lee, L.H., and Tang, S.Y. 2021 Controlled release fertilizers: A review on coating materials and mechanism of release. *Plants*, 10, p. 238.
51. Accinelli, C., Abbas, H.K., Shier, W.T., Vicari, A., Little, N.S., Aloise, M.R., and Giacomini, S. 2019 Degradation of microplastic seed film-coating fragments in soil. *Chemosphere*, 226, p. 645.
52. Buta, M., Hubeny, J., Zielínski, W., Harnisz, M., and Korzeniewska, E. 2021 Sewage sludge in agriculture—the effects of selected chemical pollutants and emerging genetic resistance determinants on the quality of soil and crops—a review. *Ecotoxicology and Environmental Safety*, 214, p. 112070.
53. Habib, R., Thiemann, T., and al Kindi, R. 2020 Microplastics and wastewater treatment plants. *Journal of Water Resource Protection (JWARP)*, 2020, p. 1.
54. Zhang, Z. and Chen, Y. 2020 Effects of microplastics on wastewater and sewage sludge treatment and their removal: A review. *Chemical Engineering Journal*, 382, p. 122955.
55. Rolsky, C., Kelkar, V., Driver, E., and Halden, R.U. 2020 Municipal sewage sludge as a source of microplastics in the environment. *Current Opinion in Environmental Science & Health*, 14, p. 16.
56. Hernández-Arenas, R., Beltrán-Sanahuja, A., Navarro-Quirant, P., and Sanz-Lazaro, C. 2021 The effect of sewage sludge containing microplastics on growth and fruit development of tomato plants. *Environmental Pollution*, 268B, p. 115779.
57. Hassan, F., Prasetya, K.D., Hanun, J.N., Bui, H.M., Rajendran, S., Kataria, N., Khoo, K.S., Wang, Y.F., You, S.J., and Jiang, J.J. 2023 Microplastic contamination in sewage sludge: Abundance, characteristics, and impacts on the environment and human health. *Environmental Technology & Innovation*, 31, p. 103176.
58. Kasirajan, S. and Ngouajio, M. 2012 Polyethylene and biodegradable mulches for agricultural applications: A review. *Agronomy for Sustainable Development*, 32, p. 501.

59. Wang, J., Lv, S., Zhang, M., Chen, G., Zhu, T., Zhang, S., Teng, Y., Christie, P., and Luo, Y. 2016 Effects of plastic film residues on occurrence of phthalates and microbial activity in soils. *Chemosphere*, 151, p. 171.
60. Liu, E.K., He, W.Q., and Yan, C.R. 2014 "White revolution" to "white pollution"—agricultural plastic film mulch in China. *Environmental Research*, 9, p. 1.
61. Lamont, W.J. 2005 Plastics: Modifying the microclimate for the production of vegetable crops. *Hort Technology*, 15, p. 477.
62. Espi, E., Salmeron, A., Fontecha, A., Garcia, Y., and Real, A.I. 2006 Plastic films for agricultural applications. *Journal of Plastic Film and Sheeting*, 22, p. 85.
63. Halley, P., Rutgers, R., Coombs, S., Kettels, J., Gralton, J., Christie, G., Jenkins, M., Beh, H., Griffin, K., Jayasekara, R., and Lonergan, G. 2001 Developing biodegradable mulch films from starch-based polymers. *Starch*, 53, p. 362.
64. Wang, J., Luo, Y.M., Teng, Y., Ma, W.T., Christie, P., and Li, Z.G. 2013 Soil contamination by phthalate esters in Chinese intensive vegetable production system with different modes of use of plastic film. *Environmental Pollution*, 180, p. 265.
65. Fleck-Arnold, J.E. 2000 Plastic mulch films—additives and their effects. *Proceedings of the 15th National Agricultural Plastics Congress*, 29, p. 310.
66. US EPA. 2012 *Guidelines for using mulches and weed barriers*. US EPA.
67. Bläsing, M. and Amelung, W. 2018 Plastics in soil: Analytical methods and possible sources. *Science of the Total Environment*, 612, p. 422.
68. Lwanga, E.H., Beriot, N., Corradini, F., Silva, V., Yang, X., Baartman, J., Rezaei, M., van Schalk, L., Riksen, M., and Geissen, V. 2022 Review of microplastic sources, transport pathways and correlations with other soil stressors: A journey from agricultural sites into the environment. *Chemical and Biological Technologies in Agriculture*, 9, p. 20.
69. Plastic Europe. 2020 *Plastics—the facts 2020: An analysis of European plastics production, demand and waste data*. Published online at PlasticsEurope.org. www.plasticseurope.org/es/resources/publications/4312-plastics-facts-2020.
70. Sommer, F., Dietze, V., Baum, A., Sauer, J., Gilge, S., Maschowski, C., and Gieré, R. 2018 Tire abrasion as a major source of microplastics in the environment. *Aerosol Air Quality Research*, 18, p. 2014.
71. Luo, Z., Zhou, X., Su, Y., Wang, H., Yu, R., and Zhou, S. 2021 Environmental occurrence, fate, impact, and potential solution of tire microplastics: Similarities and differences with tire wear particles. *Science of the Total Environment*, 795, p. 148902.
72. Baensch-Baltruschat, B., Kocher, B., Stock, F., and Reifferscheid, G. 2020 Tyre and road wear particles (TRWP)—a review of generation, properties, emissions, human health risk, ecotoxicity, and fate in the environment. *Science of the Total Environment*, 733, p. 137823.
73. Duis, K. and Coors, A. 2016 Microplastics in the aquatic and terrestrial environment: sources (with a specific focus on personal care products), fate and effects. *Environmental Sciences Europe*, 28, p. 2.
74. Zylstra, E.R. 2013 Accumulation of wind-dispersed trash in desert environments. *Journal of Arid Environments*, 89, p. 13.
75. Bullard, J.E., Ockelford, A., O'Brien, P., and McKenna Neuman, C. 2021 Preferential transport of microplastics by wind. *Atmospheric Environment*, 245, p. 118038.
76. Pérez-Reverón, R., González-Sálamo, J., Hernández-Sánchez, C., González-Pleiter, M., Hernández-Borges, J., and Díaz-Peña, F.J. 2022 Recycled wastewater as a potential source of microplastics in irrigated soils from an arid-insular territory (Fuerteventura, Spain). *Science of the Total Environment*, 817, p. 152830.
77. Jiang, J.-J., Hanun, J.N., Chen, K.-Y., Hassan, F., Liu, K.-T., Hung, Y.-H., and Chang, T.-W. 2023 Current levels and composition profiles of microplastics in irrigation water, *Environmental Pollution*, 318, p. 120858.

78. Guo, S., Zhang, J., Liu, J., Guo, N., Zhang, L., Wang, S., Wang, X., Zhao, M., Zhang, B., and Chen, Y. 2023 Organic fertilizer and irrigation water are the primary sources of microplastics in the facility soil, Beijing. *Science of the Total Environment*, 895, p. 165005.
79. Nizzetto, L., Butterfield, D., Bussi, G., Whitehead, P.G., and Futter, M.N. 2016 A theoretical assessment of microplastic transport in river catchments and their retention by soils and river sediments. *Environmental Science: Processes & Impacts*, 18, p. 1050.
80. Rehm, R., Zeyer, T., Schmidt, A., and Fiener, P. 2021 Soil erosion as transport pathway of microplastic from agriculture soils to aquatic ecosystems. *Science of the Total Environment*, 795, p. 148774.
81. He, L., Ou, Z., Fan, J., Zeng, B., and Guan, W. 2022 Research on the non-point source pollution of microplastics. *Frontiers in chemistry, sec. Nanochemistry*, 10, p. 956547.
82. Helmberger, M.S., Tiemann, L.K., and Grieshop, M.J. 2019 Towards an ecology of soil microplastics. *Functional Ecology*, 34, p. 550.
83. McGechan, M.B. and Lewis, D.R. 2002. SW—soil and water: Transport of particulate and colloid-sorbed contaminants through soil, part 1: General principles. *Biosystems Engineering*, 3, p. 255.
84. McGechan, M.B. 2002 SW-soil and water: Transport of particulate and colloid-sorbed contaminants through soil, part 2: Trapping processes and soil pore geometry. *Biosystems Engineering*, 4, p. 387.
85. Wang, D., Zhang, W., Hao, X., and Zhou, D. 2013. Transport of biochar particles in saturated granular media-effects of pyrolysis temperature and particle size. *Environmental Science & Technology*, 2, p. 821.
86. Zhang, W., Morales, V.L., Cakmak, M.E., Salvucci, A.E., Geohring, L.D., Hay, A.G., Parlange, J.-Y., and Steenhuis, T.S. 2010 Colloid transport and retention in unsaturated porous media: Effect of colloid input concentration. *Environmental Science & Technology*, 13, p. 4965.
87. Hartland, A., Lead, J.R., Slaveykova, V., O'Carroll, D., and Valsami-Jones, E. 2013 The environmental significance of natural nanoparticles. *Nature Education Knowledge*, 8, p. 7.
88. Büks, F. and Kaupenjohann, M. 2020 Global concentrations of microplastics in soils—a review. *Soil*, 6, p. 649.
89. Allen, N.S., Edge, M., Mohammadian, M., and Jones, K. 1991 Hydrolytic degradation of poly(ethylene terephthalate): Importance of chain scission versus crystallinity. *European Polymer Journal*, 27, p. 1373.
90. Bullard, J.E., Zhou, Z., Davis, S., and Fowler, S. 2023 Breakdown and modification of microplastic beads by Aeolian abrasion. *Environmental Science & Technology*, 57, p. 76.
91. An, R., Liu, C., Wang, J., and Jia, P. 2023 Recent advances in degradation of polymer plastics by insects inhabiting microorganisms. *Polymers (Basel)*, 5, p. 1307.
92. Wu, W.-M. and Criddle, C.S. 2021 Chapter five—characterization of biodegradation of plastics in insect larvae. *Methods in Enzymology*, 648, p. 95.
93. Siddiqui, S.A., Manap, A.S.A., Kolobe, S.D., Monnye, M., Yudhistra, B., and Fernando, I. 2024 Insects for plastic biodegradation—a review. *Process Safety and Environmental Protection*, in press.
94. Albertsson, A.-C. 1980 The shape of the biodegradation curve for low and high density polyethenes in prolonged series of experiments. *European Polymer Journal*, 7, p. 623.
95. Arkatkar, A., Arutchelvi, J., Bhaduri, S., Uppara, P.V., and Doble, M. 2009 Degradation of unpretreated and thermally pretreated polypropylene by soil consortia. *International Biodeterioration & Biodegradation*, 1, p. 106.
96. Otake, Y., Kobayashi, T., Asabe, H., Murakami, N., and Ono, K. 1995 Biodegradation of lowdensity polyethylene, polystyrene, polyvinyl chloride, and urea formaldehyde resin buried under soil for over 32 years. *Journal of Applied Polymer Science*, 13, p. 1789.

97. Ali, M.I., Ahmed, S., Robson, G., Javed, I., Ali, N., Atiq, N., and Hameed, A. 2014 Isolation and molecular characterization of polyvinyl chloride (PVC) plastic degrading fungal isolates. *Journal of Basic Microbiology*, 1, p. 18.
98. Santana, V.T., Gonçalves, S.P.C., Agnelli, J.A.M., and Martins-Franchetti, S.M. 2012 Biodegradation of a polylactic acid/polyvinyl chloride blend in soil. *Journal of Applied Polymer Science*, 1, p. 536.
99. Zhou, J., Wen, Y., Marshall, M.R., Zhao, J., Gui, H., Yang, Y., Zeng, Z., Jones, D.L., and Zang, H. 2021. Microplastics as an emerging threat to plant and soil health in agroecosystems. *Science of the Total Environment*, 787, p. 147444.
100. de Souza Machado, A.A., Lau, C.W., Till, J., Kloas, K., Lehmann, A., Becker, R., and Rillig, M.C. 2018 Impacts of microplastics on the soil biophysical environment. *Environmental Science & Technology*, 52, p. 9656.
101. Rillig, M.C., Lehmann, A., de Souza Machado, A.A., and Yang, G. 2019 Microplastic effects on plants. *New Phytologist*, 223, p. 1066.
102. Huang, Y., Zhao, Y., Wang, J., Zhang, M., Jia, W., and Qin, X. 2019 LDPE microplastic films alter microbial community composition and enzymatic activities in soil. *Environmental Pollution*, 254, p. 112983.
103. Fei, Y., Huang, S., Zhang, H., Tong, Y., Wen, D., Xia, X., Wang, H., Luo, Y., and Barceló, D. 2020 Response of soil enzyme activities and bacterial communities to the accumulation of microplastics in an acid cropped soil. *Science of the Total Environment*, 707, p. 13563.
104. Qi, Y., Ossowicki, A., Yang, X., Lwanga, E.H., Dini-Andreote, F., Geissen, V., and Garbeva, P. 2020 Effects of plastic mulch film residues on wheat rhizosphere and soil properties. *Journal of Hazardous Materials*, 387, p. 121711.
105. Boots, B., Russell, C.W., and Green, D.S. 2019. Effects of microplastics in soil ecosystems: Above and below ground. *Environmental Science & Technology*, 19, p. 11496.
106. Rousk, J., Brookes, P.C., and Bååth, E. 2009 Contrasting soil pH effects on fungal and bacterial growth suggest functional redundancy in carbon mineralization. *Applied Environmental Microbiology*, 75, p. 1589.
107. Jones, D.L., Cooledge, E.C., Hoyle, F.C., Griffiths, R.I., and Murphy, D.V. 2019 pH and exchangeable aluminum are major regulators of microbial energy flow and carbon use efficiency in soil microbial communities. *Soil Biology and Biochemistry*, 138, p. 107584.
108. Wan, Y., Wu, C., Xue, Q., and Hui, X. 2019 Effects of plastic contamination on water evaporation and desiccation cracking in soil. *Science of the Total Environment*, 654, p. 5.
109. Sajjad, M., Huang, Q., Khan, S., Khan, M.A., Liu, Y., Wang, J., Lian, F., Wang, Q., and Gao, G. 2022 Microplastics in the soil environment: A critical review. *Environmental Technology & Innovation*, 27, p. 102408.
110. Lin, D., Yang, G., Dou, P., Qian, S., Zhao, L., Yang, Y., and Fanin, N. 2020 Microplastics negatively affect soil fauna but stimulate microbial activity: Insights from a field-based microplastic addition experiment. *Proceedings of the Royal Society B*, 287, p. 20201268.
111. Huang, X., Zhao, Y., Wang, J., Zhang, M., Jia, W., and Qin, X. 2019 LDPE microplastic films alter microbial community composition and enzymatic activities in soil. *Environmental Pollution*, 254, p. 112983.
112. Zhou, B., Wang, J., Zhang, H., Shi, H., Fei, Y., Huang, S., Tong, Y., Wen, D., Luo, Y., and Barceló, D. 2020. Microplastics in agricultural soils on the coastal plain of Hangzhou Bay, East China: Multiple sources other than mulching film. *Journal of Hazardous Materials*, 388, p. 121814.
113. Luo, S., Wang, S., Zhang, H., Zhang, J., and Tian, C. 2022 Plastic film mulching reduces microbial interactions in black soil of Northeastern China. *Applied Soil Ecology*, 169, p. 104187.
114. Hou, J., Xu, X., Yu, H., Xi, B., and Tan, W. 2021 Comparing the long-term responses of soil microbial structures and diversities to polyethylene microplastics in different aggregate fractions. *Environment International*, 149, p. 106398.

115. Shi, J., Sun, Y., Wang, X., and Wang, J. 2022 Microplastics reduce soil microbial network complexity and ecological deterministic selection. *Environmental Microbiology*, 24, 2157.
116. Zhu, D., Ma., J., Li, G., Rillig, M.C., and Zhu, Y.G. 2022 Soil plastispheres as hotspots of antibiotic resistance genes and potential pathogens. *ISME Journal*, 16, p. 521.
117. Fu, X.W. and Du, Q.Z. 2011 Uptake of di(2-ethylhexyl)phthalate of vegetables from plastic greenhouses. *Journal of Agriculture and Food Chemistry*, 59, p. 11585.
118. Wang, J., Chen, G.C., Christie, P., Zhang, M.Y., Luo, Y.M., and Teng, Y. 2015 Occurrence and risk assessment of phthalate esters (PAEs) in vegetables and soils of suburban plastic film greenhouses. *Science of the Total Environment*, 523, p. 129.
119. Thiemann, T. 2021 Isolation of phthalates and terephthalates from plant material—natural products or contaminants? *Open Chemistry Journal (Bentham)*, 8, p. 1.
120. Arrigo, F., Impellitteri, F., Piccione, G., and Faggio, C. 2023 Phthalates and their effects on human health: Focus on erythrocytes and the reproductive system. *Comparative Biochemistry and Physiology Part C: Toxicology & Pharmacology*, 270, p. 109645.
121. Hauser, R., Meeker, J.D., Duty, S., Silva, M.J., and Calafat, A.M. 2006 Altered semen quality in relation to urinary concentrations of phthalate monoester and oxidative metabolites. *Epidemiology*, 17, p. 682.
122. Chen, F.-P. and Chien, M.-H. 2014 Lower concentrations of phthalates induce proliferation in human breast cancer cells. *Climacteric*, 17, p. 377.
123. Xie, H.J., Shi, Y.J., Zhang, J., Cui, Y., Teng, S.C., Wang, S.G., and Zhao, R. 2010 Degradation of phthalate esters (PAEs) in soil and the effects of PAEs on soil microcosm activity. *Journal of Chemical Technology & Biotechnology*, 85, p. 1108.
124. Cartwright, C.D., Thompson, I.P., and Burns, R.G. 2000 Degradation and impact of phthalate plasticizers on soil microbial communities. *Environmental Toxicology and Chemistry*, 19, p. 1523.
125. Hahladakis, J.N., Velis, C.A., Weber, R., Iacovidou, E., and Purnell, P. 2018 An overview of chemical additives present in plastics: migration, release, fate and environmental impact during their use, disposal and recycling. *Journal of Hazardous Materials*, 344, p. 179.
126. Zhang, W., Chen, L., An, S., Liu, K., Lin, K., and Zhao, L. 2014 Toxic effects of the joint exposure of decabromodiphenyl ether (BDE209) and tetrabromobisphenol A (TBBPA) on soil microorganism and enzyme activity. *Pharmacology*, 38, p. 586.
127. Rothenbacher, K.P. and Pecquet, A.M. 2018 Summary of historical terrestrial toxicity data for the brominated flame retardant tetrabromobisphenol A (TBBPA): Effects on soil microorganisms, earthworms, and seedling emergence. *Environmental Science and Pollution Research*, 25, p. 17268.
128. Godoy, V., Blázquez, G., Calero, M., Quesada, L., and Martín-Lara, M.A. 2019 The potential of microplastics as carriers of metals. *Environmental Pollution*, 255, p. 113363.
129. Lan, T., Cao, F., Cao, L., Wang, T., Yu, C., and Wang, F. 2022 A comparative study on the adsorption behavior and mechanism of pesticides on agricultural film microplastics and straw degradation products. *Chemosphere*, 303, p. 135058.
130. Puckowski, A., Cwięk, W., Mioduszewska, K., Stepnowski, P., and Białk-Bielińska, A. 2021 Sorption of pharmaceuticals on the surface of microplastics. *Chemosphere*, 263, p. 127976.
131. Maguire, L.W. and Gardner, C.M. 2023 Fate and transport of biological microcontaminants bound to microplastics in the soil environment. *Science of the Total Environment*, 892, p. 164439.
132. Rillig, M.C., Ziersch, L., and Hempel, S. 2017 Microplastic transport in soil by earthworms. *Scientific Reports*, 7, 1362.
133. Cui, W., Gao, P., Zhang, M., Wang, L., Sun, H., and Liu, C. 2022 Adverse effects of microplastics on earthworms: A critical review. *Science of the Total Environment*, 850, p. 158041.

134. Zhu, D., Bi, Q.-F., Xiang, Q., Chen, Q.-L., Christie, P., Ke, X., Wu, L.-H., and Zhu, Y.-G. 2018 Trophic predator-prey relationships promote transport of microplastics compared with the single Hypoaspis aculeifer and Folsomia candida. *Environmental Pollution*, 235, p. 150.
135. Wang, L., Peng, Y., Xu, Y., Zhang, J., Liu, C., Tang, X., Lu, Y., and Sun, H. 2022 Earthworms' degradable bioplastic diet of polylactic acid: Easy to break down and slow to excrete. *Environmental Science & Technology*, 54, p. 11220.
136. Boughattas, I., Zitouni, N., Hattab, S., Mkhinini, M., Missawi, O., Helaoui, S., Mokni, M., Bousserrhine, N., and Banni, M. 2022 Interactive effects of environmental microplastics and 2,4-dichlorophenoxyacetic acid (2,4-D) on the earthworm *Eisenia andrei*. *Journal of Hazardous Materials*, 424, p. 127578.
137. Boughattas, I., Hattab, S., Zitouni, N., Mkhinini, M., Missawi, O., Bousserrhine, N., and Banni, M. 2021 Assessing the presence of microplastic particles in tunisian agriculture soils and their potential toxicity effects using *Eisenia andrei* as bioindicator. *Science of the Total Environment*, 796, p. 148959.
138. Prendergast-Miller, M.T., Katsiamides, A., Abbass, M., Sturzenbaum, S.R., Thorpe, K.L., and Hodson, M.E. 2019 Polyester-derived microfibre impacts on the soil-dwelling earthworm *Lumbricus terrestris*. *Environmental Pollution*, 251, p. 453.
139. Cao, D., Wang, X., Luo, X., Liu, G., and Zheng, H. 2017 Effects of polystyrene microplastics on the fitness of earthworms in an agricultural soil. *IOP Conference Series: Earth and Environmental Science*, 61, p. 012148.
140. Ding, W., Li, Z., Qi, R., Jones, D.L., Liu, Q., Liu, Q., and Yan, C. 2021 Effect thresholds for the earthworm *Eisenia fetida*: Toxicity comparison between conventional and biodegradable microplastics. *Science of the Total Environment*, 781, p. 146884.
141. Besseling, E. Wegner, A., Foekema, E.M., van den Heuvel-Greve, M.J., and Koelmans, A.A. 2013 Effects of microplastic on fitness and PCB bioaccumulation by the lugworm *Arenicola marina* (L.). *Environmental Science & Technology*, 47, p. 593.
142. Lahive, E., Walton, A., Horton, A.A., Spurgeon, D.J., and Svendsen, C. 2019 Microplastic particles reduce reproduction in the terrestrial worm *Enchytraeus crypticus* in a soil exposure. *Environmental Pollution*, 255, p. 113174.
143. Li, B., Song, W., Cheng, Y., Zhang, K., Tian, H., Du, Z., Wang, J., Wang, J., Zhang, W., and Zhu, L. 2021 Ecotoxicological effects of different size ranges of industrial-grade polyethylene and polypropylene microplastics on earthworms *Eisenia fetida*. *Science of the Total Environment*, 783, p. 147007.
144. Chen, Y., Liu, X., Leng, Y., and Wang, J. 2020 Defense responses in earthworms (*Eisenia fetida*) exposed to low-density polyethylene microplastics in soils. *Ecotoxicology and Environmental Safety*, 187, p. 109788.
145. Wang, H.T., Ding, J., Xiong, C., Zhu, D., Li, G., Jia, X.Y., Zhu, Y.G., and Xue, X.M. 2019 Exposure to microplastics lowers arsenic accumulation and alters gut bacterial communities of earthworm *Metaphire californica*. *Environmental Pollution*, 251, p. 110.
146. Rodríguez-Seijo, A., da Costa, J.P., Rocha-Santos, T., Duarte, A.C., and Pereira, R. 2018 Oxidative stress, energy metabolism and molecular responses of earthworms (*Eisenia fetida*) exposed to low-density polyethylene microplastics. *Environmental Science and Pollution Research*, 25, p. 33599.
147. Zhu, D., Chen, Q.-L., An, Q.-L., Yang, X.-R., Christie, P., Ke, X., Wu, L.-H., and Zhu, Y.-G. 2018 Exposure of soil collembolans to microplastics perturbs their gut microbiota and alters their isotopic composition. *Soil Biology and Biochemistry*, 116, p. 302. (*Corrigendum: Soil Biology and Biochemistry*, 124, p. 277)
148. Kim, S.W. and An, Y.-J. 2020 Edible size of polyethylene microplastics and their effects on springtail behavior. *Environmental Pollution*, 266, p. 115255.
149. Bosker, T., Bouwman, L.J., Brun, N.R., Behrens, P., and Vijver, M.G. 2019 Microplastics accumulate on pores in seed capsule and delay germination and root growth of the terrestrial vascular plant Lepidium sativum. *Chemosphere*, 226, p. 774.

150. Bao, Y., Pan, C., Li, D., Guo, A., and Dai, F. 2022 Stress response to oxytetracycline and microplastic-polyethylene in wheat (Triticum aestivum L.) during seed germination and seedling growth stages. *Science of the Total Environment*, 806, p. 150553.
151. Balestri, E., Menicagli, V., Ligorini, V., Fulignati, S., Galletti, A.M.R., and Lardicci, C. 2019. Phytotoxicity assessment of conventional and biodegradable plastic bags using seed germination test. *Ecological Indicators*, 102, p. 569.
152. Zhang, S.L., Wang, J.Q., Yan, P.K., Hao, X.H., Xu, B., Wang, W., and Aurangzeib, M. 2020 Non-biodegradable microplastics in soils: A brief review and challenge. *Journal of Hazardous Materials*, 409, p. 124525.
153. Allouzi, M.M.A., Tang, D.Y.Y., Chew, K.W., Rinklebe, J., Bolan, N., Allouzi, S.M.A., Show, P.L. 2021 Micro (nano) plastic pollution: The ecological influence on soil-plant system and human health. *Science of the Total Environment*, 788, p. 147815.
154. Ayala, F., Zeta-Flores, M., Ramos-Baldárrago, S., Tume-Ruiz, J., Rangel-Vega, A., Reyes, E., Quinde, E., De-la-Torre, G.E., Lajo-Salazar, L., and Cárdenas-Alayza, S. 2023 Terrestrial mammals of the Americas and their interactions with plastic waste. *Environmental Science and Pollution Research*, 30, p. 57759.
155. Kühn, S. and van Franeker, J.A. 2020 Quantitative overview of marine debris ingested by marine megafauna. *Marine Pollution Bulletin*, 151, p. 110858.
156. Priyanka, M. and Dey, S. 2018 Ruminal impaction due to plastic materials—an increasing threat to ruminants and its impact on human health in developing countries. *Veterinary World*, 11, p. 1307.
157. Yong, C., Valiyaveettil, S., and Tang, B. 2020 Toxicity of microplastics and nanoplastics in mammalian systems. *International Journal of Environmental Research and Public Health*, 17, p. 1509.
158. Zolotova, N., Kosyreva, A., Dzhalilova, D., Fokichev, N., and Makarova, O. 2022 Harmful effects of the microplastic pollution on animal health: A literature review. *PeerJ*, 10, p. e13599503.
159. Jankowiak, L., Malecha, A.W., and Krawczyk, A.J. 2016 Garbage in the diet of carnivores in an agricultural area. *European Journal of Ecology*, 2(1), p. 81.
160. Beriot, N., Peek, J., Zornoza, R., Geissen, V., and Huerta-Lwanga, E. 2021 Low density-microplastics detected in sheep faeces and soil: A case study from the intensive vegetable farming in Southeast Spain. *Science of the Total Environment*, 755, p. 142653.
161. Ranjeet, H., Amol, R., Khajal, J., Jhanvi, C., and Abhay, S. 2019 Studies of plastic Bezoar ingestion in free range axis deer in summer. *Journal of Animal Research*, 9, p. 383.
162. Ballejo, F., Plaza, P., Speziale, K.L., Lambertucci, A.P., and Lambertucci, S.A. 2021 Plastic ingestion and dispersion by vultures may produce plastic islands in natural areas. *Science of the Total Environment*, 755, p. 142421.
163. Wu, R.T., Cai, Y.F., Chen, Y.X., Yang, Y.W., Xing, S.C., and Liao, X.-D. 2021 Occurrence of microplastic in livestock and poultry manure in South China. *Environmental Pollution*, 277, p. 116790.
164. Sheriff, I., Yusoff, M.S., Manan, T.S.B.A. and Koroma, M. 2023 Microplastics in manure: Sources, analytical methods, toxicodynamic, and toxicokinetic endpoints in livestock and poultry. *Environmental Advances*, 12, p. 100372.
165. Urli, S., Corte Pause, F., Crociati, M., Baufeld, A., Monaci, M., Stradaioli, G. 2023 Impact of microplastics and nanoplastics on livestock health: An emerging risk for reproductive efficiency. *Animals*, 13, p. 1132.
166. Da Costa Filho, P.A., Andrey, D., Eriksen, B., Peixoto, R.P., Carreres, B.M., Ambühl, M.E., Descarrega, J.B., Dubascoux, S., Zbinden, P., Panchaud, A. and Poitevin, E. 2021 Detection and characterization of small-sized microplastics (≥5 µm) in milk products. *Scientific Reports*, 11, p. 24046.
167. Liu, Q., Chen, Z., Chen, Y., Yang, F., Yao, W. and Xie, Y. 2022 Microplastics contamination in eggs: Detection, occurrence and status. *Food Chemistry*, 397, p. 133771.

168. Bahrani, F., Mohammadi, A., Dobaradaran, S., De-la-Torre, G.E., Arfaeinia, H., Ramavandi, B., Saeedi, R. and Tekle-Röttering, A. 2024 Occurrence of microplastics in edible tissues of livestock (cow and sheep). *Environmental Science and Pollution Research*, 31, p. 22145.
169. Tao, Z., Cao, X., Luo, X., Li, X.D., and Zhou, Z. 2012 Reponses of three enzyme activities to lower molecular weight polyethylene added in pot-cultured horse bean soil. *Chinese Journal of Soil Science*, 43, p. 1104.
170. Zhang, Z., Luo, X., Fan, Y., and Wu, Q. 2015 Cumulative effects of powders of degraded PE mulching-films on chemical properties of soil. *Environmental Science & Technology*, 38, p. 115.
171. Moreno, M.M., González-Mora, S., Villena, J., Campos, J.A., and Moreno, C. 2017 Deterioration pattern of six biodegradable, potentially low-environmental impact mulches in field conditions. *Journal of Environmental Management*, 200, p. 490.
172. Ren, X. 2003 Biodegradable plastics: A solution or a challenge? *Journal of Cleaner Production*, 11, p. 27.
173. Sintim, H.Y. and Flury, M. 2017 Is biodegradable plastic mulch the solution to agriculture's plastic problem? *Environmental Science and Technology*, 51, p. 1068.
174. Yang, N., Sun, Z.-X., Feng, L.-S., Zheng, M.-Z., Chi, D.-C., and Meng, W.-Z. 2014. Plastic film mulching for water-efficient agricultural applications and degradable films materials development research. *Materials and Manufacturing Processes*, 30, p. 143.
175. Goldberger, J.R., Jones, R.E., Miles, C.A., Wallace, R.W. and Inglis, D.A. 2015 Barriers and bridges to the adoption of biodegradable plastic mulches for US specialty crop production. *Renewable Agriculture and Food Systems*, 30, p. 143.
176. Qi, Y., Yang, X., Pelaez, A.M., Lwanga, E.H., Beriot, N., Gertsen, H., Garbeva, P., and Geissen, V. 2018 Macro- and micolastics in soil-plant system: Effects of plastic mulch film residues on wheat (*Triticium aestivum*) growth. *Science of the Total Environment*, 645, p. 1048.
177. Van den Oever, M., Molenveld, K., van der Zee, M., and Bos, H. 2017 Bio-based and bio-degradable plastics: Facts and figures: Focus on food packaging in the Netherlands. *Wageningen Food & Biobased Research.*
178. Huerta Lwanga, E. and Santos-Echeandía, J. 2022. Soil remediation under microplastics pollution. In: Rocha-Santos, T., Costa, M.F., Mouneyrac, C. (eds) *Handbook of microplastics in the environment*. Springer.

7 Airborne Microplastics
Sources, Characteristics, Dynamics, and Health Implications

Vishnu S. Moorchilot, Usha K. Aravind, and Charuvila T. Aravindakumar

7.1 INTRODUCTION

The concern regarding the contamination of the environment with fine plastic particulates was brought to light nearly five decades ago. However, it was not until the early 21st century that the term "microplastics" (MPs) gained prominence [1]. MPs are fine plastic particles having sizes spanning from 1 μm to 5 mm. They are broadly categorized into primary and secondary MPs based on their origin. Primary MPs are intentionally manufactured within the specified size range, while secondary MPs result from fragmentation and deterioration from larger plastics [1].

MPs are emitted from plastic products into the environments at various points in their life cycle, which includes production, packaging, transportation, usage, and waste management [2, 3]. These fine plastic particles have been identified in diverse environmental matrices (lakes, rivers, oceans, air, precipitation, beach sands, sediments, swamps, dust, and particulate matter) [4–6]. While initial research predominantly focused on aquatic environments, of late, there has been growing interest in detecting and quantifying airborne MPs [4]. In outdoor environments, the atmosphere serves as a significant channel for the global transport of MPs and the deposition of these particles in sensitive ecosystems. This has led to serious environmental concerns [7, 8].

Similarly, in indoor spaces, the heightened concentration of MPs in air and dust particulates poses potential risks to human health [9]. MPs are increasingly recognized as both a repository and a source for numerous persistent organic pollutants and emerging contaminants. Since many of these pollutants are semi-volatile and volatile organic compounds, they can very well diffuse into surrounding matrices [10]. However, comprehensive insights into the dynamics of their diffusion remain limited.

This chapter aims to provide a comprehensive overview of airborne MPs, covering their sources, characteristics, transport, sampling, and analysis as well as their

DOI: 10.1201/9781003468141-7

associated health risks. By delving into these aspects, our objective is to provide valuable insights into the understanding of this emerging environmental challenge.

7.2 SOURCES

7.2.1 Outdoor

In outdoor environments, the emission of MPs in the air arises from a multitude of sources. Tire MPs are one such significant contributor to airborne MPs. This encompasses tire wear particles, recycled tire crumb (RTC), and tire repair-polished debris [11]. The wear particles result from the abrasion of tire treads on road surfaces, releasing minute rubber particles into the environment. RTC, utilized in applications like artificial turf and recreational areas, is produced by the mechanical grinding or shredding of tires. Tire repair-polished debris is formed through the mechanical polishing of the inner liner during tire repairs, particularly in cases involving punctures [12].

Construction and demolition (C&D) activities also contribute to MP emissions [13, 14]. Construction materials containing plastic components, such as paints, sealants, insulation, and synthetic fabrics, release MP particles during wear and tear in C&D processes. Friction from cutting or breaking down plastic-based materials generates airborne MPs. Additionally, improper disposal of C&D waste containing plastic items may lead to their breakdown, releasing MPs [14].

Road channeling materials, often overlooked, can contribute to outdoor MP load [15]. The friction between tires and road surfaces, particularly in high-traffic zones, leads to the breakdown of these materials, generating minuscule plastic particles released into the air. Maintenance activities such as repainting road markings or replacing reflective beads use plastic-based materials, contributing to MP emissions. Exposure to environmental factors like rain, sunlight, and snow may accelerate the degradation of road channeling materials, causing them to fragment into smaller particles over time [16].

Another source of airborne MPs is air drying laundry. Fabrics, especially synthetic ones like polyester and nylon, release tiny fibers when exposed to movement, heat, and airflow during the drying process [10]. Subsequently, these liberated microfibers become airborne and disperse into the surrounding environment.

7.2.2 Indoor

Indoor spaces house a unique set of contributors to airborne MPs, with textiles and furniture in living areas emerging as significant sources [17]. The release of microfibers is attributed to friction and wear on these fabrics. In some regions, wrapping/packaging materials are combusted along with cooking fuels, exacerbating indoor MPs [10, 18]. When plastics are burned, either intentionally or unintentionally, they undergo combustion, generating airborne particulate matter that may include MPs. The incomplete combustion of plastics can produce smaller fragments that become suspended in the indoor air, posing potential inhalation risks to individuals in the vicinity [10].

An unanticipated outcome concerning the SARS-CoV-2 outbreak is the extensive utilization of disposable face masks (DFM) [19]. The fine fibers released during the use of DFM, especially when improperly disposed of, have emerged as an additional source of airborne MPs [20]. Moreover, the growing practice of reusing face masks after washing can weaken the fibers, leading to the substantial release of MPs into the surrounding matrices [21]. This unintended consequence underscores the need for proper disposal practices and heightened awareness regarding the environmental impact of mask usage during the ongoing global health crisis.

7.3 PHYSICOCHEMICAL CHARACTERISTICS

7.3.1 Shape, Color, and Polymer Type

Airborne MPs display a variety of shapes, such as beads, films, fragments, fibers, flakes, and foams (Figure 7.1) [22, 23]. These distinct shapes not only highlight the multitude of sources contributing to their presence but also emphasize the dynamic interplay between human activities, material degradation, and atmospheric conditions that shape the composition of airborne MPs [24]. Primary MPs typically undergo minimal changes in shape, as they are often directly released from their sources. In contrast, the various processes involved in generating secondary MPs, such as weathering, UV radiation, mechanical abrasion, chemical reactions, and biological degradation, contribute to a more extensive range of shapes than the relatively stable shapes of primary MPs [25].

Examining the edges of MPs has become a crucial factor in understanding their residence time in the atmosphere [23]. When MPs are initially emitted, they frequently have sharp edges, which signal the recent fragmentation or breakdown of larger plastic items; various factors, including mechanical abrasion, weathering, and other physical forces, contribute to the formation of these sharp edges. MPs with sharp edges are regarded as being in a more pristine state, implying that they have not undergone extensive aging or degradation in the environment [26]. In contrast, MPs with smooth edges have experienced prolonged exposure to environmental conditions. Over time, as MPs undergo weathering, UV radiation, and various other forces, the sharp edges wear down, resulting in smoother contours [23, 26].

The compositional profile regarding the shapes of airborne MPs is noted to vary in different regions. In Germany, for example, fragments (>90%) were the most common form, whereas in Paris, fibers (>90%) dominated the total composition, and in Iran, granules emerged as the most abundant form (60%) [27–29]. Local sources,

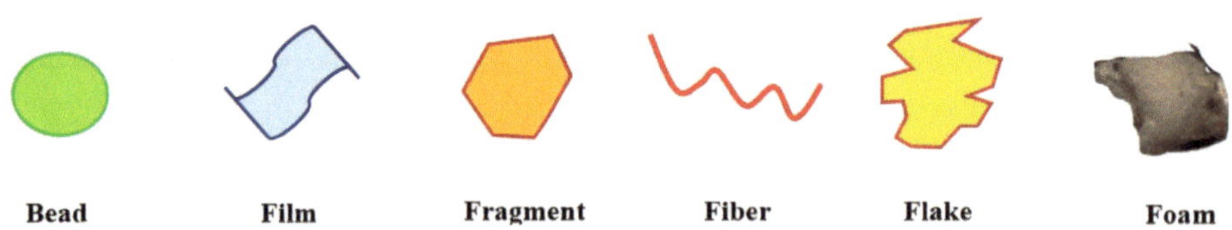

FIGURE 7.1 Different shapes of airborne MPs.

including industrial activities, transportation, and waste management practices, can significantly influence the types of MPs present in the air. Additionally, inhabitants' behaviors, atmospheric conditions such as wind patterns and precipitation can influence the dispersion and deposition of MPs, further contributing to the observed variations [22].

Color functions as an additional morphological element in the recognition and classification of MPs. Airborne MPs display a spectrum of colors, such as black, transparent, yellow, grey, blue, and red [23]. While color can be beneficial for classifying MPs, its effectiveness faces challenges. The sampling process and the impact of photo-aging can introduce complexities and undermine the accuracy of color-based identification. During sampling, factors like air pollution, environmental conditions, and the presence of other particles can influence the color of MPs. Moreover, exposure to sunlight and other weathering processes can change the appearance of MPs over time, diminishing the reliability of color as an indicator [30].

The polymer profiles of atmospheric MPs exhibit significant variation across different geographic locations. In China, studies have highlighted the prevalence of polyethylene terephthalate (PET), polyvinylchloride (PVC), polyethylene (PE), polystyrene (PS), and polycarbonate (PC) [31, 32], whereas in Germany, PE and EVA (ethyl vinyl acetate) MPs were the dominant ones [28]. An analysis of alpine glacier debris revealed the prevalence of polymer types like polyesters, PA (polyamide), PE, and PP (polypropylene) [33]. In contrast, rubber, PA, EVA, and PE MPs are the primary types found in Arctic snowfall [34]. This diversity in polymer types underscores the complex and region-specific nature of atmospheric MPs.

7.4 ATMOSPHERIC TRANSPORT

The movement of MPs in the atmosphere is governed by various environmental parameters such as advection, turbulence, precipitation, velocity, and wind direction [35]. These elements not only affect MP sources and sink but also contribute to their presence in diverse environments. The intricate cycle of MPs is highlighted by the finding that MPs originating from land can make their way into water sources and then be released into the air [36]. Subsequently, these particles can be deposited back onto land/water [37], emphasizing the complex interplay between land, water, and air in the transport of MPs. This phenomenon is particularly evident in the Arctic and Antarctic, where MP influx occurs due to their long-range transport facilitated by air currents [38]. Figure 7.2 graphically represents MP transport across land, water, and air. Understanding these pathways is essential for developing effective strategies to mitigate the environmental impact of MPs.

Research has unveiled several correlations between weather conditions and the deposition of MPs. For instance, MP deposition increases with increased rain or snowfall [37, 39], suggesting a crucial role for precipitation in removing MPs from the atmosphere. Additionally, wind speed correlates negatively with airborne levels of MPs, with higher wind speeds correlating with reduced concentrations of atmospheric MPs. This association is linked to heightened turbulence and dispersion of MPs in the air, resulting in decreased concentration [40]. Additionally, cold air masses at lower altitudes may impede atmospheric transport of MPs, causing their

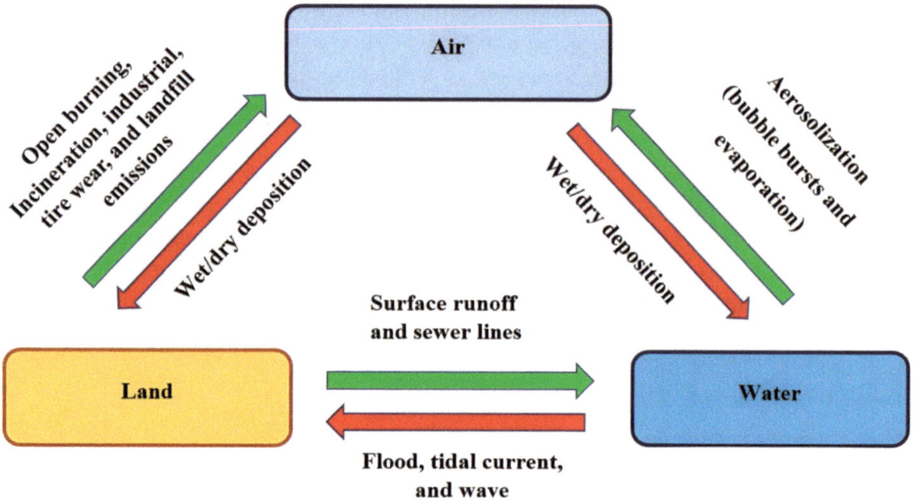

FIGURE 7.2 MP transport across land, water, and air.

accumulation closer to the ground [41]. Furthermore, both MPs and nanoplastics (NPs) have been recognized as effective in condensing water vapor in the atmosphere and creating droplets or ice crystals [42]. This condensation process plays a role in removing MPs from the air as they become incorporated into precipitation events.

The deposition flux of MPs, the rate at which these particles settle onto surfaces, is influenced not only by meteorological factors but also by anthropogenic activities. For instance, urban areas, with their higher population densities, are often associated with elevated MP levels, and even more specifically, in a comparison of fiber deposition flux between Shanghai and Paris, Shanghai exhibited higher levels [43]. In general, areas with higher population densities experience increased MP deposition flux due to the greater human activity [44], and regions characterized by significant industrialization witness increased levels of atmospheric MPs as a result of particle release during manufacturing processes, material handling, and other industrial activities [45].

The shape of MPs also influences their atmospheric transport: Films, characterized by their larger surface area than that of fragments, can be transported to more distant locations in air transport models [46]. This insight becomes crucial in predicting the potential reach of MPs and comprehending their distribution patterns on a global scale. By understanding how different MP shapes behave during transport, it is possible to refine their transport models and assessments, leading to more accurate predictions of the diffusion and settling of airborne MPs.

Despite progress in understanding the transport mechanisms of MPs, several challenges persist. The variability in particle size, shape, and composition presents difficulties in developing universal models that can accurately predict their movement. MPs exhibit diverse characteristics, and these variations impact how they behave in different environments. Furthermore, the influence of meteorological factors and human activities on transport of MPs adds complexity to the prediction models.

Continuous monitoring and research are essential to overcoming these challenges and refining predictive models. Ongoing studies will offer further insights into the factors governing the atmospheric transport of MPs, enabling researchers to develop more accurate and adaptable models.

7.5 CO-POLLUTANTS/ADDITIVES

MPs act as carriers for a diverse array of compounds, serving as critical hosts for numerous heavy metals and organic pollutants within the atmosphere [47]. The binding of these pollutants to MPs takes place through a process known as sorption, where the pollutants are adsorbed onto the MP surfaces. Subsequently, the MPs release these pollutants through a process called desorption [48]. This interaction between MPs and pollutants has significant implications for environmental pollution: MPs, acting as carriers, have the potential to transport these harmful pollutants over long distances, affecting ecosystems far from their original sources.

7.5.1 Sorption Mechanisms

Sorption can manifest as either physisorption or chemisorption [48]. Physisorption is characterized by weak and reversible interactions occurring at different rates on the diverse MP surfaces; during this process, contaminants are attracted to the MP surfaces through relatively weak forces, allowing for reversible binding. In contrast, chemisorption entails the development of strong chemical bonds between MPs and organic pollutants [49]. This establishes a more enduring and irreversible association between MPs and contaminants. Chemisorption creates robust chemical connections that contribute to the stability of the MP–contaminant complex. Different sorption mechanisms between MPs with co-pollutants are presented in Figure 7.3.

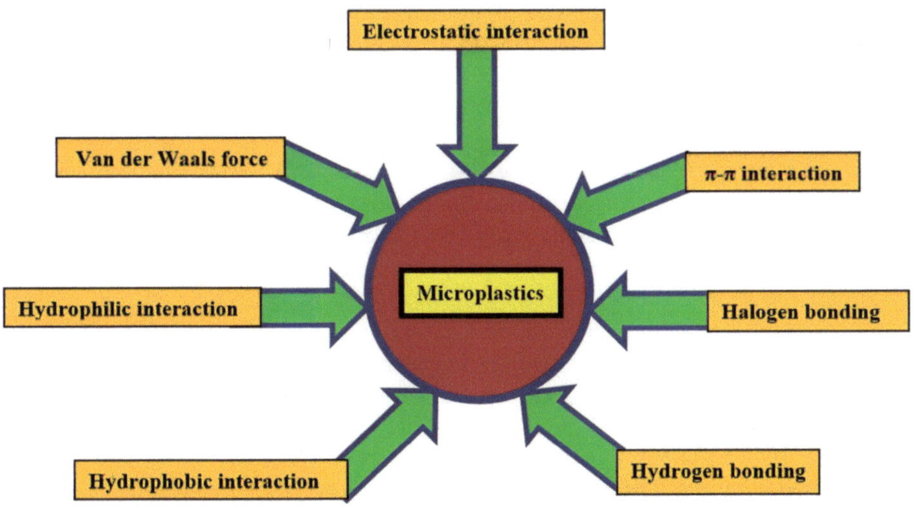

FIGURE 7.3 Sorption mechanisms between MPs and co-pollutants.

7.5.1.1 Electrostatic Interaction

Electrostatic interactions can have an important influence on the relationship between MPs and contaminants in the atmosphere. The nature of the electrostatic forces depends on the specific charges carried by MPs and contaminants. If they bear opposite charges, attractive forces come into play, causing them to be drawn closer together. Conversely, if MPs and contaminants carry similar charges, repulsive forces dominate, and they move away from each other [50].

7.5.1.2 π-π Interaction

In molecules featuring conjugated double bonds, electrons undergo delocalization, forming a π-electron cloud both above and below the molecular plane. When two molecules with these conjugated systems come close, their π-electron clouds can interact through attractive forces referred to as π-π interactions. Compounds featuring aromatic benzene rings with π bonds commonly exhibit this type of interaction [51]. These π-π interactions can occur between aromatic rings in organic pollutants and analogous structures on MP surfaces. This stacking of aromatic rings facilitates MP binding to organic molecules, creating stable associations [49].

7.5.1.3 Van der Waals Interaction

Van der Waals interactions can result in the multilayer adsorption of contaminants onto the surfaces of MPs [49]. These forces, characterized as weak and attractive, emerge from fluctuations in electron distribution within molecules, causing transient positive and negative charges. As pollutants come into proximity with MP surfaces, van der Waals forces draw them in, enabling the sequential development of layers [50].

7.5.1.4 Hydrogen Bonding

In nature, intermolecular hydrogen bonding is a prevalent mechanism. It arises from the dipole–dipole interaction wherein a hydrogen atom covalently bonded to a highly electronegative element such as nitrogen, oxygen, or fluorine is attracted to another atom that has lone pairs of electrons [52]. In the context of MPs, this mechanism binds organic pollutants to MPs, enhancing their sorption capacity [53].

7.5.1.5 Halogen Bonding

In organic compounds, the benzene ring possesses π electrons that constitute the double bonds within the ring structure. These π electrons are relatively loosely bound and can be influenced by electron acceptors. Chlorine atoms along the polymer chain serve as electron acceptors. As a chlorine atom approaches the benzene ring, it forms a halogen bond with the π electrons [50]. In this interaction, the chlorine atom partially draws the π electrons toward itself, resulting in a stabilizing effect. This bonding process shares similarities with hydrogen bonding, but it differs by involving a halogen atom as the acceptor, like the role of hydrogen in hydrogen bonding.

7.5.1.6 Hydrophobic Interaction

Hydrophobic interactions are crucial due to the hydrophobic nature of many MPs. This characteristic leads to the adsorption of hydrophobic contaminants onto MP surfaces [54]. The degree of hydrophobicity can vary based on the polymer types.

PP, PE, PS, and PVC exhibit large surface areas and display a high degree of hydrophobicity [55]. This hydrophobic nature of MPs renders them appealing hosts for organic pollutants, as many of these organic pollutants share similar hydrophobic characteristics.

7.5.1.7 Hydrophilic Interaction

Hydrophilic interactions involve the attraction of water-soluble contaminants to the hydrophilic regions of MP surfaces. This type of interaction is relevant for contaminants that are more soluble in water and may be attracted to specific regions on the MPs that exhibit hydrophilic properties [51]. While the association of organic pollutants with MPs in aquatic environments is well documented, the understanding of these interactions in the context of airborne MPs remains a relatively unexplored area. To address the growing concerns, there is a need for further research to elucidate the role of wetlands as sinks for hazardous compounds.

7.6 SAMPLING AND ANALYSIS

7.6.1 Sample Collection

Airborne MPs can be collected using either active or passive sampling methods, each offering unique advantages and considerations. Active sampling involves the use of specialized samplers equipped with motorized pumps to actively draw in air and capture suspended particulates (Figure 7.4). These collected particles are then impacted onto a suitable filter, facilitating subsequent MP analysis and characterization [56].

In contrast, passive sampling utilizes natural processes such as wet and dry deposition to collect particulates. In this method, a common technique involves attaching a funnel to glass containers to capture the deposition of airborne particles over time [22]. Techniques like vacuuming and sweeping are employed for collecting fine settled dust (Figure 7.4), proving practical for gathering dust from surfaces in both indoor and outdoor environments [10, 22].

While active sampling is generally considered more robust and capable of providing real-time data, passive sampling stands out for its cost-effectiveness. Since it doesn't require a power supply, passive sampling is suitable for remote or resource-constrained locations [22]. However, it is crucial to acknowledge certain limitations associated with passive sampling, including the potential alteration of samples due to meteorological factors and the risk of contamination from biological activity [22, 56]. The choice between active and passive sampling ultimately depends on the specific objectives of the monitoring program, the desired level of detail in the collected data, and the available resources. Careful consideration of these factors is essential for selecting the most appropriate sampling approach based on the study's requirements and the environmental context.

7.6.2 Sample Processing

The assessment of MPs can face challenges due to the presence of organic contaminants on their surfaces. This issue can be addressed by employing oxidizing

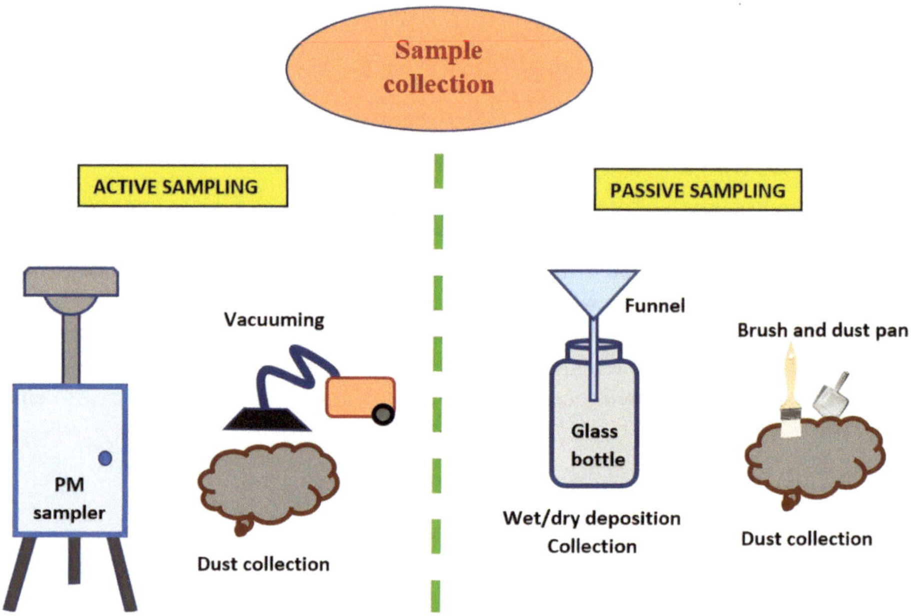

FIGURE 7.4 Different techniques used for sampling airborne MPs.

agents such as hydrogen peroxide, sodium hypochlorite, nitric acid, and Fenton reagents [57, 58]. However, certain polymers like PA and PET, when subjected to acidic conditions, may undergo chemical changes that affect their structural integrity, potentially leading to the deterioration of MPs [59]. Furthermore, the temperature set in oven for drying samples should not exceed 60 °C to prevent thermal degradation [60].

Ultrasonic treatment is another effective method of removing organic matter from MPs [61]. This technique involves the use of ultrasonic waves, which generate high-frequency sound waves in a liquid medium. The cavitation effect generated due to ultrasonic waves can disrupt and dislodge organic contaminants from the MP surface. However, it's worth noting that ultrasonic treatment may also result in the physical breakdown of MPs into finer fragments, potentially complicating the analysis process [62].

Septic tank cleaners have demonstrated high efficiency, surpassing 90%, in removing organic matter from MPs. These cleaners, likely containing a combination of chemicals, prove effective in breaking down and removing organic substances adhered to MPs [63]. While this method is efficient in organic matter removal, it is crucial to carefully consider its potential impact on the structural integrity of MPs and their subsequent analysis. MP separation after sample digestion is a critical step in the analysis process. This separation is typically carried out through a series of methods, with density separation followed by vacuum filtration being common practices.

i) **Density separation**
Density separation involves the use of a highly saturated salt solution such as NaCl (sodium chloride), $ZnCl_2$ (zinc chloride), $CaCl_2$ (calcium chloride), NaI (sodium iodide), $ZnBr_2$ (zinc bromide), and sodium bromide [22]. The principle behind density separation is that MPs, having lower density than the solution, will float while denser materials sink. This enables the isolation of MPs for further analysis.

ii) **Vacuum filtration**
Following density separation, vacuum filtration is often employed to capture the MPs on a suitable filter membrane (e.g., polytetrafluoroethylene or glass fiber filter). This process involves applying a vacuum to draw the solution through the filter, leaving the MPs behind [22]. The filter membrane can then be collected and analyzed for the presence of MPs.

These separation techniques are crucial for ensuring the accuracy and reliability of analyses in the assessment of MPs in various environmental samples. They help isolate MPs from other materials, allowing for a more focused and precise examination of their characteristics.

It is important to note that in determining the additives/co-pollutants attached to MPs, the digestion of samples with an oxidizing agent is omitted to prevent potential degradation of organic compounds [15]. Consequently, the samples undergo only density separation, and the isolated MPs are subsequently extracted using an appropriate solvent. Some of the extraction techniques employed in earlier reports involve liquid–solid extraction, ultrasound solvent extraction, or accelerated solvent extraction [64].

7.6.3 Analytical Techniques for MPs

7.6.3.1 Microscopy

Different microscopic techniques employed for analyzing MPs include stereomicroscopy, florescence microscopy, polarized optical microscopy (POM), scanning electron microscopy (SEM), and atomic force microscopy (AFM).

Stereomicroscopy is an optical method that offers a three-dimensional perspective of a sample. This technique is frequently employed for facilitating the initial identification and sorting of MP particles based on their morphology [65]. Fluorescence microscopy utilizes fluorescence to generate images. By applying specific dyes or fluorophores to the sample, fluorescence is emitted when the sample is illuminated with light of a particular wavelength, producing an image. This method is advantageous for identifying and quantifying MPs labeled with fluorescent markers, enhancing visibility and enabling specific identification [65]. POM utilizes polarized light to enhance contrast and visibility, particularly for materials with crystalline structures. It is used to differentiate MPs from natural particles based on their birefringence, providing insights into their material composition [66].

SEM uses electron beams to scan the surface of a sample, producing high-resolution images with detailed topographical information. This technique is valuable for studying the surface morphologies of MPs at a nanoscale level, aiding in the characterization of their structures and textures [67]. AFM involves scanning a sharp probe over the sample surface and measuring the interaction forces between the probe and the material. This provides high-resolution topographical information, making it particularly useful for examining the nanoscale features of MPs, including their surface roughness and mechanical properties. AFM offers insights into the structural characteristics of MPs [67].

7.6.3.2 Spectroscopy

Raman spectroscopy and Fourier transform Infrared (FT-IR) spectroscopy find wide application in MP analysis. Raman spectroscopy relies on the inelastic scattering of laser light, where the resulting frequency shift, or Raman shift, corresponds to molecular vibrations [68]. This shift creates a unique signature for each material, making Raman spectroscopy an excellent, nondestructive tool for differentiating MP types [69].

FT-IR spectroscopy measures the absorption of infrared radiation, revealing information about the functional groups in a material and providing high-resolution spectra that are sensitive to the chemical structures of MPs. Additionally, FT-IR also provides insights into the physical and chemical weathering of these particles through the analysis of their oxidation state [56]. This technique is widely employed for identifying MPs. In summary, Raman and FT-IR spectroscopy offer complementary insights, with Raman being more adept at distinguishing between similar polymers and FT-IR excelling in identifying specific functional groups, and the combination of Raman and FT-IR spectroscopy enhances the precision and reliability of MP identification, providing detailed information crucial for understanding the presence and impact of MPs in diverse environmental samples.

7.6.3.3 Thermal Analysis

Thermogravimetric analysis (TGA) monitors the fluxes in a sample's weight relative to temperature, offering insights into processes such as decomposition, degradation, and volatilization [70]. In the context of MPs, TGA serves as a valuable tool for assessing their thermal stability. Each polymer has a unique decomposition temperature, and TGA helps identify MP types based on the temperatures at which weight loss occurs. This information aids in distinguishing between different polymers present in environmental samples [71]. Moreover, TGA provides insights into the thermal characteristics of MPs, helping us to understand their behaviors under varying temperatures.

Differential scanning calorimetry (DSC) measures the heat flow linked with thermal transitions in a material as a function of temperature. It is particularly sensitive to phase transitions such as melting and crystallization [72]. In the context of MPs, DSC is a valuable tool for identifying these particles based on their thermal behaviors because different polymers exhibit distinct melting points and thermal profiles. By subjecting MP samples to controlled temperature changes, DSC provides information about the thermal properties of the material. The observed heat flow during phase transitions offers insights into the specific polymer composition of the MP [72], aiding researchers in characterizing and distinguishing between different MP types.

Pyrolysis gas chromatography–mass spectrometry (Py-GC-MS) involves the thermal decomposition (pyrolysis) of a material in the absence of oxygen. During pyrolysis, the material is heated, leading to the release of volatile products that are analyzed using gas chromatography–mass spectrometry (GC-MS) [73]. Py-GC-MS is highly efficient for identifying MPs, offering detailed insights into polymer composition and specific chemical compounds within the material [74]. The technique is well suited for analyzing complex mixtures, which is common in environmental samples containing various types of MPs. While DSC and TGA are well suited for analyzing bulk samples, offering insights into the overall thermal behavior of a material, Py-GC-MS is versatile and can be applied to both bulk samples and individual MP particles. This dual capability enables Py-GC-MS to provide detailed chemical information at different scales, making it a valuable tool for comprehensive MP analysis.

7.6.4 Analytical Techniques for Monitoring Co-pollutants/Additives

Investigations into co-pollutants/additives associated with MPs have gained momentum lately. The majority of existing literature focuses on aquatic ecosystems [56], and this focus on aquatic systems arises from concerns about ecological consequences and the potential transfer of pollutants up the food chain. Predominantly among the analytical techniques, GC-based techniques have been employed to investigate MP co-pollutants and additives. GC-MS has demonstrated its versatility and potency for detecting a broad spectrum of organic compounds associated with MPs. It has been utilized for monitoring organic compounds such as polychlorinated biphenyls (PCBs), polybrominated diphenyl ethers (PBDEs), organochlorine pesticides, polycyclic aromatic hydrocarbons (PAHs), organophosphorus esters, and phthalic acid esters (PAEs) [75]. Another GC technique, electron capture detector (GC-ECD), has been employed to identify hexachlorocyclohexanes, PCBs, DDT, and PBDEs associated with MPs [75].

In addition to GC techniques, liquid chromatography (LC) techniques, such as LC-MS (liquid chromatography-mass spectrometry), are valuable for analyzing co-pollutants and additives in MPs. LC-MS has been employed to identify per- and polyfluoroalkyl substances in plastic pellets [76]. Another study showcased the use of High-performance liquid chromatography with fluorescence detection for monitoring bisphenol A (BPA) and alkylphenols [77]. LC offers a particularly effective alternative approach for nonvolatile and polar compounds, thereby expanding the scope of detectable co-pollutants in MPs.

Of late, there has been a noticeable shift toward exploring other ecosystems, including air matrices. This shift is driven by the recognition that MPs are not confined to water bodies and can be present in the air, posing potential risks to air quality and human health [78]. GC–MS was effectively utilized to monitor different types of plastic additives (dioctyl adipate, 2-ethylhexyl diphenyl phosphate, di-ethyl hexyl phthalate (DEHP), 2-ethylhexyl 2-cyano-3,3-diphenylacrylate, and 2-hydroxy-4-(octyloxy) benzophenone) in MPs isolated from road dust samples [15]. The combination of techniques such as thermal desorption, pyrolysis, and DART-MS was able to screen a wide range of additives in MPs [78]. Another study demonstrated the use of LC quadrupole time-of-flight mass spectrometry in identifying DEHP, monobutyl phthalate, 1,2-dihydro-2,2,4-trimethylquinoline, benzisothiazolone, di-butyl amine, metoprolol, and dichloroaniline (3,4-DCA) in MPs present in indoor dust samples [10].

In summary, the analysis of co-pollutants and additives in MPs has seen a diversification of techniques beyond the traditional focus on gas chromatography. While GC-MS remains a widely used and effective method, the inclusion of LC-based approaches broadens the analytical toolkit, allowing researchers to explore a wider range of substances associated with MPs in various environmental settings.

7.7 HUMAN EXPOSURE

Humans can be exposed to airborne MPs via pathways like inhalation, ingestion, and dermal adsorption [79]. Inhalation involves the uptake of airborne MPs present in the atmosphere that can be inhaled into the respiratory system. Ingestion refers to the consumption of food and water containing MPs, leading to their entry into the gastrointestinal tract. Dermal adsorption involves MP absorption through the skin, particularly in cases where individuals come into direct contact with surfaces or products containing these particles [40, 79]. Each pathway poses potential health risks, and understanding these exposure routes is crucial for predicting the overall effect of MPs on human health. Among these routes, inhalation, and ingestion of MPs accounts for most of the risk due to the direct entry of these particles into the respiratory and digestive systems [80].

Once MPs enter the human body, they promote the activity of antioxidant enzymes, like superoxide dismutase, glutathione peroxidases, catalase, and glutathione [81]. Antioxidant enzymes are a part of the body's defense mechanism against oxidative stress. However, high antioxidant enzyme activity produces reactive oxidative species (ROS), which contains highly reactive molecules like superoxide radicals, hydrogen peroxide, and hydroxyl radicals [81]. The accumulation of ROS results in oxidative stress that exceeds the body's capacity to detoxify. Due to their high reactivity, ROS can interact with and damage cellular structures [81]. Extended exposure to oxidative stress has been linked with neurodegenerative disorders and cardiovascular diseases [82, 83].

Of late, it is suggested that MPs can have a dual nature, functioning both as sources and sinks in diverse environments, adding complexity to our understanding of their impact [10]. This dual functionality of MPs has the potential to complicate the assessment of their effects on human health. Consequently, it is crucial for future researchers to meticulously consider these aspects.

7.8 CONCLUSION

In conclusion, the issue of MPs and plastic pollution presents a highly intricate challenge that demands a comprehensive understanding and strategic approach. The complexity arises not only from the diverse sources and transport pathways of MPs but also from their interactions with co-pollutants. Moving forward, focus should be placed on estimating and representing the fluxes and pools, deciphering the movement of airborne MPs across various ecosystem compartments. These estimates and models, in turn, become invaluable tools in supporting the identification of MPs, elucidating important pathways and mechanisms, and advancing scientific understanding. Furthermore, advancing scientific understanding lays the groundwork for effective policymaking that addresses these issues in a holistic manner. By weaving together local efforts, recognizing global implications, and leveraging scientific advancements, an effective response can be chalked out to counter the various challenges presented by MPs. This integrated approach is essential for fostering a sustainable and resilient environment for generations to come.

DECLARATION OF COMPETING INTEREST

The authors declare that they do not have any known competing financial benefits or personal relationships that could have appeared to affect the work described in this chapter.

ACKNOWLEDGEMENT

MVS is thankful to UGC for the research (UGC-SRF) fellowship.

REFERENCES

1. Napper, I.E. and Thompson, R.C., 2020. Plastic debris in the marine environment: history and future challenges. *Global Challenges*, *4*(6), p. 1900081.
2. Sobhani, Z., Lei, Y., Tang, Y., Wu, L., Zhang, X., Naidu, R., Megharaj, M. and Fang, C., 2020. Microplastics generated when opening plastic packaging. *Scientific Reports*, *10*(1), p. 4841.
3. Ryberg, M.W., Hauschild, M.Z., Wang, F., Averous-Monnery, S. and Laurent, A., 2019. Global environmental losses of plastics across their value chains. *Resources, Conservation and Recycling*, *151*, p. 104459.
4. Duis, K. and Coors, A., 2016. Microplastics in the aquatic and terrestrial environment: sources (with a specific focus on personal care products), fate and effects. *Environmental Sciences Europe*, *28*(1), p. 2.
5. Abbasi, S., Keshavarzi, B., Moore, F., Turner, A., Kelly, F.J., Dominguez, A.O. and Jaafarzadeh, N., 2019. Distribution and potential health impacts of microplastics and microrubbers in air and street dusts from Asaluyeh County, Iran. *Environmental Pollution*, *244*, pp. 153–164.
6. Abbasi, S., 2021. Microplastics washout from the atmosphere during a monsoon rain event. *Journal of Hazardous Materials Advances*, *4*, p. 100035.
7. Yang, H., He, Y., Yan, Y., Junaid, M. and Wang, J., 2021. Characteristics, toxic effects, and analytical methods of microplastics in the atmosphere. *Nanomaterials*, *11*(10), p. 2747.
8. Bullard, J.E., Ockelford, A., O'Brien, P. and Neuman, C.M., 2021. Preferential transport of microplastics by wind. *Atmospheric Environment*, *245*, p. 118038.
9. Dewika, M., Markandan, K., Irfan, N.A., Abdah, M.A.A.M., Ruwaida, J.N., Sara, Y.Y. and Khalid, M., 2023. Review of microplastics in the indoor environment: Distribution, human exposure and potential health impacts. *Chemosphere*, *324*, p. 138270.
10. Moorchilot, V.S., Aravind, U.K. and Aravindakumar, C.T., 2023. Occurrence of azo-dyes, plasticizers, and PAH-bound microplastics: an emerging source and sink for hazardous compounds in indoor environments? *Air Quality, Atmosphere & Health*, pp. 1–14.

11. Zhou, X., Luo, Z., Wang, H., Luo, Y., Yu, R., Zhou, S., Wang, Z., Hu, G. and Xing, B., 2023. Machine learning application in forecasting tire wear particles emission in China under different potential socioeconomic and climate scenarios with tire microplastics context. *Journal of Hazardous Materials*, *441*, p. 129878.
12. Luo, Z., Zhou, X., Su, Y., Wang, H., Yu, R., Zhou, S., Xu, E.G. and Xing, B., 2021. Environmental occurrence, fate, impact, and potential solution of tire microplastics: similarities and differences with tire wear particles. *Science of the Total Environment*, *795*, p. 148902.
13. Yuk, H., Jo, H.H., Nam, J., Kim, Y.U. and Kim, S., 2022. Microplastic: a particulate matter (PM) generated by deterioration of building materials. *Journal of Hazardous Materials*, *437*, p. 129290.
14. Prasittisopin, L., Ferdous, W. and Kamchoom, V., 2023. Microplastics in construction and built environment. *Developments in the Built Environment*, p. 100188.
15. Kitahara, K.I. and Nakata, H., 2020. Plastic additives as tracers of microplastic sources in Japanese road dusts. *Science of the Total Environment*, *736*, p. 139694.
16. Burghardt, T.E., Pashkevich, A., Babić, D., Mosböck, H., Babić, D. and Żakowska, L., 2022. Microplastics and road markings: the role of glass beads and loss estimation. *Transportation Research Part D: Transport and Environment*, *102*, p. 103123.
17. Kacprzak, S. and Tijing, L.D., 2022. Microplastics in indoor environment: Sources, mitigation and fate. *Journal of Environmental Chemical Engineering*, *10*(2), p. 107359.
18. Kumar, R., Verma, A., Shome, A., Sinha, R., Sinha, S., Jha, P.K., Kumar, R., Kumar, P., Shubham, Das, S. and Sharma, P., 2021. Impacts of plastic pollution on ecosystem services, sustainable development goals, and need to focus on circular economy and policy interventions. *Sustainability*, *13*(17), p. 9963.
19. Jiang, H., Luo, D., Wang, L., Zhang, Y., Wang, H. and Wang, C., 2023. A review of disposable facemasks during the COVID-19 pandemic: a focus on microplastics release. *Chemosphere*, *312*, p. 137178.
20. Chen, X., Chen, X., Liu, Q., Zhao, Q., Xiong, X. and Wu, C., 2021. Used disposable face masks are significant sources of microplastics to environment. *Environmental Pollution*, *285*, p. 117485.
21. De Felice, B., Antenucci, S., Ortenzi, M.A. and Parolini, M., 2022. Laundering of face masks represents an additional source of synthetic and natural microfibers to aquatic ecosystems. *Science of the Total Environment*, *806*, p. 150495.
22. Ahmad, M., Chen, J., Khan, M.T., Yu, Q., Phairuang, W., Furuuchi, M., Ali, S.W., Nawab, A. and Panyametheekul, S., 2023. Sources, analysis, and health implications of atmospheric microplastics. *Emerging Contaminants*, p. 100233.
23. Zhang, Y., Kang, S., Allen, S., Allen, D., Gao, T. and Sillanpää, M., 2020. Atmospheric microplastics: a review on current status and perspectives. *Earth-Science Reviews*, *203*, p. 103118.
24. Jahandari, A., 2023. Microplastics in the urban atmosphere: Sources, occurrences, distribution, and potential health implications. *Journal of Hazardous Materials Advances*, *12*, p. 100346.
25. Huber, M., Archodoulaki, V.M., Pomakhina, E., Pukánszky, B., Zinöcker, E. and Gahleitner, M., 2022. Environmental degradation and formation of secondary microplastics from packaging material: a polypropylene film case study. *Polymer Degradation and Stability*, *195*, p. 109794.
26. Acharya, S., Rumi, S.S., Hu, Y. and Abidi, N., 2021. Microfibers from synthetic textiles as a major source of microplastics in the environment: a review. *Textile Research Journal*, *91*(17–18), pp. 2136–2156.
27. Dris, R., Gasperi, J., Rocher, V., Saad, M., Renault, N. and Tassin, B., 2015. Microplastic contamination in an urban area: a case study in Greater Paris. *Environmental Chemistry*, *12*(5), pp. 592–599.

28. Klein, M. and Fischer, E.K., 2019. Microplastic abundance in atmospheric deposition within the Metropolitan area of Hamburg, Germany. *Science of the Total Environment*, *685*, pp. 96–103.
29. Dehghani, S., Moore, F. and Akhbarizadeh, R., 2017. Microplastic pollution in deposited urban dust, Tehran metropolis, Iran. *Environmental Science and Pollution Research*, *24*, pp. 20360–20371.
30. Vithanage, M. and Prasad, M.N.V. eds., 2023. *Microplastics in the Ecosphere: Air, Water, Soil, and Food*. John Wiley & Sons.
31. Qian, Z.H.O.U., ChongGuo, T.I.A.N. and YongMing, L.U.O., 2017. Various forms and deposition fluxes of microplastics identified in the coastal urban atmosphere. *Chinese Science Bulletin*, *62*(33), pp. 3902–3909.
32. Liu, C., Li, J., Zhang, Y., Wang, L., Deng, J., Gao, Y., Yu, L., Zhang, J. and Sun, H., 2019. Widespread distribution of PET and PC microplastics in dust in urban China and their estimated human exposure. *Environment International*, *128*, pp. 116–124.
33. Ambrosini, R., Azzoni, R.S., Pittino, F., Diolaiuti, G., Franzetti, A. and Parolini, M., 2019. First evidence of microplastic contamination in the supraglacial debris of an alpine glacier. *Environmental Pollution*, *253*, pp. 297–301.
34. Bergmann, M., Mützel, S., Primpke, S., Tekman, M.B., Trachsel, J. and Gerdts, G., 2019. White and wonderful? Microplastics prevail in snow from the Alps to the Arctic. *Science Advances*, *5*(8), p.eaax1157.
35. Kumar, R., Sharma, P., Verma, A., Jha, P.K., Singh, P., Gupta, P.K., Chandra, R. and Prasad, P.V., 2021. Effect of physical characteristics and hydrodynamic conditions on transport and deposition of microplastics in riverine ecosystem. *Water*, *13*(19), p. 2710.
36. Masry, M., Rossignol, S., Roussel, B.T., Bourgogne, D., Bussière, P.O., R'mili, B. and Wong-Wah-Chung, P., 2021. Experimental evidence of plastic particles transfer at the water-air interface through bubble bursting. *Environmental Pollution*, *280*, p. 116949.
37. Sun, J., Peng, Z., Zhu, Z.R., Fu, W., Dai, X. and Ni, B.J., 2022. The atmospheric microplastics deposition contributes to microplastic pollution in urban waters. *Water Research*, *225*, p. 119116.
38. Cunningham, E.M., Rico Seijo, N., Altieri, K.E., Audh, R.R., Burger, J.M., Bornman, T.G., Fawcett, S., Gwinnett, C., Osborne, A.O. and Woodall, L.C., 2022. The transport and fate of microplastic fibres in the Antarctic: The role of multiple global processes. *Frontiers in Marine Science*, *9*, p. 1056081.
39. Wen, H., Xu, H., Ma, Y., Zhang, C., Zhang, D. and Wang, X., 2024. Diverse and high pollution of microplastics in seasonal snow across Northeastern China. *Science of the Total Environment*, *907*, p. 167923.
40. Bhat, M.A., Gedik, K. and Gaga, E.O., 2023. Atmospheric micro (nano) plastics: future growing concerns for human health. *Air Quality, Atmosphere & Health*, *16*(2), pp. 233–262.
41. López, A.P.A., Trilleras, J., Arana, V.A., Garcia-Alzate, L.S. and Grande-Tovar, C.D., 2023. Atmospheric microplastics: exposure, toxicity, and detrimental health effects. *RSC Advances*, *13*(11), pp. 7468–7489.
42. Aeschlimann, M., Li, G., Kanji, Z.A. and Mitrano, D.M., 2022. Potential impacts of atmospheric microplastics and nanoplastics on cloud formation processes. *Nature Geoscience*, *15*(12), pp. 967–975.
43. Haque, F. and Fan, C., 2023. Fate and impacts of microplastics in the environment: hydrosphere, pedosphere, and atmosphere. *Environments*, *10*(5), p. 70.
44. Zhu, J., Xu, A., Shi, M., Su, Y., Liu, W., Zhang, Y., She, Z., Xing, X. and Qi, S., 2024. Atmospheric deposition is an important pathway for inputting microplastics: insight into the spatiotemporal distribution and deposition flux in a mega city. *Environmental Pollution*, *341*, p. 123012.

45. Li, Y., Shao, L., Wang, W., Zhang, M., Feng, X., Li, W. and Zhang, D., 2020. Airborne fiber particles: types, size and concentration observed in Beijing. *Science of the Total Environment, 705*, p. 135967.
46. Long, X., Fu, T.M., Yang, X., Tang, Y., Zheng, Y., Zhu, L., Shen, H., Ye, J., Wang, C., Wang, T. and Li, B., 2022. Efficient atmospheric transport of microplastics over Asia and adjacent oceans. *Environmental Science & Technology, 56*(10), pp. 6243–6252.
47. Verla, A.W., Enyoh, C.E., Verla, E.N. and Nwarnorh, K.O., 2019. Microplastic–toxic chemical interaction: a review study on quantified levels, mechanism and implication. *SN Applied Sciences, 1*(11), pp. 1–30.
48. Sahai, H., Valverde, M.G., Morales, M.M., Hernando, M.D., Del Real, A.M.A. and Fernández-Alba, A.R., 2023. Exploring sorption of pesticides and PAHs in microplastics derived from plastic mulch films used in modern agriculture. *Chemosphere, 333*, p. 138959.
49. Agboola, O.D. and Benson, N.U., 2021. Physisorption and chemisorption mechanisms influencing micro (nano) plastics-organic chemical contaminants interactions: a review. *Frontiers in Environmental Science, 9*, p. 678574.
50. Li, W., Zu, B., Yang, Q., Guo, J. and Li, J., 2023. Sources, distribution, and environmental effects of microplastics: a systematic review. *RSC Advances, 13*(23), pp. 15566–15574.
51. Atugoda, T., Vithanage, M., Wijesekara, H., Bolan, N., Sarmah, A.K., Bank, M.S., You, S. and Ok, Y.S., 2021. Interactions between microplastics, pharmaceuticals and personal care products: implications for vector transport. *Environment International, 149*, p. 106367.
52. Szalewicz, K., 2003. Hydrogen bond. In: R.A. Meyers (Ed.), *Encyclopedia of Physical Science and Technology* (Third Edition). Academic Press, pp. 505–538.
53. Sun, Q., Liu, L., Gong, Y. and Liu, P., 2024. Adsorption behavior and interaction mechanism of microplastics with typical hydrophilic pharmaceuticals and personal care products. *Environmental Research, 244*, p. 117897.
54. Fu, L., Li, J., Wang, G., Luan, Y. and Dai, W., 2021. Adsorption behavior of organic pollutants on microplastics. *Ecotoxicology and Environmental Safety, 217*, p. 112207.
55. Joo, S.H., Liang, Y., Kim, M., Byun, J. and Choi, H., 2021. Microplastics with adsorbed contaminants: mechanisms and treatment. *Environmental Challenges, 3*, p. 100042.
56. Shao, L., Li, Y., Jones, T., Santosh, M., Liu, P., Zhang, M., Xu, L., Li, W., Lu, J., Yang, C.X. and Zhang, D., 2022. Airborne microplastics: A review of current perspectives and environmental implications. *Journal of Cleaner Production, 347*, p. 131048.
57. Liu, Y., Wang, B., Pileggi, V. and Chang, S., 2022. Methods to recover and characterize microplastics in wastewater treatment plants. *Case Studies in Chemical and Environmental Engineering, 5*, p. 100183.
58. Monteiro, S.S., Rocha-Santos, T., Prata, J.C., Duarte, A.C., Girão, A.V., Lopes, P., Cristovão, T. and da Costa, J.P., 2022. A straightforward method for microplastic extraction from organic-rich freshwater samples. *Science of the Total Environment, 815*, p. 152941.
59. Pfeiffer, F. and Fischer, E.K., 2020. Various digestion protocols within microplastic sample processing—evaluating the resistance of different synthetic polymers and the efficiency of biogenic organic matter destruction. *Frontiers in Environmental Science, 8*, p. 572424.
60. Nayebi, B., Khurana, P., Pulicharla, R., Karimpour, S. and Brar, S.K., 2023. Preservation, storage, and sample preparation methods for freshwater microplastics–a comprehensive review. *Environmental Science: Advances, 2*, pp. 1060–1081.
61. Goli, V.S.N.S. and Singh, D.N., 2023. Extraction and characterization of microplastics in landfill-mined-soil-like-fractions: a novel methodology. *Chemical Engineering Journal, 452*, p. 139217.

62. Von der Esch, E., Lanzinger, M., Kohles, A.J., Schwaferts, C., Weisser, J., Hofmann, T., Glas, K., Elsner, M. and Ivleva, N.P., 2020. Simple generation of suspensible secondary microplastic reference particles via ultrasound treatment. *Frontiers in Chemistry*, *8*, p. 169.
63. Lavoy, M. and Crossman, J., 2021. A novel method for organic matter removal from samples containing microplastics. *Environmental Pollution*, *286*, p. 117357.
64. Martín, J., Santos, J.L., Aparicio, I. and Alonso, E., 2022. Microplastics and associated emerging contaminants in the environment: analysis, sorption mechanisms and effects of co-exposure. *Trends in Environmental Analytical Chemistry*, *35*, p.e00170.
65. Shim, W.J., Hong, S.H. and Eo, S.E., 2017. Identification methods in microplastic analysis: a review. *Analytical Methods*, *9*(9), pp. 1384–1391.
66. Sierra, I., Chialanza, M.R., Faccio, R., Carrizo, D., Fornaro, L. and Pérez-Parada, A., 2020. Identification of microplastics in wastewater samples by means of polarized light optical microscopy. *Environmental Science and Pollution Research*, *27*, pp. 7409–7419.
67. Mariano, S., Tacconi, S., Fidaleo, M., Rossi, M. and Dini, L., 2021. Micro and nanoplastics identification: classic methods and innovative detection techniques. *Frontiers in Toxicology*, *3*, p. 636640.
68. Moorchilot, V.S., Aravind, U.K., Menacherry, S.P.M. and Aravindakumar, C.T., 2022. Single-particle analysis of atmospheric aerosols: applications of Raman spectroscopy. *Atmosphere*, *13*(11), p. 1779.
69. Fang, C., Awoyemi, O.S., Saianand, G., Xu, L., Niu, J. and Naidu, R., 2024. Characterising microplastics in indoor air: insights from Raman imaging analysis of air filter samples. *Journal of Hazardous Materials*, *464*, p. 132969.
70. Lup, A.N.K., Abnisa, F., Daud, W.M.A.W. and Aroua, M.K., 2018. Delayed volatiles release phenomenon at higher temperature in TGA via sample encapsulation technique. *Fuel*, *234*, pp. 422–429.
71. Pandey, B., Pathak, J., Singh, P., Kumar, R., Kumar, A., Kaushik, S. and Thakur, T.K., 2022. Microplastics in the ecosystem: an overview on detection, removal, toxicity assessment, and control release. *Water*, *15*(1), p. 51.
72. Shabaka, S.H., Marey, R.S., Ghobashy, M., Abushady, A.M., Ismail, G.A. and Khairy, H.M., 2020. Thermal analysis and enhanced visual technique for assessment of microplastics in fish from an urban harbor, Mediterranean Coast of Egypt. *Marine Pollution Bulletin*, *159*, p. 111465.
73. Dris, R., Imhof, H.K., Löder, M.G., Gasperi, J., Laforsch, C. and Tassin, B., 2018. Microplastic contamination in freshwater systems: Methodological challenges, occurrence and sources. In: *Microplastic Contamination in Aquatic Environments*. Elsevier, pp. 51–93.
74. Picó, Y. and Barceló, D., 2020. Pyrolysis gas chromatography-mass spectrometry in environmental analysis: focus on organic matter and microplastics. *TrAC Trends in Analytical Chemistry*, *130*, p. 115964.
75. Jiménez-Skrzypek, G., Hernández-Sánchez, C., Ortega-Zamora, C., González-Sálamo, J., González-Curbelo, M.Á. and Hernández-Borges, J., 2021. Microplastic-adsorbed organic contaminants: analytical methods and occurrence. *TrAC Trends in Analytical Chemistry*, *136*, p. 116186.
76. Llorca, M., Farré, M., Karapanagioti, H.K. and Barceló, D., 2014. Levels and fate of perfluoroalkyl substances in beached plastic pellets and sediments collected from Greece. *Marine Pollution Bulletin*, *87*(1–2), pp. 286–291.
77. Da Costa, J.P., Avellan, A., Mouneyrac, C., Duarte, A. and Rocha-Santos, T., 2023. Plastic additives and microplastics as emerging contaminants: mechanisms and analytical assessment. *TrAC Trends in Analytical Chemistry*, *158*, p. 116898.
78. Salthammer, T., 2022. Microplastics and their additives in the indoor environment. *Angewandte Chemie*, *134*(32), p.e202205713.

79. Choudhury, A., Simnani, F.Z., Singh, D., Patel, P., Sinha, A., Nandi, A., Ghosh, A., Saha, U., Kumari, K., Jaganathan, S.K. and Kaushik, N.K., 2023. Atmospheric microplastic and nanoplastic: the toxicological paradigm on the cellular system. *Ecotoxicology and Environmental Safety*, *259*, p. 115018.
80. Ramsperger, A.F., Bergamaschi, E., Panizzolo, M., Fenoglio, I., Barbero, F., Peters, R., Undas, A., Purker, S., Giese, B., Lalyer, C.R. and Tamargo, A., 2023. Nano-and microplastics: a comprehensive review on their exposure routes, translocation, and fate in humans. *NanoImpact*, *29*, p. 100441.
81. Hu, M. and Palić, D., 2020. Micro-and nano-plastics activation of oxidative and inflammatory adverse outcome pathways. *Redox Biology*, *37*, p. 101620.
82. Uttara, B., Singh, A.V., Zamboni, P. and Mahajan, R., 2009. Oxidative stress and neurodegenerative diseases: a review of upstream and downstream antioxidant therapeutic options. *Current Neuropharmacology*, *7*(1), pp. 65–74.
83. D'Oria, R., Schipani, R., Leonardini, A., Natalicchio, A., Perrini, S., Cignarelli, A., Laviola, L. and Giorgino, F., 2020. The role of oxidative stress in cardiac disease: from physiological response to injury factor. *Oxidative Medicine and Cellular Longevity, 2020*, p. 5732956.

8 Household Microplastic Pollutants

A Potential Contributor to Indoor Microplastics Pollution

Dipika Narula and Sushma Negi

8.1 INTRODUCTION

Originating from highly durable and desirable plastics, microplastics (MPs) (diameter size 100 nm–5 mm) have emerged as fibers, fragments, or pellets formed by the degradation of plastics into small pieces by biotic and abiotic factors [1]. Thompson [2] introduced the idea of MPs, and their sources range from huge industrial products to common household items. Based on their sources, the subdivision of MPs into primary and secondary MPs has been well demonstrated [3]. Primary MPs (microbeads) are purposefully designed to be small (microsized) for use in personal care commodities, cosmetics, scrubbers (in air-blasting), and medicines (as drug vectors), whereas secondary MPs emerge as a result of chemical, physical, or biological degradation of bigger plastic objects [4].

The toxicity of MPs is attributed to their ability to release chemicals into the environment since a variety of additives (plasticizers, flame retardants, heat, and UV stabilizers) are employed in the plastic production on industrial scale [5]. The biotoxicity of MPs has a profound dependence on the MP surface area, size, functional groups, and composition, which varies according to polymer type: polyethylene (PE), polypropylene (PP), polystyrene (PS), polyethylene terephthalate, polyvinyl chloride (PVC), low-density PE (LDPE), and high-density (HDPE) [6–8]. The elevated surface area, high adsorption capacity and electrostatic properties along with hydrophobic and lipophilic character associated with MPs make them carriers of several heavy metals (Pb, Zn, Ni, Cd), toxic materials, organic pollutants (PCBs, organochlorine pesticides, PAHs), and pathogenic microorganisms by their adsorption, thereby changing their biological/environmental toxicities [5, 9].

MPs, rightly referred to as "cocktail of contaminants", and their adsorbed contaminants have the potential to travel long distances and cause detrimental health risks to living organisms and ecology by their bioaccumulation in the food chain, subsequently affecting humans [9]. Following their entry, the interaction of MPs

DOI: 10.1201/9781003468141-8

with the biological barriers in organisms causes reproductive, neuro-, immuno- and development toxicities, affecting intestines and metabolic functions, and inhaling MPs leads to respiratory illness and asthma [9]. Inflammation, skin irritations, and endocrine disruption have been reported upon dermal exposure to MPs [3, 9]. Other health ailments include disturbed circadian rhythms and production of melatonin; disturbed sleep; and increased risk of developing diabetes, obesity, and cancer [3]. These hazards reveal MPs to be a significant concern in health and ecosystems together owing to their ubiquitous presence in the environment.

In addition to their presence outdoors, the existence of MPs indoors cannot be overlooked since humans spend almost 90% of their lifetime indoors (includes educational institutes, offices, household, surgical rooms etc.) [10]. Reports reveal greater indoor than outdoor MP pollution [11]. Sources that contribute to indoor MP pollution include wear and tear of architectural materials, indoor furniture, washing and drying of synthetic clothes and laundry, sloughing and flaking of textile fibers, plastic waste incineration, release from plastic packaging and storage materials, electronics, plastic kitchen tools, indoor paint, personal care products, medical supplies, masks, and others [1, 12]. In addition, the exact composition and abundance of indoor MPs is impacted by the kind of human activities carried out in the nearby areas [13].

Since people spend almost 60% of their time within their homes, it seems necessary to thoroughly understand the impacts of household exposure to indoor MPs. In this regard, this chapter gives a deep insight into the occurrence, types, human exposure, and health impacts of indoor household MPs pollutants, which deserves immediate attention. More knowledge on indoor household MP pollution and its severity could assist in the creation of environmental health policies for its control and prevention at household sites.

8.2 TYPES OF INDOOR HOUSEHOLD MICROPLASTIC POLLUTANTS

Human health is seriously threatened by indoor contaminants, and investigating their origin is an essential first step in the study of indoor pollution [14]. In 2021, Soltani et al. [15] reported that indoor MPs cause higher risks to human health than outdoor MPs. The release of MPs in the indoor environment is often related to mechanical stresses and strong abrasive agitations. Here, we discuss the various types of indoor household MPs pollutants.

8.2.1 Indoor Dust

Numerous exposure pathways allow MPs to enter the human body, one route being breathing dust loaded with MPs [16]. Particulate matter is deposited as dust indoors from many sources. Walking and crawling on carpets as well as vacuuming in homes has accounted for the presence of MPs such as PE, polyacrylic, polyester, PS, and polyamide fibers in the indoor air [15]. It is estimated that the annual intake of MP particles from dust can reach 1,000 for adults and 3,000 for children. Children are more exposed to dust inside because they are closer to the ground [17].

8.2.2 Cosmetics and Personal Care Products

According to a Cosmetics Europe survey, 714–793 tons of microbeads were utilized in cosmetics and personal hygiene products in the European Union in 2015 [18]. In the UK, 680 million microbeads are reportedly used annually in cosmetics and personal hygiene products [19]. The presence of MPs in cosmetics and personal care products (PCPs) like face scrubs, [20] shower gels, shampoos, creams, eye shadows, deodorants, blush powders, make-up foundations, skin creams (as exfoliators, emulsifiers, binding agents, opacifying agents, anti-static agents and film-forming agents), powders, glitters, concealers, mascaras, eye shadows, eye pencils, eyeliners, lip care, sun care, hair care (like leave-on conditioners, dry shampoos), consumer goods, nail polish, etc. has been well reported [21]. Glitter sticks to human skin because of its tiny size, dermal oils, or even merely static electrical force [22]. PE, PP, PET, and PMMA [23] are the common MPs used in these PCPs, but big business cosmetic brands like Unilever and L'Oréal have either eliminated MPs from their products or replaced them with alternatives like chitin, microbeads made of cellulose, and bio-based polymers [24].

8.2.3 Drinks

Humans directly inhale, swallow, and consume MPs from consuming bottled water and alcoholic beverages (such as beers, wines, and spirits), nonalcoholic beverages (such as milk, chocolate, tea, coffee, soft drinks, carbonated and non-carbonated sweetened drinks). In one research study, researchers examined a group of beverages and found 80% MP content in cold tea, 40% ± 24.53% in soft drinks and 14% ± 5.79% in energy drinks [25]. As a result, there is growing concern throughout the world about MPs as a significant source of anthropogenic contamination in drinks. PP was found to be the major MPs pollutant in drinks, suggesting its origin from plastic packaging bottles [26].

Around the world, disposable cups are frequently used for take-out coffee, milk, tea, etc. [27]. Commercial disposable cups typically come in two varieties: paper cups with a plastic coating and cups made of plastic [28]. The amount of MPs released from disposable cups into beverages that contain them should be of concern because MPs can directly enter the human body through beverage consumption.

8.2.4 Household Laundry

The use of household washing machines is increasing at an alarming rate, and therefore, characterizing MPs in laundry wastewater has become imperative. Washing domestic laundry made of nylon, polyester, and acrylic fibers leads to substantial MP emission at the household level [29]. Additionally, washing a variety of different fabric structures in one load releases different amounts and types of MPs into laundry wastewater [30].

8.2.5 Dishwashers

Researchers have experimentally examined how MPs are released into urban wastewater by dishwashers [31]. This release of MPs from a dishwasher into the sewage system can be attributed to the degradation of the dishwasher plastic components/materials during the wash cycle. Therefore, dishwashing can be regarded as a significant household source of MP contamination, much like laundry [32]. It has also been reported that using a detergent causes a 35%–54% increase in the deterioration of plastic dishwashing components [31].

8.2.6 Chopping Board

Fang et al. studied the release of MPs by cutting/chopping on plastic chopping boards [33]. The chopping methods used to extract MPs debris from the chopping board involve pressing the knife to chop (in perpendicular direction to the chopping board surface), pushing the knife in a scratching motion (for transporting food from one chopping premise to another), and slicing the knife (in horizontal direction and parallel to the chopping board surface). Analytical studies indicated the release of 100–300 MPs/mm/cut along the grove produced on the chopping board and 3,000/mm_2/cut in the scratched chopping board area. Henceforth, the MPs find an easy route for human ingestion during food preparation utilizing plastic chopping boards. To avoid this MP contamination, wooden chopping boards are a better choice.

8.2.7 Toothpaste and Dental Care Products

It has been speculated that a number of toothpastes, including Sensodyne, Prodent, Oral-B, Aquafresh, and Zendium, contain PE microbeads, which are presumably being phased out as we speak. These microbeads are visibly spherical plastic particles (<5 mm) that assist in peeling, scrubbing, and exfoliating [34]. However, in toothpaste, these MPs can stick to the gums, trap bacteria, and result in gingivitis. Over time, the gum infection may spread into the tooth-supporting bone, which could cause gum bleeding; the presence of MPs such as PVC, PP, and cellophane in dental care products has been well recognized [35].

8.2.8 Food Packaging

Packaging is required to protect food from the environment from the time of its production till it is consumed. The shelf life of packaged food may range from a few days to more than a year [36]. Because plastic packaging is so inexpensive, it has supplanted other materials in the food sector; its other benefits are that plastics are noncorrosive, durable, and lightweight.

For instance, because of PS's extended shelf life, optical translucency, and insulating qualities, it is widely used in throwaway food packaging. However, despite these advantageous properties, PS presents a health risk to humans due to the leaching of styrene monomer into the edible contents; in one research study, PS food containers exhibited significantly more styrene leaching (up to 6 μg/mL) than did PS cups (up to 0.110 μg/mL) [37]. Moreover, Sobhani et al. reported on producing larger fragmented plastics and fine plastic particles by opening packaged plastic materials with a knife [38].

8.3 CONCLUSIONS

So far, there are inadequate awareness of and resources for the issue of indoor contamination by household microplastic pollutants as well as ineffective laws. Actions must be taken immediately to address the growing risks that MPs present to the environment as a whole and to human health. Furthermore, additional investigation and development work are required to more fully associate, connect, and comprehend the effects of microplastics on humans.

REFERENCES

1. Tiwari, B.R., Lecka, J., Pulicharla, R. and Brar, S.K., 2023. Microplastic pollution and associated health hazards: Impact of COVID-19 pandemic. *Current Opinion in Environmental Science & Health*, p. 100480.
2. Thompson, R.C., Olsen, Y., Mitchell, R.P., Davis, A., Rowland, S.J., John, A.W., McGonigle, D. and Russell, A.E., 2004. Lost at sea: Where is all the plastic? *Science*, 304(5672), pp. 838–838.
3. Ghosh, S., Sinha, J.K., Ghosh, S., Vashisth, K., Han, S. and Bhaskar, R., 2023. Microplastics as an emerging threat to the global environment and human health. *Sustainability*, 15(14), p. 10821.
4. Karbalaei, S., Hanachi, P., Walker, T.R. and Cole, M., 2018. Occurrence, sources, human health impacts and mitigation of microplastic pollution. *Environmental Science and Pollution Research*, 25, pp. 36046–36063.
5. Street, M.E. and Bernasconi, S., 2021. Microplastics, environment and child health. *Italian Journal of Pediatrics*, 47(1), p. 75.
6. Cao, G. and Cai, Z., 2023. Getting health hazards of inhaled nano/microplastics into focus: Expectations and challenges. *Environmental Science & Technology*, 57(9), pp. 3461–3463.
7. Sabilillah, A.M., Palupi, F.R., Adji, B.K. and Nugroho, A.P., 2023. Health risk assessment and microplastic pollution in streams through accumulation and interaction by heavy metals. *Global Journal of Environmental Science and Management*, 9(4), pp. 719–740.
8. Dewika, M., Markandan, K., Irfan, N.A., Abdah, M.A.A.M., Ruwaida, J.N., Sara, Y.Y. and Khalid, M., 2023. Review of microplastics in the indoor environment: Distribution, human exposure and potential health impacts. *Chemosphere*, p. 138270.
9. Zhu, L., Xie, C., Chen, L., Dai, X., Zhou, Y., Pan, H. and Tian, K., 2023. Transport of microplastics in the body and interaction with biological barriers, and controlling of microplastics pollution. *Ecotoxicology and Environmental Safety*, 255, p. 114818.
10. Klepeis, N.E., Nelson, W.C., Ott, W.R., Robinson, J.P., Tsang, A.M., Switzer, P., Behar, J.V., Hern, S.C. and Engelmann, W.H., 2001. The national human activity pattern survey (NHAPS): A resource for assessing exposure to environmental pollutants. *Journal of Exposure Science & Environmental Epidemiology*, 11(3), pp. 231–252.
11. Dris, R., Gasperi, J., Mirande, C., Mandin, C., Guerrouache, M., Langlois, V. and Tassin, B., 2017. A first overview of textile fibers, including microplastics, in indoor and outdoor environments. *Environmental Pollution*, 221, pp. 453–458.
12. Bahrina, I., Syafei, A.D., Satoto, R., Jiang, J.J., Nurasrin, N.R., Assomadi, A.F., Boedisantoso, R., Hermana, J. and Nasir, M., 2020. An occupant-based overview of microplastics in indoor environments in the City of Surabaya, Indonesia. *Journal of Ecological Engineering*, 21(8).
13. Kacprzak, S. and Tijing, L.D., 2022. Microplastics in indoor environment: sources, mitigation and fate. *Journal of Environmental Chemical Engineering*, 10(2), p. 107359.

14. Yang, D.L., Zhang, Z.N., Liu, H., Yang, Z.Y., Liu, M.M., Zheng, Q.X., Chen, W. and Xiang, P., 2023. Indoor air pollution and human ocular diseases: Associated contaminants and underlying pathological mechanisms. *Chemosphere*, 311, p. 137037.
15. Soltani, N.S., Taylor, M.P. and Wilson, S.P., 2021. Quantification and exposure assessment of microplastics in Australian indoor house dust. *Environmental Pollution*, 283, p. 117064.
16. Sangkham, S., Faikhaw, O., Munkong, N., Sakunkoo, P., Arunlertaree, C., Chavali, M., Mousazadeh, M. and Tiwari, A., 2022. A review on microplastics and nanoplastics in the environment: Their occurrence, exposure routes, toxic studies, and potential effects on human health. *Marine Pollution Bulletin*, 181, p. 113832.
17. Ramasamy, E.V. and Harit, A.K., 2023. *Microplastics in Human Consumption*. CRC Press.
18. Shanker, R., Khan, D., Hossain, R., Islam, M.T., Locock, K., Ghose, A., Sahajwalla, V., Schandl, H. and Dhodapkar, R., 2022. Plastic waste recycling: Existing Indian scenario and future opportunities. *International Journal of Environmental Science and Technology*, pp. 1–18.
19. Habib, R.Z., Aldhanhani, J.A., Ali, A.H., Ghebremedhin, F., Elkashlan, M., Mesfun, M., Kittaneh, W., Al Kindi, R. and Thiemann, T., 2022. Trends of microplastic abundance in personal care products in the United Arab Emirates over the period of 3 years (2018–2020). *Environmental Science and Pollution Research*, 29(59), pp. 89614–89624.
20. Fendall, L.S. and Sewell, M.A., 2009. Contributing to marine pollution by washing your face: Microplastics in facial cleansers. *Marine Pollution Bulletin*, 58(8), pp. 1225–1228.
21. Singh, A. and Mishra, B.K., 2023. Microplastic in personal care products: An overlooked environmental concern. *Journal of Cleaner Production*, p. 139082.
22. Lei, K., Qiao, F., Liu, Q., Wei, Z., Qi, H., Cui, S., Yue, X., Deng, Y. and An, L., 2017. Microplastics releasing from personal care and cosmetic products in China. *Marine Pollution Bulletin*, 123(1–2), pp. 122–126.
23. Šunta, U., Trebše, P. and Kralj, M.B., 2021. Simply applicable method for microplastics determination in environmental samples. *Molecules*, 26(7), p. 1840.
24. Berpotensi, A., MRahim, N.A.S., Islahudin, F., Tahrim, N.A. and Jasamai, M., 2022. Microplastics in cosmetics and personal care products: Impacts on aquatic life and rodents with potential alternatives. *Sains Malaysiana*, 51(8), pp. 2495–2506.
25. Shruti, V.C., Pérez-Guevara, F., Elizalde-Martínez, I. and Kutralam-Muniasamy, G., 2020. First study of its kind on the microplastic contamination of soft drinks, cold tea and energy drinks-Future research and environmental considerations. *Science of the Total Environment*, 726, p. 138580.
26. Mortensen, N.P., Fennell, T.R. and Johnson, L.M., 2021. Unintended human ingestion of nanoplastics and small microplastics through drinking water, beverages, and food sources. *NanoImpact*, 21, p. 100302.
27. Liu, G., Wang, J., Wang, M., Ying, R., Li, X., Hu, Z. and Zhang, Y., 2022. Disposable plastic materials release microplastics and harmful substances in hot water. *Science of the Total Environment*, 818, p. 151685.
28. Ranjan, V.P., Joseph, A. and Goel, S., 2021. Microplastics and other harmful substances released from disposable paper cups into hot water. *Journal of Hazardous Materials*, 404, p. 124118.
29. Praveena, S.M., Syahira Asmawi, M. and Chyi, J.L.Y., 2021. Microplastic emissions from household washing machines: Preliminary findings from Greater Kuala Lumpur (Malaysia). *Environmental Science and Pollution Research*, 28, pp. 18518–18522.
30. Lin, Q., Pang, L., Ngo, H.H., Guo, W., Zhao, S., Liu, L., Chen, L. and Li, F., 2023. Occurrence of microplastics in three types of household cleaning products and their estimated emissions into the aquatic environment. *Science of the Total Environment*, 902, p. 165903.

31. Sol, D., Menéndez-Manjón, A., Carrasco, S., Crisóstomo-Miranda, J., Laca, A., Laca, A. and Díaz, M., 2023. Contribution of household dishwashing to microplastic pollution. *Environmental Science and Pollution Research*, 30(15), pp. 45140–45150.
32. Galvão, A., Aleixo, M., De Pablo, H., Lopes, C. and Raimundo, J., 2020. Microplastics in wastewater: Microfiber emissions from common household laundry. *Environmental Science and Pollution Research*, 27, pp. 26643–26649.
33. Luo, Y., Chuah, C., Al Amin, M., Khoshyan, A., Gibson, C.T., Tang, Y., Naidu, R. and Fang, C., 2022. Assessment of microplastics and nanoplastics released from a chopping board using Raman imaging in combination with three algorithms. *Journal of Hazardous Materials*, 431, p. 128636.
34. Kavery, C.S., Rao, A., Aparna, K.S., Jodalli, P.S. and Shenoy, K.R., 2023. Microplastic content of over-the-counter toothpastes-a systematic review. *F1000Research*, 12.
35. Madhumitha, C.T., Karmegam, N., Biruntha, M., Arun, A., Al Kheraif, A.A., Kim, W. and Kumar, P., 2022. Extraction, identification, and environmental risk assessment of microplastics in commercial toothpaste. *Chemosphere*, 296, p. 133976.
36. Macena, M.W., Carvalho, R., Cruz-Lopes, L.P. and Guiné, R.P., 2022. Perceptions and knowledge regarding quality and safety of plastic materials used for food packaging. *Open Agriculture*, *7*(1), pp. 132–146.
37. Ajaj, A., J'Bari, S., Ononogbo, A., Buonocore, F., Bear, J.C., Mayes, A.G. and Morgan, H., 2021. An insight into the growing concerns of styrene monomer and poly(styrene) fragment migration into food and drink simulants from poly(styrene) packaging. *Foods*, 10(5), p. 1136.
38. Sobhani, Z., Lei, Y., Tang, Y., Wu, L., Zhang, X., Naidu, R., Megharaj, M. and Fang, C., 2020. Microplastics generated when opening plastic packaging. *Scientific Reports*, 10(1), p. 4841.

9 Adverse Impacts of Microplastics on Animal Health

Kushal Thakur, Danish Mahajan, Navya Rana, Kanika Choudhary, Amit Kumar Sharma, Dixit Sharma, Hishani Kumari, Sunil Kumar, and Rakesh Kumar

9.1 INTRODUCTION

The widespread presence of microplastics (MPs) in our environment has raised significant concern over their influence on the health of organisms in both terrestrial and aquatic ecosystems. These minute particles, usually less than five millimeters in size, have infiltrated several ecosystems, posing a significant danger to species worldwide [1]. In order to comprehensively grasp the effects of MPs on animal health, researchers need to thoroughly examine MP sources, pathways of exposure, harmful processes, and consequences on different species [2]. MPs can originate from diverse sources and processes, including from the fragmentation of larger plastic objects, the inclusion of microbeads in personal care products, the shedding of synthetic fibers from fabrics, and the activities that take place in industrial environments [3]. These MP particles accumulate in soil, sediment, and aquatic habitats, and the organisms inhabiting these habitats, both small invertebrates and massive marine and terrestrial mammals, are exposed to these plentiful pollutants [4].

MPs have both physical and chemical qualities and have shown ill effects on animal health. These particles adsorb and concentrate hazardous compounds that are prevalent in the environment such as heavy metals and persistent organic pollutants [5]; when creatures consume these potentially dangerous substances, they can leak into their tissues, leading to bioaccumulation and biomagnification throughout the food chain [6]. Furthermore, the ingestion of MPs can result in mechanical injury, obstruction of digestive tracts, and changes in feeding behavior in animals, which affects their nutrient intake and their general health [7].

The small size of MP particles makes it easier for a wide variety of organisms to take them in. These organisms include filter feeders, plankton, fish, birds, and even terrestrial animals, which further exacerbates the ubiquitous nature of this problem [8]. MP exposure has a wide range of physiological effects on animal health, and these effects can present themselves in a variety of different ways. Instances of these include immunosuppression, disruptions to reproductive functions, organ damage,

DOI: 10.1201/9781003468141-9

and heightened susceptibility to infections [9]. In recent years, there has been growing concern among researchers and conservationists about the consequences of prolonged chronic exposure to these minute particles for the viability and sustainability of populations within ecosystems [10]. Interdisciplinary research efforts that include biology, ecology, toxicology, and environmental science are required to address the nuanced and complicated nature of the influence that MPs have on the health of animals [11].

9.2 MP EFFECTS ON DIFFERENT FAUNA

MP pollution poses a significant threat to various vertebrate species across terrestrial and aquatic ecosystems. This chapter highlights the adverse effects of MPs on different animal species in both terrestrial and aquatic ecosystems.

9.2.1 MP Impacts on Insects

Pollutants like MPs are widely distributed in the food and agriculture industries, posing a threat to the health of plants and animals [12]. Insects play an important role in ecosystem services and are widespread in many environments. Insects' survival, reproduction, development, and gut microbiota can be negatively affected by MPs [13]. House cricket (*Gryllodes sigillatus*) was used as a dietary model in one study, and MP ingestion affected specific organs; additionally, changes in organ size were more pronounced than the body size changes in general [14]. Separately, MP effects appear to be sex specific: MP ingestion caused damage to reproductive organs (inhibiting spermatogenesis) in male earthworms (*Eisenia andrei*), but there was no such damage in the reproductive tissue of female earthworms [15].

MPs also have negative effects on the ecological functions performed by insects like mayflies, dragonflies, midges, and mosquitoes whose development occurs exclusively in freshwater environments [16]. MPs can be dispersed in river ecosystems by trophic transfer to other aquatic animals, affecting higher trophic levels in the food chain. MPs can be eaten by mosquito larvae because they cannot distinguish between food and MPs [17]. *Culex pipiens* larvae retained most of their MPs as they grew into adults, supporting the hypothesis that MPs are transferred to adults [18]. In various toxicological studies, *Drosophila melanogaster* is used as a model for studying the effects of short-term exposure to different concentrations polystyrene MPs [19]. Polyethylene terephthalate (PET) MPs also had sex-specific effects on the flies, decreasing egg production in females [20] and decreasing lipid, glucose, and starvation resistance in males [21].

Successful pollination requires an interaction between plant and pollinator. The health of plants and pollinators is affected by MP pollution, which disrupts their communication [22]. In most aquatic and terrestrial ecosystems, insects play a critical role. MP pollution is highly prevalent in most ecosystems due to insects' ubiquitous distribution. Larvae of black solider fly (*Hermetia illucens*) were found to be less likely to survive in plastic-contaminated environments [23]. There has been evidence that MPs are present in honey [24] and in the cuticles of honeybees [25]. Through the cuticle or by ingestion, *Apis mellifera* can absorb MPs from the environment and transfer them to other parts of the hive, including larvae, wax, and honey

[26]. Researchers established that the long-horned beetle (order Coleoptera, family Cerambycidae), which makes up around 25% of all animal species, contains MPs in the majority of its population [27].

9.2.2 MP Impacts on Aquatic Organisms

Increasing MP concentrations in the aquatic environment have drawn the attention of many researchers in the last few years. The presence of MPs in almost all freshwater and marine aquatic ecosystems indicates their potential negative effects on the global health of aquatic ecosystems [28]. A vast array of organisms easily consume these particles because of their abundance and small size, ranging from microscopic zooplankton to large filter feeders and from filter feeders to top predators. Several negative impacts can result from this ingestion on individual organisms as well as populations and, in the end, entire aquatic ecosystems [29]. Numerous studies on aquatic species have revealed the detrimental impacts of MPs, including decreased fertility and eating behavior, slowed larval growth and development, increased oxygen use, and structural damage to organs.

The fate of benthic invertebrates, whether they live in freshwater or marine habitats, is of particular concern because it is estimated that up to 90% of the biomass of fish prey comes from them [30]. In both aquatic ecosystems, MP particles in the sediments expose benthic organisms to plastics [31, 32]. The ingestion of MPs by benthic organisms adversely affects their health, feeding, development, mortality, behavior, reproduction, and other aspects of their physiological functioning (Table 9.1) [33, 34].

These MPs have a significant influence on benthic animals, which include both invertebrates and vertebrates. These minute particles penetrate the sediment beds, changing the habitats of the benthic organisms [35]. Mollusks, crabs, and worms are examples of invertebrates that are likely to experience severe consequences due to the presence of MPs in water because of their filter-feeding or deposit-feeding behaviors. These organisms mistake MPs for food and ingest them, and the MPs accumulate in their digestive systems, which in turn causes bodily injury, decreased feeding efficiency, and disturbance in the absorption of nutrients [36]. In addition, these plastics contain absorbed toxins that affect invertebrate physiology and reproductive capacity, which affects population dynamics and ecological balance [37].

Researchers have made significant advancements in investigating MP pollution in bivalve organisms. As a result of their widespread distribution, filtering capabilities, and close association with human health, bivalves are excellent bioindicators of MP pollution. They are affected directly and indirectly by MPs, including their physiology, habitat structure, food source, and persistence of organic pollutants [34, 38].

Vertebrates that live in benthic settings, such as fish and organisms that feed on the bottom, are also subject to the negative effects of MPs. In the case of fish, for example, they mistake MPs for prey, which results in the accumulation and intensification of toxins in their tissues as these plastics migrate up the food chain [39]. It seems that visual eaters are more likely to ingest MP particles that resemble natural prey objects [5]. The bioaccumulation of MPs impairs the immune system, increases susceptibility to infections, and hinders reproductive capacity, while also posing a potential hazard to human health when consuming infected seafood [40].

TABLE 9.1
Effects of MP Particles on Invertebrates' Fauna [33]

Taxa	Species	Material	Size	Parameter	Effect	References
Annelida	*Arenicola Marina*	PS (Polystyrene)	400–1,300 μm	Behavior	Significant, dose-dependent decrease in feeding activity	[77]
		PVC (Polyvinylchloride)	63–500 μm	Egestion	Significantly prolonged gut residence time	[78]
				Behavior	Significantly reduced feeding activity	
				Cellular response	Significantly increased phagocytic activity	
	Perinereis aibuhitensis	PS	8–12 μm	Mortality	Increased mortality significantly	[79]
				Regeneration	Significantly reduced regeneration rate	
	Hediste diversicolor	PVC	250 μm	Cellular response	Significantly increased phagocytic activity	[80]
	Eisenia fetida	PE (Polyethylene)	(≤50 μm		Neurotoxicity, energy metabolism disruption, inflammation, and oxidative stress	[81]
Arthropoda	*Chironomus tepperi*	PE	1–4, 10–27, and 43–54 μm;	Mortality	Significantly increased mortality at 1–4, 10–27, and 43–54 μm	[82]
				Development	Significantly decreased body length at 1–4, 10–27, and 43–54 μm	
					Significantly decreased length of head capsule at 10–27 μm	
					Reduced development of head capsule, mouth parts, and antenna at 10–27 μm	
					Development delay at 10–27 μm	
				Emergence	Significantly lower emerging rates for all size ranges	
	Gammarus fossarum	PA (polyamide)	20 × 500 μm	Assimilation	Significantly decreased assimilation	[83]
	Gammarus fossarum	PMMA (polymethyl methacrylate)	32–64 μm	Development	Significantly decreased wet weight	[84]
			125–250 μm	Assimilation	Significantly decreased assimilation	
		PHB (Polyhydroxybutyrate)	32–64 μm	Development	Significantly decreased wet weight	
	Gammarus pulex	PS	20–500 μm	Development	Significantly reduced growth	[85]

(Continued)

TABLE 9.1 *(Continued)*
Effects of MP Particles on Invertebrates' Fauna [33]

Taxa	Species	Material	Size	Parameter	Effect	References
	Hyalella azteca	PE	10–27 μm	Mortality	Significant, dose-dependent increase in mortality	[86]
		PP (Polypropylene)	20–75 × 20 μm	Reproduction	Significantly decreased reproduction	
	Calanus helgolandicus	PS	20 μm	Behavior	Significantly lower feeding rate	[87]
				Development	Significantly smaller eggs	
	Palaemonetes pugio	PE, PS, PP	35–165 μm, 30 and 75 μm, 34 and 93 μm	Mortality	Significantly increased mortality	[88]
	Parvocalanus crassirostris	PET (Polyethylene terephthalate)	5–10 μm	Reproduction	Significantly reduced egg production	[89]
					Significantly reduced relative population size	
Mollusca	*Dreissena polymorpha*	PS	1 and 10 μm	Cellular response	CAT and gpx activity significantly affected	[90]
	Atactodea striata	PS	63–250 μm	Behavior	Significantly reduced feeding rate	[91]
	Mytilus galloprovincialis	PE	<100 μm	Cellular response	Significantly reduced granulocyte/hyalinocyte ratio	[92]
Nematoda	*Caenorhabditis elegans*	PA, PP	~70 μm	Mortality	Significantly increased mortality	[93]
				Development	Significantly decreased body length	
				Reproduction	Significantly fewer embryos	
				Cellular Response	Significantly lower calcium level in intestines	
	Caenorhabditis elegans	PS	0.1 μm	Behavior	Significant, dose-dependent increase in defecation cycle length	[94]
					Significant, dose-dependent decrease of locomotion (head trash, body bend)	
				Reproduction	Significant, dose-dependent decrease in reproduction (brood size)	
Rotifera	*Brachionus koreanus*	PS	0.05 μm	Reproduction	Significantly longer reproduction time	[95]
			0.5 μm		Fecundity and population growth significantly retarded	
			6 μm	Life Span	Life span significantly affected	
Echinodermata	*Paracentrotus lividus*	PS	10 μm	Development	Significant effects on body length, body weight	[96]
	Paracentrotus lividus	PS, HDPE (High-density polyethylene)		Growth	Reduced and abnormal larval growth	[97]

Around the world, fish are important nutritional and economic components of aquatic ecosystems [41], but fish meals contain MPs, with cellophane, polypropylene and polyethylene terephthalate (PET) the most common types; these are predominantly produced in China, Myanmar, and Peru [42]. Soliño et al. investigated the presence of MPs in fish species dwelling in the deep bathyal zone of the North Atlantic Sea and observed PET filaments of MPs within the digestive tracts of *Alepocephalus bairdii* and *Coryphaenoids rupestris* [43]. Fish are at risk and often die before they reach maturity as a result of MP toxicity [44]. In an electrophysiological study, Liang et al. found that fish–MP interactions caused increased cough rates, decreased intake, and augmented dynamic mucus secretion during ventilation, feeding, and swimming processes in goldfish [45]. The exposure of fish to MPs resulted in several behavioral, morphological, and physiological changes as well as damage to their nervous and visual systems, and alterations in their neurological processes and immune response [46–49].

9.2.3 MP Impacts on Birds

Avian species are also susceptible to the detrimental impacts of MPs due to ingestion, entanglement, and habitat disruption [50]. An important consequence arises from birds consuming MPs, frequently mistaking them for food [51]. Since the 1960s, it has been estimated that over 78% of identified species of seabirds have consumed MPs in their digestive systems. It is projected that by 2050, over 99% of more than 300 seabird species will have ingested plastic trash [52, 53]. In a study conducted by Zhao et al., MPs were found in the digestive systems of 16 out of 17 land-dwelling bird species [54]. MPs that accumulate in birds' digestive tracts cause various adverse effects, and they reside longer in the intestine than other undigestible substances [55].

Wild seabirds show alterations in gut microbiomes due to the presence of MP concentrations. MPs have been linked to decreased commensal microbiota and increased pathogens as well as antibiotic-resistant and plastic-degrading microbes [56]. Empirically, polystyrene MPs harmed chicken lung tissue of chickens by triggering apoptotic and autophagic pathways [57], and MPs can cause severe scar tissue formation in birds, a condition known as plasticosis [58]. As a consequence of macro-plastic and MP exposure in seabird tissues, Rivers-Auty et al. observed a decrease in the number of tubular glands and rugae in the proventriculus, along with inflammation, fibrosis, and deterioration of kidney and spleen structures [59].

9.2.4 MP Impacts on Mice

Findings from studies on MP impacts on terrestrial animals, mainly laboratory mice and rats, show that these particles can lead to hepatotoxicity, neurotoxic effects, local inflammation, gastrointestinal toxicity, and reproductive disorders (Figure 9.1) [60].

Replacing mouse drinking water with different MP suspensions resulted in different impacts on the health of laboratory rodents. In male mice, 0.024 and 0.24 mg/kg/day of MPs induced considerable alterations to the makeup of the intestinal microbiota, abnormalities in lipid metabolism in the liver and in how bile acids are metabolized, decreased mucin secretion, and decreased mass of the liver and adipose

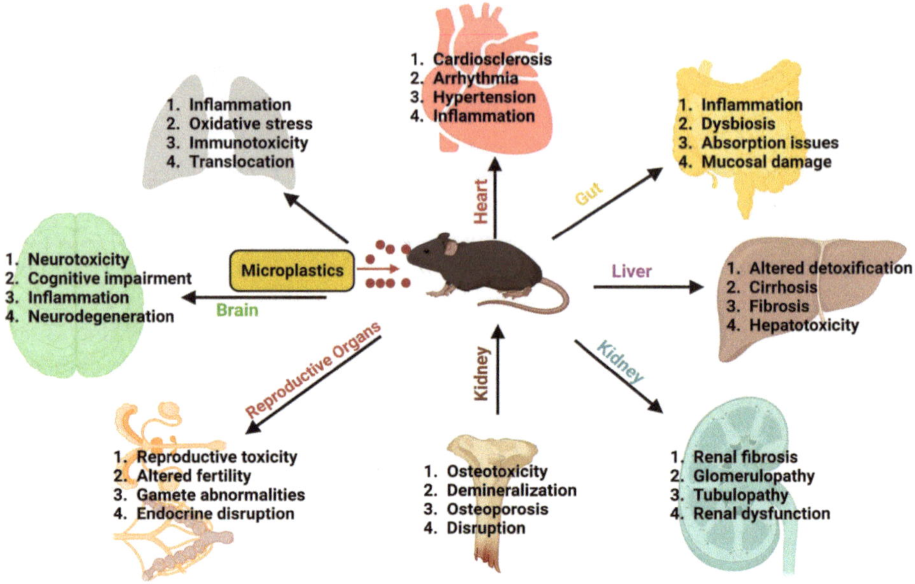

FIGURE 9.1 Harmful effects of microplastics on mice.

tissue in the colon [61]. Hou et al. revealed that MP exposure induced deleterious effects on the mouse ovary and contributed to female infertility. Long-term exposure to polystyrene (PS) MPs could lead to pyroptosis and apoptosis of granulosa cells via NLRP3/Caspase-1 signaling pathway triggered by oxidative stress [62].

Shengchen et al. studied the effect of MPs on skeletal muscle and observed that PS MPs produced excessive ROS, enhanced oxidative stress, and suppressed myogenic differentiation by lowering p38 MAPK phosphorylation [63]. MPs also produced adverse effects in kidneys as elevated levels of ER stress, as well as increased production of inflammatory markers and autophagy-associated proteins [64]. In male mice, exposure to micro-PS significantly reduced sperm motility and quantity while increasing sperm deformity rate [65]. Araujo et al. studied the trophic transfer effects of MPs on mice via tadpoles and fish, and when tadpoles were exposed to MPs and fed *tambatinga* fish, the mice displayed behavioral disorders like decreased risk assessment behaviour and a lack of defensive social aggregation [66].

9.2.5 MP Impacts on Cattle

Livestock and poultry systems have also experienced significant and intense contamination of MPs [67]. Accumulation of MPs has been reported in species like pigs and cows. Both meat and dairy products also include plastics [68]. The main routes by which MPs and nanoplastics (NPs) can access the animal body are (1) inhalation of plastic particles present in the air, due to synthetic textiles and polluted environments, (2) ingestion of food and water that are contaminated with these particles, and (3) physical contact allowing the plastic particles to penetrate the skin barrier [69].

The potential risk of MPs to cattle involves physical, chemical, and microbiological aspects. The tiny sizes of MPs pose a physical risk since they can pass through biological barriers such as the skin, stomach, haemato-encephalic, testicular, and even placental tissues, resulting in direct injury. Chemical risks result from the presence of potentially hazardous pollutants or persistent additives. The attachment of bacteria to the plastic particle surface is linked to microbiological concerns [70]. MPs with sizes less than 20 μm can infiltrate organs, while less than 10 μm can penetrate cell membranes and cross the placental barrier in exposed cells or laboratory animals [71].

The accumulation of MPs occurs in various organs, including the liver, kidneys, gut, stomach, small intestine, and mesenteric lymph nodes (Figure 9.2) [72]. Exposure of cattle to MPs induces inflammation, cytotoxicity, genotoxicity, immunotoxicity [73], translocation to tissues, physical harm, ingestion, and toxicological difficulties [74]. Several studies indicate that MPs can damage the reproductive system, including irregular offspring development, infertility, and disturbances in cattle's estrus cycle; MPs also weaken cattle's immune systems, and high levels of MP contamination make cattle more susceptible to illnesses and infections [75]. In addition, many studies indicate that MPs may cause immunotoxicity in cattle [76].

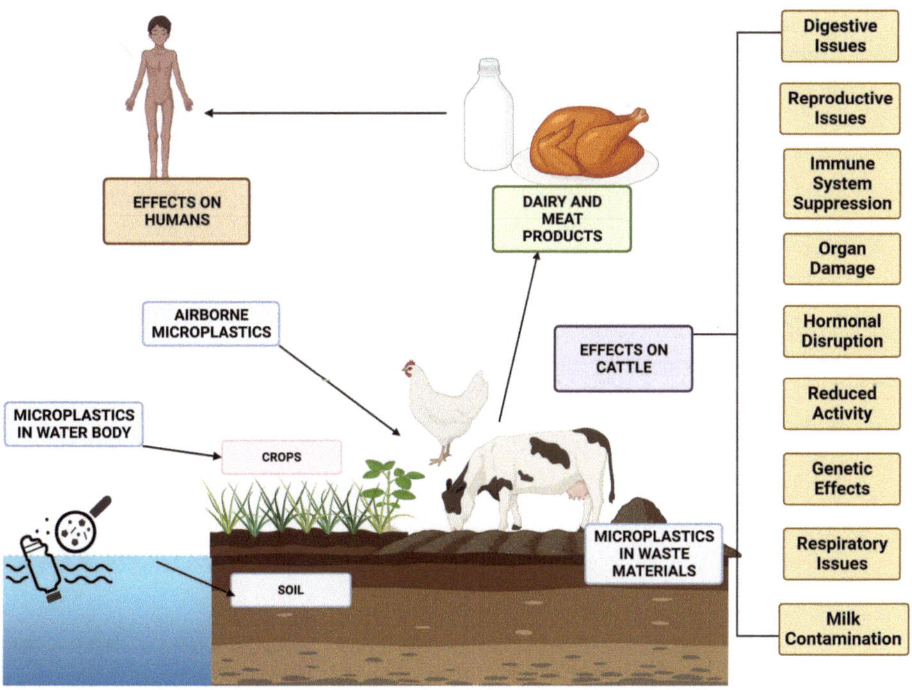

FIGURE 9.2 Impact of microplastic on cattle

9.3 CONCLUSION

Microplastics are widespread in the environment, and they pose a significant threat to the health of many animal species, including insects, aquatic invertebrates and vertebrates, birds, and livestock. The ingestion and accumulation of these MPs cause a range of harmful effects including physical impairment and disturbed reproduction in insects. Aquatic ecosystems face a significant challenge as the accumulation of these particles affects the health of fish, Mollusca, and other aquatic organisms. Plastic consumption also poses a threat to both marine and terrestrial birds, leading to internal damage, starvation, and other potential health repercussions. Similarly, cattle within agricultural systems experience health issues stemming from exposure through contaminated feed water or pasture. The interdependence of ecosystems highlights the necessity of tackling the origins of microplastics, adopting sustainable waste management methods, and promoting worldwide dedication to mitigating the extensive repercussions of this environmental problem. In order to safeguard animals from the harmful effects of microplastics, it is imperative to participate in a collaborative endeavor to maintain the equilibrium of our ecosystems and guarantee the welfare of all organisms.

REFERENCES

1. Vivekanand, A.C., Mohapatra, S. and Tyagi, V.K., 2021. Microplastics in aquatic environment: Challenges and perspectives. *Chemosphere*, 282, p. 131151.
2. Campanale, C., Massarelli, C., Savino, I., Locaputo, V. and Uricchio, V.F., 2020. A detailed review study on potential effects of microplastics and additives of concern on human health. *International Journal of Environmental Research and Public Health*, 17(4), p. 1212.
3. Anand, U., Dey, S., Bontempi, E., Ducoli, S., Vethaak, A.D., Dey, A. and Federici, S., 2023. Biotechnological methods to remove microplastics: A review. *Environmental Chemistry Letters*, pp. 1–24.
4. Issac, M.N. and Kandasubramanian, B., 2021. Effect of microplastics in water and aquatic systems. *Environmental Science and Pollution Research*, *28*, pp. 19544–19562.
5. Bhuyan, M.S., 2022. Effects of microplastics on fish and in human health. *Frontiers in Environmental Science*, *10*, p. 250.
6. Miller, M.E., Hamann, M. and Kroon, F.J., 2020. Bioaccumulation and biomagnification of microplastics in marine organisms: A review and meta-analysis of current data. *PLoS One*, *15*(10), p. e0240792.
7. Akdogan, Z. and Guven, B., 2019. Microplastics in the environment: A critical review of current understanding and identification of future research needs. *Environmental Pollution*, *254*, p. 113011.
8. Alfonso, M.B., Lindsay, D.J., Arias, A.H., Nakano, H., Jandang, S. and Isobe, A., 2023. Zooplankton as a suitable tool for microplastic research. *Science of the Total Environment*, p. 167329.
9. Lee, Y., Cho, J., Sohn, J. and Kim, C., 2023. Health effects of microplastic exposures: Current issues and perspectives in South Korea. *Yonsei Medical Journal*, *64*(5), p. 301.
10. Kumar, R., Verma, A., Shome, A., Sinha, R., Sinha, S., Jha, P.K., Kumar, R., Kumar, P., Shubham, Das, S. and Sharma, P., 2021. Impacts of plastic pollution on ecosystem services, sustainable development goals, and need to focus on circular economy and policy interventions. *Sustainability*, *13*(17), p. 9963.

11. Henderson, L. and Green, C., 2020. Making sense of microplastics? Public understandings of plastic pollution. *Marine Pollution Bulletin*, *152*, p. 110908.
12. Zurier, H.S. and Goddard, J.M., 2021. Biodegradation of microplastics in food and agriculture. *Current Opinion in Food Science*, *37*, pp. 37–44.
13. Shen, J., Liang, B. and Jin, H., 2023. The impact of microplastics on insect physiology and the indication of hormesis. *TrAC Trends in Analytical Chemistry*, p. 117130.
14. Bertram, S.M., Yaremchuk, D.D., Reifer, M.L., Villarreal, A., Muzzatti, M.J. and Kolluru, G.R., 2021. Tests of the positive and functional allometry hypotheses for sexually selected traits in the Jamaican field cricket. *Behavioural Processes*, *188*, p. 104413.
15. Kwak, J.I. and An, Y.J., 2021. Microplastic digestion generates fragmented nanoplastics in soils and damages earthworm spermatogenesis and coelomocyte viability. *Journal of Hazardous Materials*, *402*, p. 124034.
16. Oliveira, M., Ameixa, O.M. and Soares, A.M., 2019. Are ecosystem services provided by insects "bugged" by micro (nano) plastics? *TrAC Trends in Analytical Chemistry*, *113*, pp. 317–320.
17. Al-Jaibachi, R., Cuthbert, R.N. and Callaghan, A., 2018. Up and away: Ontogenic transference as a pathway for aerial dispersal of microplastics. *Biology Letters*, *14*(9), p. 20180479.
18. Malafaia, G., da Luz, T.M., Guimarães, A.T.B. and da Costa Araújo, A.P., 2020. Polyethylene microplastics are ingested and induce biochemical changes in Culex quinquefasciatus (Diptera: Culicidae) freshwater insect larvae. *Ecotoxicology and Environmental Contamination*, *15*(1), pp. 79–89.
19. El Kholy, S. and Al Naggar, Y., 2023. Exposure to polystyrene microplastic beads causes sex-specific toxic effects in the model insect Drosophila melanogaster. *Scientific Reports*, *13*(1), p. 204.
20. Liang, B., Zhang, D., Liu, X., Xu, Y., Tang, H., Li, Y. and Shen, J., 2022. Sex-specific effects of PET-MPs on Drosophila lifespan. *Archives of Insect Biochemistry and Physiology*, *110*(3), p. e21909.
21. Shen, J., Liang, B., Zhang, D., Li, Y., Tang, H., Zhong, L. and Xu, Y., 2021. Effects of PET microplastics on the physiology of Drosophila. *Chemosphere*, *283*, p. 131289.
22. Shah, S., Ilyas, M., Li, R., Yang, J. and Yang, F.L., 2023. Microplastics and nanoplastics effects on plant–pollinator interaction and pollination biology. *Environmental Science & Technology*, *57*(16), pp. 6415–6424.
23. Lee, M., Simakova, A.V., Babkina, I.B., Bagirov, R.T. and Frank, Y.A., 2023. Interaction of microplastics and terrestrial and aquatic insects (bioaccumulation, degradation, ecotoxicological effects). *Acta Biologica Sibirica*, *9*, pp. 549–564.
24. Diaz-Basantes, M.F., Conesa, J.A. and Fullana, A., 2020. Microplastics in honey, beer, milk and refreshments in Ecuador as emerging contaminants. *Sustainability*, *12*(14), p. 5514.
25. Deng, Y., Jiang, X., Zhao, H., Yang, S., Gao, J., Wu, Y., Diao, Q. and Hou, C., 2021. Microplastic polystyrene ingestion promotes the susceptibility of honeybee to viral infection. *Environmental Science & Technology*, *55*(17), pp. 11680–11692.
26. Alma, A.M., de Groot, G.S. and Buteler, M., 2023. Microplastics incorporated by honeybees from food are transferred to honey, wax and larvae. *Environmental Pollution*, *320*, p. 121078.
27. Zhu, J., Wu, P., Zhao, N., Jiang, S., Zhu, H. and Jin, H., 2023. Microplastics in terrestrial insects, long-horned beetles (Coleoptera: Cerambycidae), from China. *Science of The Total Environment*, *888*, p. 164197.
28. Ma, H., Pu, S., Liu, S., Bai, Y., Mandal, S. and Xing, B., 2020. Microplastics in aquatic environments: Toxicity to trigger ecological consequences. *Environmental Pollution*, *261*, p. 114089.

29. De Sá, L.C., Oliveira, M., Ribeiro, F., Rocha, T.L. and Futter, M.N., 2018. Studies of the effects of microplastics on aquatic organisms: What do we know and where should we focus our efforts in the future? *Science of the Total Environment, 645*, pp. 1029–1039.
30. Schindler, D.E. and Scheuerell, M.D., 2002. Habitat coupling in lake ecosystems. *Oikos, 98*(2), pp. 177–189.
31. Galloway, T.S., Cole, M. and Lewis, C., 2017. Interactions of microplastic debris throughout the marine ecosystem. *Nature Ecology & Evolution, 1*(5), p. 0116.
32. Neelavannan, K. and Sen, I.S., 2023. Microplastics in freshwater ecosystems of India: Current trends and future perspectives. *ACS Omega, 8*(38), pp. 34235–34248.
33. Haegerbaeumer, A., Mueller, M.T., Fueser, H. and Traunspurger, W., 2019. Impacts of micro-and nano-sized plastic particles on benthic invertebrates: A literature review and gap analysis. *Frontiers in Environmental Science, 7*, p. 17.
34. Khanjani, M.H., Sharifinia, M. and Mohammadi, A.R., 2023. The impact of microplastics on bivalve mollusks: A bibliometric and scientific review. *Marine Pollution Bulletin, 194*, p. 115271.
35. Mehra, S., Sharma, K., Sharma, G., Singh, M. and Chadha, P., 2020. Sources, fate, and impact of microplastics in aquatic environment. In *Emerging Contaminants*. A. Nuro, Ed., IntechOpen, 2021. doi: 10.5772/intechopen.93805.
36. Doyle, D., Sundh, H. and Almroth, B.C., 2022. Microplastic exposure in aquatic invertebrates can cause significant negative effects compared to natural particles-A meta-analysis. *Environmental Pollution*, p. 120434.
37. De Sá, L.C., Oliveira, M., Ribeiro, F., Rocha, T.L. and Futter, M.N., 2018. Studies of the effects of microplastics on aquatic organisms: What do we know and where should we focus our efforts in the future? *Science of the Total Environment, 645*, pp. 1029–1039.
38. Wang, R., Mou, H., Lin, X., Zhu, H., Li, B., Wang, J., Junaid, M. and Wang, J., 2021. Microplastics in mollusks: Research progress, current contamination status, analysis approaches, and future perspectives. *Frontiers in Marine Science, 8*, p. 759919.
39. Ziani, K., Ioniţă-Mîndrican, C.B., Mititelu, M., Neacşu, S.M., Negrei, C., Moroşan, E., Drăgănescu, D. and Preda, O.T., 2023. Microplastics: A real global threat for environment and food safety: A state of the art review. *Nutrients, 15*(3), p. 617.
40. Smith, M., Love, D.C., Rochman, C.M. and Neff, R.A., 2018. Microplastics in seafood and the implications for human health. *Current Environmental Health Reports, 5*, pp. 375–386.
41. Thakur, K., Kumar, R. and Brar, B., 2021. A review on freshwater fish diversity of India and concept of DNA barcoding in fish identification. *Egyptian Journal of Aquatic Biology and Fisheries, 25*(3), pp. 667–693.
42. Thiele, C.J., Hudson, M.D., Russell, A.E., Saluveer, M. and Sidaoui-Haddad, G., 2021. Microplastics in fish and fishmeal: An emerging environmental challenge? *Scientific Reports, 11*(1), p. 2045.
43. Soliño, L., Vidal-Liñán, L., Pérez, P., García-Barcelona, S., Baldó, F. and Gago, J., 2022. Microplastic occurrence in deep-sea fish species *Alepocephalus bairdii* and *Coryphaenoides rupestris* from the Porcupine Bank (North Atlantic). *Science of The Total Environment, 834*, p. 155150.
44. Bhuyan, M.S., 2022. Effects of microplastics on fish and in human health. *Frontiers in Environmental Science, 10*, p. 250.
45. Liang, W., Li, B., Jong, M.C., Ma, C., Zuo, C., Chen, Q. and Shi, H., 2023. Process-oriented impacts of microplastic fibers on behavior and histology of fish. *Journal of Hazardous Materials, 448*, p. 130856.
46. Critchell, K. and Hoogenboom, M.O., 2018. Effects of microplastic exposure on the body condition and behaviour of planktivorous reef fish (Acanthochromis polyacanthus). *PLoS One, 13*(3), p. e0193308.

47. Ahrendt, C., Perez-Venegas, D.J., Urbina, M., Gonzalez, C., Echeveste, P., Aldana, M., Pulgar, J. and Galbán-Malagón, C., 2020. Microplastic ingestion cause intestinal lesions in the intertidal fish *Girella laevifrons*. *Marine Pollution Bulletin*, *151*, p. 110795.
48. Alomar, C., Sureda, A., Capó, X., Guijarro, B., Tejada, S. and Deudero, S., 2017. Microplastic ingestion by *Mullus surmuletus* Linnaeus, 1758 fish and its potential for causing oxidative stress. *Environmental Research*, *159*, pp. 135–142.
49. Huang, J.N., Wen, B., Xu, L., Ma, H.C., Li, X.X., Gao, J.Z. and Chen, Z.Z., 2022. Micro/nano-plastics cause neurobehavioral toxicity in discus fish (*Symphysodon aequifasciatus*): Insight from brain-gut-microbiota axis. *Journal of Hazardous Materials*, *421*, p. 126830.
50. Wang, L., Nabi, G., Yin, L., Wang, Y., Li, S., Hao, Z. and Li, D., 2021. Birds and plastic pollution: Recent advances. *Avian Research*, *12*(1), pp. 1–9.
51. Mendez-Sanhueza, S., Torres, M., Pozo, K., Del Aguila, G., Hernandez, F., Jacobsen, C. and Echeverry, D., 2023. Microplastics in seabird feces from coastal areas of central Chile. *Animals*, *13*(18), p. 2840.
52. Wilcox, C., Van Sebille, E. and Hardesty, B.D., 2015. Threat of plastic pollution to seabirds is global, pervasive, and increasing. *Proceedings of the National Academy of Sciences*, *112*(38), pp. 11899–11904.
53. Basto, M.N., Nicastro, K.R., Tavares, A.I., McQuaid, C.D., Casero, M., Azevedo, F. and Zardi, G.I., 2019. Plastic ingestion in aquatic birds in Portugal. *Marine Pollution Bulletin*, *138*, pp. 19–24.
54. Zhao, S., Zhu, L. and Li, D., 2016. Microscopic anthropogenic litter in terrestrial birds from Shanghai, China: Not only plastics but also natural fibers. *Science of the Total Environment*, *550*, pp. 1110–1115.
55. Carlin, J., Craig, C., Little, S., Donnelly, M., Fox, D., Zhai, L. and Walters, L., 2020. Microplastic accumulation in the gastrointestinal tracts in birds of prey in central Florida, USA. *Environmental Pollution*, *264*, p. 114633.
56. Fackelmann, G., Pham, C.K., Rodríguez, Y., Mallory, M.L., Provencher, J.F., Baak, J.E. and Sommer, S., 2023. Current levels of microplastic pollution impact wild seabird gut microbiomes. *Nature Ecology & Evolution*, *7*(5), pp. 698–706.
57. Lu, H., Yin, K., Su, H., Wang, D., Zhang, Y., Hou, L., Li, J.B., Wang, Y. and Xing, M., 2023. Polystyrene microplastics induce autophagy and apoptosis in birds lungs via PTEN/PI3K/AKT/mTOR. *Environmental Toxicology*, *38*(1), pp. 78–89.
58. Charlton-Howard, H.S., Bond, A.L., Rivers-Auty, J. and Lavers, J.L., 2023. 'Plasticosis': Characterising macro-and microplastic-associated fibrosis in seabird tissues. *Journal of Hazardous Materials*, *450*, p. 131090.
59. Rivers-Auty, J., Bond, A.L., Grant, M.L. and Lavers, J.L., 2023. The one-two punch of plastic exposure: Macro-and micro-plastics induce multi-organ damage in seabirds. *Journal of Hazardous Materials*, *442*, p. 130117.
60. da Silva Brito, W.A., Mutter, F., Wende, K., Cecchini, A.L., Schmidt, A. and Bekeschus, S., 2022. Consequences of nano and microplastic exposure in rodent models: The known and unknown. *Particle and Fibre Toxicology*, *19*(1), pp. 1–24.
61. Jin, Y., Lu, L., Tu, W., Luo, T. and Fu, Z., 2019. Impacts of polystyrene microplastic on the gut barrier, microbiota and metabolism of mice. *Science of the Total Environment*, *649*, pp. 308–317.
62. Hou, J., Lei, Z., Cui, L., Hou, Y., Yang, L., An, R., Wang, Q., Li, S., Zhang, H. and Zhang, L., 2021. Polystyrene microplastics lead to pyroptosis and apoptosis of ovarian granulosa cells via NLRP3/Caspase-1 signaling pathway in rats. *Ecotoxicology and Environmental Safety*, *212*, p. 112012.
63. Shengchen, W., Jing, L., Yujie, Y., Yue, W. and Shiwen, X., 2021. Polystyrene microplastics-induced ROS overproduction disrupts the skeletal muscle regeneration by converting myoblasts into adipocytes. *Journal of Hazardous Materials*, *417*, p. 125962.

64. Wang, J., Li, Y., Lu, L., Zheng, M., Zhang, X., Tian, H., Wang, W. and Ru, S., 2019. Polystyrene microplastics cause tissue damages, sex-specific reproductive disruption and transgenerational effects in marine medaka (*Oryzias melastigma*). *Environmental Pollution*, *254*, p. 113024.
65. Xie, X., Deng, T., Duan, J., Xie, J., Yuan, J. and Chen, M., 2020. Exposure to polystyrene microplastics causes reproductive toxicity through oxidative stress and activation of the p38 MAPK signaling pathway. *Ecotoxicology and Environmental Safety*, *190*, p. 110133.
66. da Costa Araujo, A.P. and Malafaia, G., 2021. Microplastic ingestion induces behavioral disorders in mice: A preliminary study on the trophic transfer effects via tadpoles and fish. *Journal of Hazardous Materials*, *401*, p. 123263.
67. Wu, R.T., Cai, Y.F., Chen, Y.X., Yang, Y.W., Xing, S.C. and Liao, X.D., 2021. Occurrence of microplastic in livestock and poultry manure in South China. *Environmental Pollution*, *277*, p. 116790.
68. van-der Veen, I., van Mourik, L.M., Van-Velzen, M.J.M., Groenewoud, Q.R. and Leslie, H.A., 2022. Plastic particles in livestock feed, milk, meat and blood. *Environmental Health, Report EH22-01*, pp. 1–48.
69. Shen, M., Song, B., Zeng, G., Zhang, Y., Huang, W., Wen, X. and Tang, W., 2020. Are biodegradable plastics a promising solution to solve the global plastic pollution? *Environmental Pollution*, *263*, p. 114469.
70. Hwang, J., Choi, D., Han, S., Choi, J. and Hong, J., 2019. An assessment of the toxicity of polypropylene microplastics in human derived cells. *Science of the Total Environment*, *684*, pp. 657–669.
71. Kannan, K. and Vimalkumar, K., 2021. A review of human exposure to microplastics and insights into microplastics as obesogens. *Frontiers in Endocrinology*, *12*, p. 724989.
72. Allouzi, M.M.A., Tang, D.Y.Y., Chew, K.W., Rinklebe, J., Bolan, N., Allouzi, S.M.A. and Show, P.L., 2021. Micro (nano) plastic pollution: The ecological influence on soil-plant system and human health. *Science of the Total Environment*, *788*, p. 147815.
73. Poma, A., Vecchiotti, G., Colafarina, S., Zarivi, O., Aloisi, M., Arrizza, L., Chichiriccò, G. and Di Carlo, P., 2019. In vitro genotoxicity of polystyrene nanoparticles on the human fibroblast Hs27 cell line. *Nanomaterials*, *9*(9), p. 1299.
74. Sheriff, I., Yusoff, M.S., Abd Manan, T.S.B. and Koroma, M., 2023. Microplastics in manure: Sources, analytical methods, toxicodynamic, and toxicokinetic endpoints in livestock and poultry. *Environmental Advances*, p. 100372.
75. Urli, S., Corte Pause, F., Crociati, M., Baufeld, A., Monaci, M. and Stradaioli, G., 2023. Impact of microplastics and nanoplastics on livestock health: An emerging risk for reproductive efficiency. *Animals*, *13*(7), p. 1132.
76. Zolotova, N., Kosyreva, A., Dzhalilova, D., Fokichev, N. and Makarova, O., 2022. Harmful effects of the microplastic pollution on animal health: A literature review. *PeerJ*, *10*, p. e13503.
77. Besseling, E., Wegner, A., Foekema, E.M., Van Den Heuvel-Greve, M.J. and Koelmans, A.A., 2013. Effects of microplastic on fitness and PCB bioaccumulation by the lugworm *Arenicola marina* (L.). *Environmental Science & Technology*, *47*(1), pp. 593–600.
78. Wright, S.L., Rowe, D., Thompson, R.C. and Galloway, T.S., 2013. Microplastic ingestion decreases energy reserves in marine worms. *Current Biology*, *23*(23), pp. R1031–R1033.
79. Leung, J. and Chan, K.Y.K., 2018. Microplastics reduced posterior segment regeneration rate of the polychaete *Perinereis aibuhitensis*. *Marine Pollution Bulletin*, *129*(2), pp. 782–786.
80. Gomiero, A., Strafella, P., Pellini, G., Salvalaggio, V. and Fabi, G., 2018. Comparative effects of ingested PVC micro particles with and without adsorbed benzo (a) pyrene vs. spiked sediments on the cellular and sub cellular processes of the benthic organism Hediste diversicolor. *Frontiers in Marine Science*, *5*, p. 99.

81. Yang, X., Zhang, X., Shu, X., Gong, J., Yang, J., Li, B., Lin, J., Chai, Y. and Liu, J., 2023. The effects of polyethylene microplastics on the growth, reproduction, metabolic enzymes, and metabolomics of earthworms *Eisenia fetida*. *Ecotoxicology and Environmental Safety*, *263*, p. 115390.
82. Ziajahromi, S., Kumar, A., Neale, P.A. and Leusch, F.D., 2018. Environmentally relevant concentrations of polyethylene microplastics negatively impact the survival, growth and emergence of sediment-dwelling invertebrates. *Environmental Pollution*, *236*, pp. 425–431.
83. Blarer, P. and Burkhardt-Holm, P., 2016. Microplastics affect assimilation efficiency in the freshwater amphipod *Gammarus fossarum*. *Environmental Science and Pollution Research*, *23*, pp. 23522–23532.
84. Straub, S., Hirsch, P.E. and Burkhardt-Holm, P., 2017. Biodegradable and petroleum-based microplastics do not differ in their ingestion and excretion but in their biological effects in a freshwater invertebrate *Gammarus fossarum*. *International Journal of Environmental Research and Public Health*, *14*(7), p. 774.
85. Redondo-Hasselerharm, P.E., Falahudin, D., Peeters, E.T. and Koelmans, A.A., 2018. Microplastic effect thresholds for freshwater benthic macroinvertebrates. *Environmental Science & Technology*, *52*(4), pp. 2278–2286.
86. Au, S.Y., Bruce, T.F., Bridges, W.C. and Klaine, S.J., 2015. Responses of *Hyalella azteca* to acute and chronic microplastic exposures. *Environmental Toxicology and Chemistry*, *34*(11), pp. 2564–2572.
87. Cole, M. and Galloway, T.S., 2015. Ingestion of nanoplastics and microplastics by Pacific oyster larvae. *Environmental Science & Technology*, *49*(24), pp. 14625–14632.
88. Gray, A.D. and Weinstein, J.E., 2017. Size-and shape-dependent effects of microplastic particles on adult daggerblade grass shrimp (Palaemonetes pugio). *Environmental Toxicology and Chemistry*, *36*(11), pp. 3074–3080.
89. Heindler, F.M., Alajmi, F., Huerlimann, R., Zeng, C., Newman, S.J., Vamvounis, G. and van Herwerden, L., 2017. Toxic effects of polyethylene terephthalate microparticles and Di (2-ethylhexyl) phthalate on the calanoid copepod, *Parvocalanus crassirostris*. *Ecotoxicology and Environmental Safety*, *141*, pp. 298–305.
90. Magni, S., Gagné, F., André, C., Della Torre, C., Auclair, J., Hanana, H., Parenti, C.C., Bonasoro, F. and Binelli, A., 2018. Evaluation of uptake and chronic toxicity of virgin polystyrene microbeads in freshwater zebra mussel *Dreissena polymorpha* (Mollusca: Bivalvia). *Science of the Total Environment*, *631*, pp. 778–788.
91. Xu, X.Y., Lee, W.T., Chan, A.K.Y., Lo, H.S., Shin, P.K.S. and Cheung, S.G., 2017. Microplastic ingestion reduces energy intake in the clam *Atactodea striata*. *Marine Pollution Bulletin*, *124*(2), pp. 798–802.
92. Avio, C.G., Gorbi, S., Milan, M., Benedetti, M., Fattorini, D., d'Errico, G., Pauletto, M., Bargelloni, L. and Regoli, F., 2015. Pollutants bioavailability and toxicological risk from microplastics to marine mussels. *Environmental Pollution*, *198*, pp. 211–222.
93. Lei, L., Wu, S., Lu, S., Liu, M., Song, Y., Fu, Z., Shi, H., Raley-Susman, K.M. and He, D., 2018. Microplastic particles cause intestinal damage and other adverse effects in zebrafish *Danio rerio* and nematode *Caenorhabditis elegans*. *Science of the Total Environment*, *619*, pp. 1–8.
94. Zhao, L., Qu, M., Wong, G. and Wang, D., 2017. Transgenerational toxicity of nanopolystyrene particles in the range of μg L− 1 in the nematode *Caenorhabditis elegans*. *Environmental Science: Nano*, *4*(12), pp. 2356–2366.
95. Jeong, C.B., Won, E.J., Kang, H.M., Lee, M.C., Hwang, D.S., Hwang, U.K., Zhou, B., Souissi, S., Lee, S.J. and Lee, J.S., 2016. Microplastic size-dependent toxicity, oxidative stress induction, and p-JNK and p-p38 activation in the monogonont rotifer (*Brachionus koreanus*). *Environmental Science & Technology*, *50*(16), pp. 8849–8857.

96. Messinetti, S., Mercurio, S., Parolini, M., Sugni, M. and Pennati, R., 2018. Effects of polystyrene microplastics on early stages of two marine invertebrates with different feeding strategies. *Environmental Pollution*, *237*, pp. 1080–1087.
97. Martínez-Gómez, C., León, V.M., Calles, S., Gomáriz-Olcina, M. and Vethaak, A.D., 2017. The adverse effects of virgin microplastics on the fertilization and larval development of sea urchins. *Marine Environmental Research*, *130*, pp. 69–76.

10 Novel Remediation Pathways for Mitigating Microplastic Pollution

Prerna Attri, Preeti Garg, and Ganga Ram Chaudhary

10.1 INTRODUCTION

Plastics are well-defined as "polymeric material that may contain other substances to improve performance and/or reduce costs", the authoritative explanation given by the International Union of Pure and Applied Chemistry [1]. Plastic is a manmade polymer made up of long hydrocarbon chains that are generated using petrochemicals, an unrenewable resource. Because it is readily available, affordable, stable, and extremely durable, plastics can be easily molded into infinite essential products. Polystyrene (PS), polyethylene terephthalate (PET), polyethylene (PE), polyvinyl chloride (PVC), polypropylene (PP), and other common polymers found in the market have been used extensively [2]. Around the world, 9.1 billion tons of plastic garbage have been produced in the previous 50 years, growing at an average yearly rate of 8.7% [3].

Asia is the world's greatest plastics manufacturer as well as the greatest consumer. China makes up the largest share (30%) of the world's plastic output, followed by Europe (17%), and the US, Canada, and Mexico (18%) [4]. It is concerning that every year, from 4.8 to 12.8 million metric tons of plastic waste is dumped into water bodies due to the nonexistence of sustainable waste management strategies, and this statistic will continue to grow in the future. Plastic materials play a significant role in the global economy, and the difficulties related to their extensive use should not be neglected. Broken tiny plastic fragments degraded into the environment are receiving increasing attention because of their danger; the majority of these plastics are nonbiodegradable and will persist in the environment for a lengthy period. Based on size, these particles are classified as nanoplastics (NPs) and microplastics (MPs) [5]. These are the foremost contributors to the plastic contamination in ecosystems such as agroecosystems, marine habitats, and freshwater environments.

MP wastes that persist and accumulate are threatening to ecosystems and people's well-being. Current research indicates that cells are becoming distorted and NPs are aggregating in organs (liver, kidney, and gut). Plastic waste's robust polymeric composition prevents it from biodegrading in open atmospheres. Instead, via physicochemical processes, it deteriorates into MPs and NPs [6]. These tiny plastic units are limitless in their ability to travel and bring a variety of dangerous substances

DOI: 10.1201/9781003468141-10

and infections to a broad range of environments, where they have adverse effects on many ecosystems and many biological systems. Typically, hydrogen bonds, van der Waals forces, and ionic and electrostatic interactions are how the broken pieces of plastic interact with the other contaminants [7]. Considering that MPs are difficult to break down, effective waste management technology is required to combat these negative consequences. In this chapter, we discuss impacts on the environment of novel plastic pollutants that are smaller than microscale, along with novel methods of cleanup.

10.2 SOURCES OF MP POLLUTANTS

MPs are categorized as primary or secondary MPs. Primary MPs result from home goods, polymeric raw materials, cosmetics, and other products of plastics manufacturing such as PS, PE, and PP. Secondary MPs emerge as an outcome of the fragmentation of huge amounts of plastic debris caused by abiotic processes such as ultraviolet (UV) radiations, air pressure, temperature, and biological degradation [8]. Radical reactive species such as ozone and physical stress are also important in polymer breakdown. (Figure 10.1).

Physical erosion of plastic materials, surprisingly, can result in secondary MPs due to oxidation [9]. Because of their chemical structure, polymer materials such as PP, polybutadiene (PBD), and PVC are extremely vulnerable to oxidative-thermal deterioration even at ambient temperature. When antioxidants and stabilizers are added, these materials are capable of degrading. Physical erosion on these polymer compounds also breaks molecular chains' rigid chemical interactions, resulting in oxidative-thermal breakdown. Additional secondary sources are road markings, tire abrasion, marine coatings, and synthetic textiles during washing [10]. Large plastic fragments breaking down releases hazardous chemicals and greatly increases the

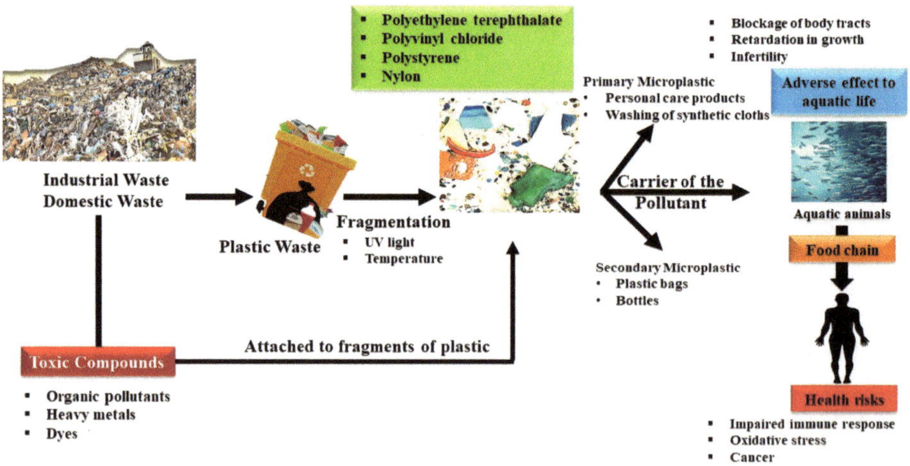

FIGURE 10.1 Schematic representation of different sources and impacts of microplastic.

plastic content that is exposed to the ecosystem. Furthermore, the larger surface area and smaller size encourage the adsorption of heavy metals and microcontaminants, increasing the rate at which MPs are ingested by biota and raising their potential risks.

10.3 MP IMPACTS ON ECOSYSTEMS

MP particles, characterized by their water insolubility and resistance to biodegradation, pose significant threats to both aquatic and terrestrial ecosystems. These particles, unable to undergo easy digestion, have been associated with detrimental effects on various organisms, manifesting in behavioral abnormalities, damage to the peripheral nervous system, oxidative stress, reproductive infertility, and liver failure [11]. The toxicity of MPs is intricately linked to manifold aspects including their physiological properties, quantity, exposure route, and chemical composition [12].

In plants, MP toxicity hinders growth and stunts development. The impacts on plant life are influenced by MP size, form, and concentration, as well as the specific plant species involved [13]. The three main ways that individuals are exposed to MPs are through their diet, lungs, and skin. Direct ingestion can happen when a person consumes water, salt, or soil tainted with plastic; indirect ingestion can happen when a person consumes marine life, shellfish, and plants [14]. According to studies, individuals absorb 4,000 particles of MPs daily from drinking water 11,000 particles from shellfish, and 37–1,000 particles from sea salt that can be eaten. Respiratory exposure results in the inhalation of almost 272 airborne MP particles every day. Importantly, the ingestion of MPs has been shown to impede the digestion of lipids, which form hetero-aggregates of MP oil droplets that inhibit digestive enzyme activities, posing a severe risk to digestive health [15].

Recent researchers have extensively explored the noxious influence of MPs on aquatic organisms, encompassing a wide range of species such as algae, seabirds, invertebrates, mollusks, phytoplankton, zooplankton, fish, and mammals [16]. MP aggregates in the gills, liver, and gut; scientists observed up to 35% plastic particles, size ranging in between 1 and 2.8 mm, in the stomachs of pelagic fish, altering gut microbioma [17]. These findings underscore the pervasive and harmful effects of MPs on both marine and terrestrial ecosystems.

10.4 NOVEL TECHNIQUES FOR MP DETERIORATION

To alleviate the presence of MPs in aquatic ecosystem, researchers have investigated advanced remediation methods in addition to preventive measures including limiting plastic manufacture and consumption. Even though plastic degradation has been studied for many years, there are still only a few specific strategies that target MPs [18]. These strategies can be generally categorized into one of four physical approaches: adsorption and membrane filtration, microbial degradation, biotechnological interventions, and advanced oxidation processes (AOPs). Figure 10.2 shows a condensed illustration of MP degradation routes.

Microbial degradation is a standout among these strategies because of its long history of reducing MPs and its minimal energy input need [19]. AOPs and heat

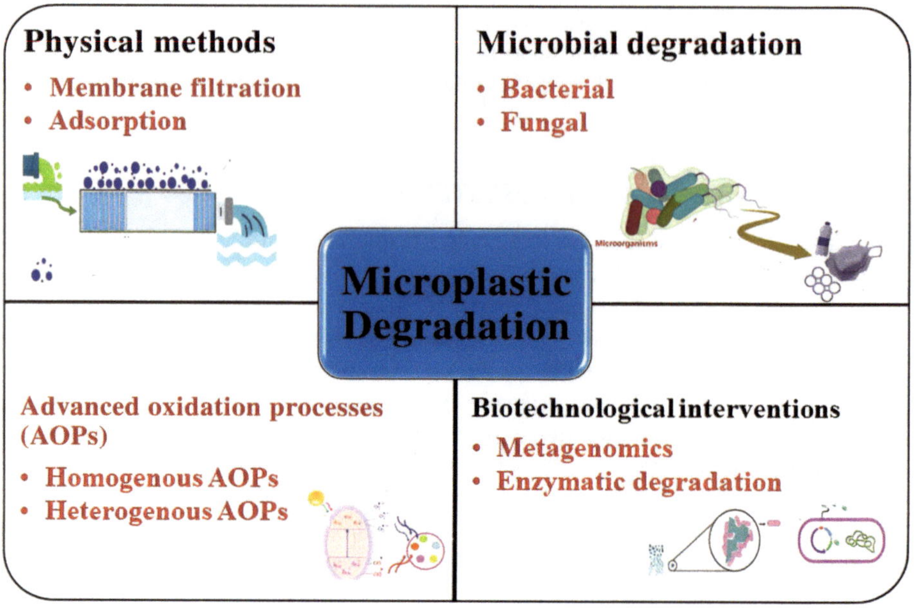

FIGURE 10.2 Different novel methods used for microplastic degradation.

treatment show promise in harnessing the hydrogen and carbon components found in MPs to manufacture fuels and in accelerating the rates of decomposition. Showcasing noteworthy research from every area, this section will provide insights into the advancements made in addressing MP pollution [20].

10.4.1 Physical Methods

10.4.1.1 Membrane Techniques

Membrane technology, which is now classified into nanofiltration, ultrafiltration (UF), and reverse osmosis, plays an important role in water purification within wastewater treatment plants [21] This encompasses methods such as sand filtration, granular activated carbon and coagulation–flocculation, as well as sedimentation and precipitation. Wastewater can be treated to remove organic materials using UF membranes [22]. Research indicates that this method exhibits a high degree of pollution separation efficiency, little energy usage, and a small plant footprint. Membranes used in water treatment have the same porosity as MPs. As a result, these membranes have great stability and selectivity, making them useful in eliminating MPs pollutants and their byproducts [22].

Ma et al. studied the dynamics of PE, a common MP component, removal from potable water using current techniques like coagulation and UF (Figure 10.3). Selecting coagulants based on iron instead of aluminum because of the latter's higher density, the authors demonstrated increased removal effectiveness for smaller PE particles. On the other hand, the conventional coagulation method showed no effect

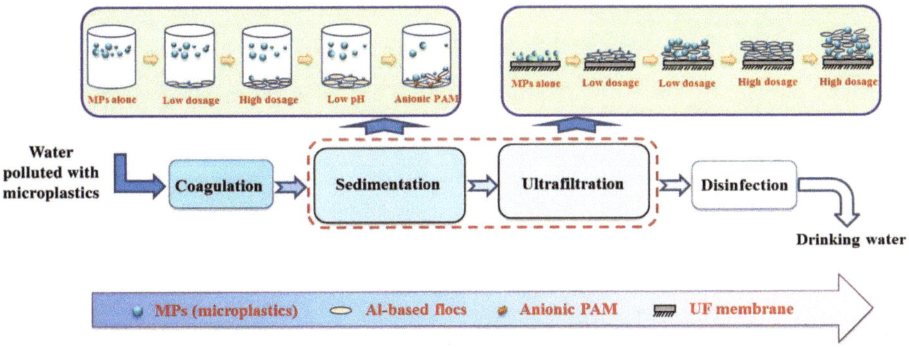

FIGURE 10.3 Schematic of microplastics during coagulation and ultrafiltration processes [23].

from water properties and continuously poor PE removal efficacy, staying below 15% [23]. The usage of polyacrylamide (PAM) was exhibited a noteworthy effect on removal efficacy. Specifically, after large dosages of anionic PAM were used, removal efficiency increased to 90.9%, outperforming the effect of solution pH.

10.4.1.2 Adsorption

Adsorption is the prevalent method employed in wastewater purification as an enhanced and final treatment for minuscule molecule contaminants in water. Adsorption has several benefits over other technologies, including ease of operation, a high purification rate, minimal energy usage, and cost-effectiveness.

Research on MP removal using adsorption is important since it is a fundamental technique for wastewater treatment. Carbon nanotubes (CNTs) [24] and titanium dioxide (TiO_2) [25] and graphene nanotubes are examples of emerging materials that have been produced and deployed by exploiting the widespread use of adsorption technologies. Tang et al. fabricated magnetic carbon nanotubes (MCNTs) as adsorbents for the elimination of MPs [26]. These MCNTs showed good adsorption to PE, PET, and polyamide (PA). Notably, employing magnetic force, all of the MPs/MCNTs composites could be readily removed from aqueous solutions. Within 300 minutes, the addition of 5 gL^{-1} of MCNTs resulted in the total elimination of target MPs (5 gL^{-1}). PE, PET, and PA maximal adsorption capabilities were determined to be 1,650, 1,400, and 1,100 mg MCNTs g^{-1}, respectively.

Similarly, Wang et al. conducted an experiment on removing MP spheres. They investigated the use of biochar as a possible low-cost material to improve the performance of sand filter systems, notably in the context of eliminating microbeads in wastewater treatment facilities (Figure 10.4). The results show that biochar filters have a remarkable capability for extracting and immobilizing MP spheres with a diameter of 10 m; over 95% of the samples were removed [27]. This capacity outperformed that of the comparably sized sand filters the authors examined. Extensive study of the samples collected following leaching experiments using environmental scanning electron microscopy (ESEM) demonstrates that MP spheres are immobilized via three morphologically regulated pathways, as conceptualized.

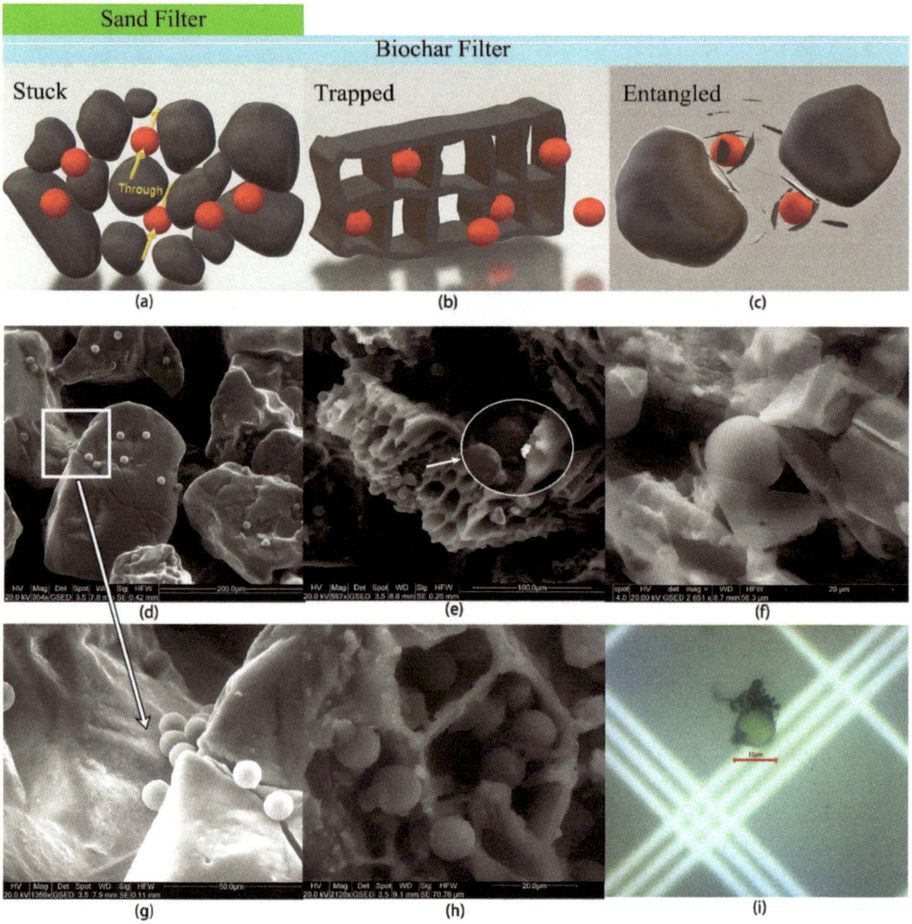

FIGURE 10.4 Three mechanisms for microplastic immobilization: stuck (a), trapped (b), and entangled (c); corresponding ESEM images: sand filter (d, g), biochar filter (e, f, g); and optical microscope image (i) [27].

10.4.2 Biodegradation Pathways

Microorganisms are often utilized as MP cleanup agents. The degradation pattern is influenced by several factors such as type of functional groups, molecular weight, chemical structure (plasticizers/additives present), and susceptibility to bacterial assault. The degradation rate reduces as molecular weight and crystallinity rise.

Plastic degradation happens in a series of phases that include assimilation, bio-fragmentation, bio-deterioration, and mineralization. It is essential to comprehend the catalytic capacity of microbial enzymes in the breakdown of these polymers, especially the active catalytic sites included in these enzymes. These active sites are made up of certain amino acids that are essential to the breakdown of MPs [28].

Extracellular enzymes—hydrolases—which break down polymers with larger molecular weight into small chains of oligomers and monomers are released by microorganisms during the decomposition process. The result of this modification is an alteration in the chemical and physical structure. Smaller chains that are produced, like monomers, are easily able to pass through the semipermeable barrier. These breakdown products then provide bacteria with carbon sources. Following that, the mineralization phase begins, during which intracellular enzymes break the shorter chains into end products [29]. Under aerobic circumstances, these end products appear as CO_2 and H_2O. In an anaerobic environment, however, mineralization produces CH_4. The process of using these final products as carbon, hydrogen, and energy sources through ingestion is referred to as bioattenuation [30].

10.4.2.1 Bacterial Degradation

Recent years have seen an increase in the identification of bacterial isolates that are capable of degrading MPs, drawing heightened attention to the properties and impact of these particles throughout the process of decomposition [31]. To determine how well *Bacillus cereus* CH_6 degrades polystyrene microplastics (PS MPs), Yuan et al. isolated the bacteria from lake sediments, and it displayed short growth times, strong metabolic activity, and quick substrate uptake [32]. The bacteria's average deterioration rate in PS MPs was 0.967 mg/d throughout the 50-day incubation period, resulting in a 10.7% fall in weight.

Similarly, H.S. Auta et al. identified eight strains of bacteria from peninsular Malaysian mangrove sediment with the goal of addressing the negative environmental effects of MPs and developing a workable cleanup strategy (Figure 10.5). PE, PET, PS, and PP were among the UV-treated MPs that these bacterial strains were tested for their ability to break down. Only *B. cereus* and *B. gottheilii*, two of the isolates, showed signs of growth on a synthetic medium in which the only carbon source was a variety of MP polymers [33].

10.4.2.2 Fungal Degradation

Fungi, similar to bacteria, have shown the ability to use and degrade MPs. Over the last few decades, scientists have discovered a new fungus with improved activity in degrading MPs [34]. It is worth mentioning, however, that investigations into fungus-mediated MP breakdown have been rather restricted, highlighting difficulties in efficiently screening fungal strains with superior MP-degrading capacities by ectopic approaches.

Zhang et al. extracted PEDX3, an *Aspergillus flavus*-like fungus, from the gastrointestinal fluids of the wax moth *Galleria mellonella*. After a 28-day incubation period, the strain PEDX3 successfully decomposed high-density polyethylene (HDPE) MP particles into lower molecular weight forms (Figure 10.6). The presence of ether groups and carbonyl groups in the degraded MP particles was verified by Fourier transform-infrared spectroscopy, giving additional evidence for PE breakdown. Using reverse transcription-polymerase chain reaction, researchers found two up-regulated laccase-like multicopper oxidase genes, AFLA_053930 and AFLA_006190, during the degradation process. These enzymes are thought to be good candidates for *A. flavus* PEDX3 remediation of PE MP particles [35].

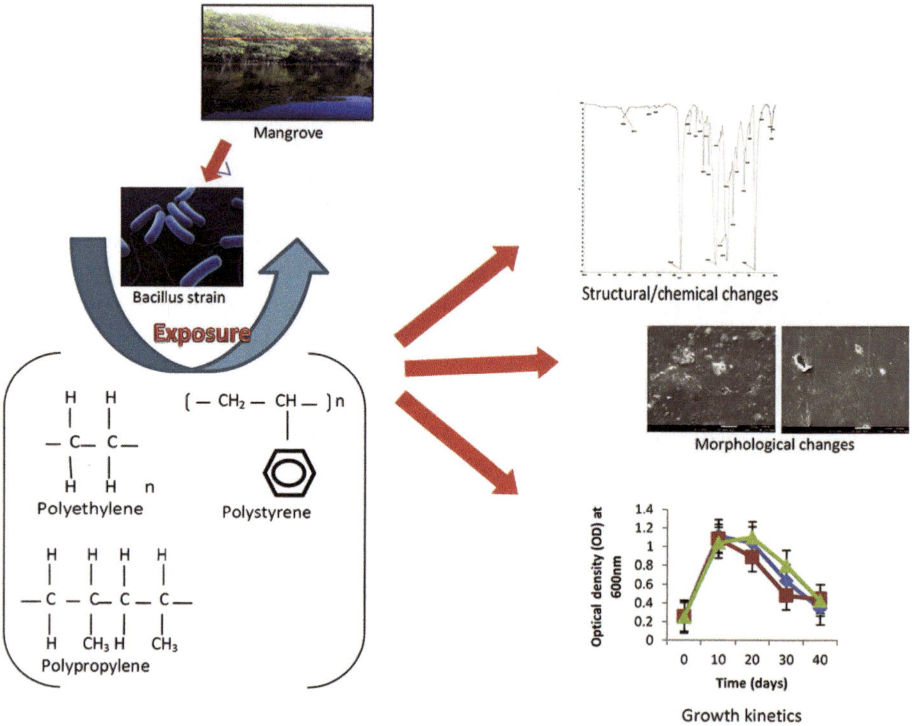

FIGURE 10.5 Graphical abstract of microplastic removal using extracted bacterial strains from peninsular Malaysian mangrove sediment [33].

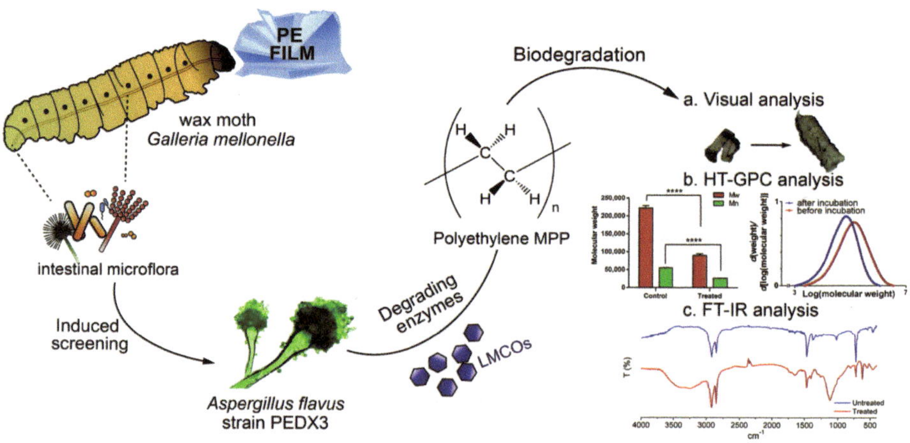

FIGURE 10.6 Schematic illustration of isolated PEDX3, an *Aspergillus flavus*-like successfully decomposing high-density polythene [35].

Note: MPP, microplastic particles; HT-GPC, high-temperature gel permeation chromatography.

10.4.3 Biotechnological Interventions

10.4.3.1 Metagenomics

Exploration of novel enzymes, metabolic pathways, and functionally enhanced metabolically active molecules through a systems biology approach, combined with the use of genetically modified organisms or customized microbial consortia, has the ability to considerably accelerate the evolution of cutting-edge biotechnological tools for plastic remediation and potential recovery [36]. These materials are generally modified using recombinant DNA (rDNA) technologies, gene editing, and gene cloning (Figure 10.7). The method entails identifying and isolating the target gene responsible for degradation, followed by polymerase chain reaction (PCR) amplification and transformation into an appropriate host organism, such as *Escherichia coli*. Peroxidases, cutinase, and laccases were efficiently segregated from *Saccharomyces cerevisiae* and *P. chryososporium* BY 4741 using rDNA technology, indicating higher efficiency in PET degradation [37].

Aside from biotechnological uses, there is a current investigation of systems biology networks to discover plastic-degrading mechanisms. Data mining is used extensively in this technique to extract, design, optimize, and create microbial consortia from various databases. BioCyc, Biodegradative Oxygenases Database (OxDBase), and MetaCyc are examples of commonly used databases. The marine bacterial strain

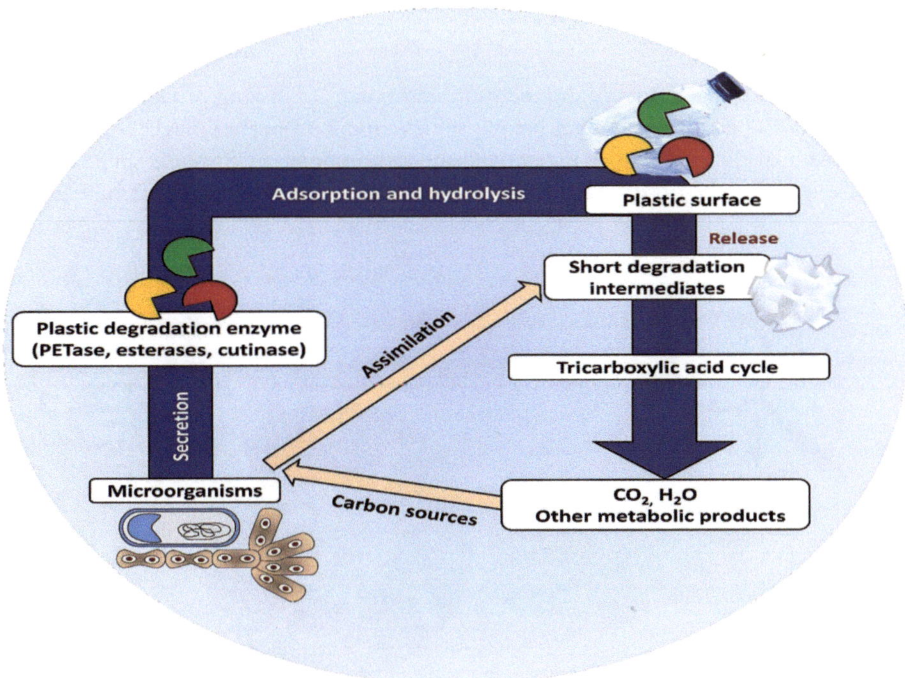

FIGURE 10.7 Overarching mechanism of microbial disintegration of microplastic degradation [36].

AIIW2 has been effectively separated using these integrated methods, illustrating the capacity for decomposing both PVC and low-density polyethylene (LDPE) [38].

10.4.3.2 Enzymatic Degradation

Lipase, cutinase, esterase, and oxygenase are enzymes that demonstrate effectiveness in breaking down MPs [39]. For instance, the enzyme alkane hydroxylase that was isolated from *Pseudomonas* sp. E4 strain showed that it could break down 30% of PE in 80 days. Over 75 days, alkane hydroxylase obtained from *E. coli*-degraded PE-based MPs with an efficiency of 20%. Interestingly PET can be broken down into mono (2-hydroxyethyl) terephthalate and terephthalic acid (TPA) by the enzymes MHETase and PETase, which were isolated from the bacteria *Ideonella sakaiensisis* [40].

10.4.4 Advanced Oxidation Processes

AOPs have been extensively explored and used for the decomposition of many tenacious environmental pollutants. The efficiency of AOPs is contingent upon the creation of reactive oxygen species (ROS), which are extremely efficient at degrading a wide range of pollutants [41]. Several AOPs have been used to decompose MPs such as ozone, UV photolysis, UV/H_2O_2, and UV/visible light-induced photocatalysis in both homogeneous and heterogeneous processes (Figure 10.8). These AOPs are effective at decomposing different types and sizes of MPs.

10.4.4.1 Homogeneous AOPs

10.4.4.1.1 Ozone

Ozone (O_3), known for its high reactivity and potent oxidizing properties, is widely used in AOPs because it changes the physicochemical properties of MPs. Researchers have extensively explored MP removal and decomposition through surface oxidation, leveraging the observed changes in their properties. Hideayaturahman and Lee

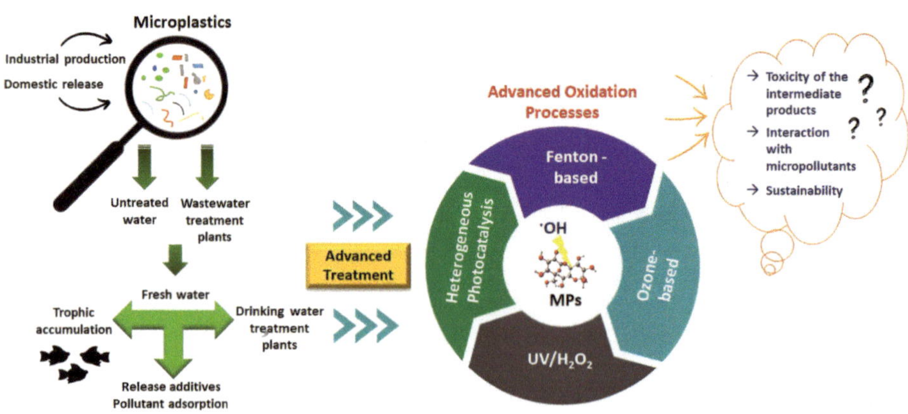

FIGURE 10.8 Graphical representation of removal of microfibers and microplastics by AOPs [41].

combined O_3 treatment with traditional wastewater treatment procedures to mitigate MPs [42]. The combination of ozone and coagulation resulted in an amazing 99.2% elimination of MPs. In this investigation, ozonation alone resulted in a MP degradation rate of 89.9%.

Zafar et al. investigated using O_3 to decompose MPs. For the periods of 60, 120, and 180 minutes, ozone decomposition was carried out using different O_3 dosages in the range of 4–7 mg/min. They measured the carbon index (CI) and hydroxyl index (HI) to track the breakdown of MPs. Greater O_3 doses increased the CI and HI readings as a result of the oxidation of PE, which showed reliance on the amount of ozone [43].

10.4.4.1.2 UV Photolysis and UV/H_2O_2

A substance's breakdown brought on by light is referred to as photolysis. Light irradiation, especially UV light, produces reactive intermediates and radicals. These include singlet oxygen ($^1O_2$), hydroxyl radicals ($^{\bullet}OH$), hydrated electrons (e^-_{aq}), triple organic matter ($^3OM^{\bullet}$), carbonate radicals ($CO_3^{\bullet-}$), organoperoxy radicals ($ROO^{\bullet}$), and superoxide radical anions ($O_2^{\bullet-}$) [44]. These species can attack and decompose various persistent contaminants.

With sunlight as its principal source, photolysis has mostly been investigated for the environmental breakdown of plastics, and there has been little research on using photolysis to remove MPs from environmental pollution treatment plants. UV radiation changes the physicochemical characteristics of a number of plastics, like PE, PVC, PS, and PET despite the complexity of the degrading pathways; specifically, UV photolysis caused fissures to grow on the surfaces of plastic samples, resulting in a much coarser texture [45]. Finally, the samples were fragmented, breaking down into tiny fragments ranging in size from nanometers to micrometers. Separately, Zhu et al. investigated MP photodegradation using PS MPs as a model. The experiment lasted up to 150 days under simulated solar light irradiation, and light exposure generated significant amounts of ROS such as $O_2^{\bullet}$, $^{\bullet}OH$, and H_2O_2 in the PS-MP solution. Environmentally persistent free radicals were produced in substantial correlation with these ROS [46].

10.4.4.2 Heterogeneous AOPs

10.4.4.2.1 UV-Induced Photocatalysis

In order to treat water and wastewater, UV light is often used as a light source for UV photolysis, UV-induced photocatalysis, and UV disinfection. UV has been widely used in the breakdown of persistent pollutants by activating different photocatalysts to create ROS such as $O_2^{\bullet}$ and $^{\bullet}OH$. UV-induced photocatalysis has been the topic of much investigation for the removal of MPs in the environment due to the efficiency with which these created ROS break down diverse contaminants.

Nabi et al. explored the photocatalytic removal of common MPs, specifically PE and PS microspheres, under UV light exposure utilizing TiO_2 nanoparticle films formulated with Triton X-100, and the films almost completely mineralized (98.40%) 400 nm PS within 12 hours [47]. The authors also investigated the degradation of PE of various sizes and observed substantial photodegradation after 36 hours, with CO_2 identified as the primary end product. Liu et al. used a sol-gel synthesized form of cryptomelane (OMS-2) and a boron-doped form of the same compound (B-OMS-2) as photocatalysts to degrade PE film when exposed to UV and visible light [48]. The

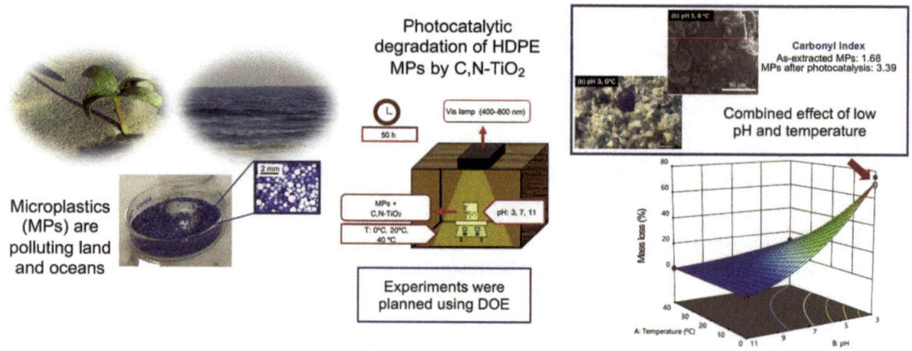

FIGURE 10.9 Schematic representation of microplastic degradation by a carbon and nitrogen-doped TiO_2 [51].

Note: DOE = design of experiments.

PE-B-OMS-2 film outperformed the PE and PE-OMS-2 composite films in terms of PE weight reduction and textural alterations. The B-OMS-2 catalyst improved the PE film photodegradation, and when exposed to UV light, B-OMS-2 displayed better catalytic performance than ordinary OMS-2.

10.4.4.2.2 Visible Light-Induced Photocatalysis

Since UV light comprises just a small fraction of the solar spectrum, roughly 5%, many efforts have been devoted to extending the photo-response range into visible light to enable the photocatalytic decomposition of different types of contaminants. Presently, researchers are intensely investigating visible light-induced photocatalysis, which is considered an environmentally beneficial method of using solar light for removing pollutants from the environment [49].

A novel method of eliminating MPs was presented by Uheida et al., who used glass fiber substrates to support the photocatalyst material and collect low-density MP particles like PP. The authors used zinc oxide (ZnO) nanorods immobilized on the glass fiber substrates in a flow-through system to demonstrate the photocatalytic destruction of spherical PP MP particles suspended in water under visible light irradiation. Following two weeks of exposure to visible light, average PP MP particle volume decreased by 65% [50].

Researchers have also extensively investigated TiO_2 for eliminating MPs. To decrease the pollution caused by MPs under visible light radiation, Ariza-Tarazona et al. designed nitrogen-doped (N-doped) and carbon-nitrogen co-doped (C-N co-doped) TiO_2. For the purpose of examining the degradation of MPs in both cases by visible light-induced photocatalysis, HDPE was used as a model MPs. After 18 hours of visible irradiation, only 6.4% of the N-doped subject's weight was lost. However, after 50 hours of visible light at pH 3.0 and 0 °C, around 72% of HDPE degraded with C-N co-doping. Interestingly, temperature and pH had impacts on the photocatalysis-induced degradation of HDPE. Considering TiO_2's zeta potential and MP breakdown, the study shown in Figure 10.9 suggests that lower pH and temperature are advantageous for breaking down HDPE [51].

Furthermore, Jiang et al. created a new hydroxy-rich ultrathin bismuth-oxychloride (BiOCl-X) photocatalyst at ambient room temperature. BiOCl-X has a substantial photocatalytic degradation capacity for MPs, with a mass loss of plastics that is 24 times more than that with BiOCl nanosheets [52]. The authors examined photocatalysis using electron paramagnetic resonance and scavenger examinations, indicating that hydroxyl on the surface of BiOCl efficiently increases the formation of hydroxyl radicals, hence improving degrading performance.

10.5 CONCLUSION

Microplastics from plastic waste are a major hazard since they are emerging toxins that affect human health and the environment. Plastic trash is a result of numerous human activities. The potential dangers associated with MPs are highlighted by their cytotoxic, neurotoxic, and genotoxic effects on humans, which can occur by oral consumption, cutaneous exposure, or inhalation. In conclusion, investigating new remediation strategies to reduce MP pollution is an essential first step in tackling the growing environmental problems these persistent pollutants are causing. This chapter covered a variety of creative ways and strategies for remediation, each with its own benefits and considerations.

Physical approaches, such as adsorption and filtration, offer realistic alternatives for eliminating microplastics from a variety of settings. The use of membranes, particularly in membrane bioreactor technology, emerges as a robust technique, displaying great efficiency in microplastic removal. Microplastics may be broken down in an environmentally beneficial way by biodegradation processes, which take advantage of microbial enzymes' catalytic capacity. A thorough analysis of microbial species and their enzymatic activity offers important new perspectives on how to use them for plastic cleanup. Plastic cleanup has expanded possibilities thanks to biotechnological treatments like customized microbial consortia and genetically engineered species. More improvements in sustainable plastic mitigation are anticipated with the identification of enzymes with improved functions and the use of system biology techniques. Chemical technologies, particularly advanced oxidation processes, are useful in degrading microplastics by producing reactive oxygen species. Ozone treatment, UV photolysis, and other chemical methods provide a variety of instruments for targeted plastic deterioration.

The combination of these novel remediation methods, together with a better understanding of microplastic sources, processes, and toxicity, results in a holistic strategy for reducing microplastic pollution. Moving forward, continuous multidisciplinary research, technological innovation, and global cooperation will be critical in realizing effective, sustainable solutions for the preservation of our ecosystems and the well-being of future generations.

REFERENCES

1. da Costa, J.P., 2018. Nanoplastics in the environment. *Plastics and the Environment*, *47*, p. 82.
2. Kankanige, D. and Babel, S., 2020. Smaller-sized micro-plastics (MPs) contamination in single-use PET-bottled water in Thailand. *Science of the Total Environment*, *717*, p. 137232.

3. He, D., Luo, Y., Lu, S., Liu, M., Song, Y. and Lei, L., 2018. Microplastics in soils: Analytical methods, pollution characteristics and ecological risks. *Trends in Analytical Chemistry*, *109*, pp. 163–172.
4. Liang, Y., Tan, Q., Song, Q. and Li, J., 2021. An analysis of the plastic waste trade and management in Asia. *Waste Management*, *119*, pp. 242–253.
5. Iyare, P.U., Ouki, S.K. and Bond, T., 2020. Microplastics removal in wastewater treatment plants: A critical review. *Environmental Science: Water Research & Technology*, *6*(10), pp. 2664–2675.
6. Reynaud, S., Aynard, A., Grassl, B. and Gigault, J., 2022. Nanoplastics: From model materials to colloidal fate. *Current Opinion in Colloid & Interface Science*, *57*, p. 101528.
7. Fred-Ahmadu, O.H., Bhagwat, G., Oluyoye, I., Benson, N.U., Ayejuyo, O.O. and Palanisami, T., 2020. Interaction of chemical contaminants with microplastics: Principles and perspectives. *Science of the Total Environment*, *706*, p. 135978.
8. Murphy, F., Ewins, C., Carbonnier, F. and Quinn, B., 2016. Wastewater treatment works (WwTW) as a source of microplastics in the aquatic environment. *Environmental Science & Technology*, *50*(11), pp. 5800–5808.
9. El Hadri, H., Gigault, J., Maxit, B., Grassl, B. and Reynaud, S., 2020. Nanoplastic from mechanically degraded primary and secondary microplastics for environmental assessments. *NanoImpact*, *17*, p. 100206.
10. Jaikumar, G., Brun, N.R., Vijver, M.G. and Bosker, T., 2019. Reproductive toxicity of primary and secondary microplastics to three cladocerans during chronic exposure. *Environmental Pollution*, *249*, pp. 638–646.
11. Gola, D., Tyagi, P.K., Arya, A., Chauhan, N., Agarwal, M., Singh, S.K. and Gola, S., 2021. The impact of microplastics on marine environment: A review. *Environmental Nanotechnology, Monitoring & Management*, *16*, p. 100552.
12. Du, H., Xie, Y. and Wang, J., 2021. Microplastic degradation methods and corresponding degradation mechanism: Research status and future perspectives. *Journal of Hazardous Materials*, *418*, p. 126377.
13. Colzi, I., Renna, L., Bianchi, E., Castellani, M.B., Coppi, A., Pignattelli, S., Loppi, S. and Gonnelli, C., 2022. Impact of microplastics on growth, photosynthesis and essential elements in Cucurbita pepo L. *Journal of Hazardous Materials*, *423*, p. 127238.
14. Yang, X., Man, Y.B., Wong, M.H., Owen, R.B. and Chow, K.L., 2022. Environmental health impacts of microplastics exposure on structural organization levels in the human body. *Science of the Total Environment*, *825*, p. 154025.
15. Prata, J.C., da Costa, J.P., Lopes, I., Andrady, A.L., Duarte, A.C. and Rocha-Santos, T., 2021. A One Health perspective of the impacts of microplastics on animal, human and environmental health. *Science of the Total Environment*, *777*, p. 146094.
16. Botterell, Z.L., Beaumont, N., Dorrington, T., Steinke, M., Thompson, R.C. and Lindeque, P.K., 2019. Bioavailability and effects of microplastics on marine zooplankton: A review. *Environmental Pollution*, *245*, pp. 98–110.
17. Bai, Z., Wang, N. and Wang, M., 2021. Effects of microplastics on marine copepods. *Ecotoxicology and Environmental Safety*, *217*, p. 112243.
18. Shen, M., Song, B., Zhu, Y., Zeng, G., Zhang, Y., Yang, Y., Wen, X., Chen, M. and Yi, H., 2020. Removal of microplastics via drinking water treatment: Current knowledge and future directions. *Chemosphere*, *251*, p. 126612.
19. Ahmed, R., Hamid, A.K., Krebsbach, S.A., He, J. and Wang, D., 2022. Critical review of microplastics removal from the environment. *Chemosphere*, *293*, p. 133557.
20. Nabi, I. and Zhang, L., 2022. A review on microplastics separation techniques from environmental media. *Journal of Cleaner Production*, *337*, p. 130458.
21. Wang, Y., Liu, Y., Yu, Y. and Huang, H., 2018. Influence of CNT-rGO composite structures on their permeability and selectivity for membrane water treatment. *Journal of Membrane Science*, *551*, pp. 326–332.

22. Khanzada, N.K., Farid, M.U., Kharraz, J.A., Choi, J., Tang, C.Y., Nghiem, L.D., Jang, A. and An, A.K., 2020. Removal of organic micropollutants using advanced membrane-based water and wastewater treatment: A review. *Journal of Membrane Science*, *598*, p. 117672.
23. Ma, B., Xue, W., Hu, C., Liu, H., Qu, J. and Li, L., 2019. Characteristics of microplastic removal via coagulation and ultrafiltration during drinking water treatment. *Chemical Engineering Journal*, *359*, pp. 159–167.
24. Athari, M., Fattahi, M., Khosravi-Nikou, M. and Hajhariri, A., 2022. Adsorption of different anionic and cationic dyes by hybrid nanocomposites of carbon nanotube and graphene materials over UiO-66. *Scientific Reports*, *12*(1), p. 20415.
25. Van Viet, P. and Tran, H.N., 2019. Adsorption and photocatalytic degradation of methylene blue by titanium dioxide nanotubes at different pH conditions. *Advances in Natural Sciences: Nanoscience and Nanotechnology*, *10*(4), p. 045011.
26. Tang, Y., Zhang, S., Su, Y., Wu, D., Zhao, Y. and Xie, B., 2021. Removal of microplastics from aqueous solutions by magnetic carbon nanotubes. *Chemical Engineering Journal*, *406*, p. 126804.
27. Wang, Z., Sedighi, M. and Lea-Langton, A., 2020. Filtration of microplastic spheres by biochar: Removal efficiency and immobilisation mechanisms. *Water Research*, *184*, p. 116165.
28. Yuan, J., Ma, J., Sun, Y., Zhou, T., Zhao, Y. and Yu, F., 2020. Microbial degradation and other environmental aspects of microplastics/plastics. *Science of the Total Environment*, *715*, p. 136968.
29. Fred-Ahmadu, O.H., Bhagwat, G., Oluyoye, I., Benson, N.U., Ayejuyo, O.O. and Palanisami, T., 2020. Interaction of chemical contaminants with microplastics: Principles and perspectives. *Science of the Total Environment*, *706*, p. 135978.
30. Danso, D., Chow, J. and Streit, W.R., 2019. Plastics: Environmental and biotechnological perspectives on microbial degradation. *Applied and Environmental Microbiology*, *85*(19), pp. e01095–19.
31. Auta, H.S., Emenike, C.U., Jayanthi, B. and Fauziah, S.H., 2018. Growth kinetics and biodeterioration of polypropylene microplastics by Bacillus sp. and Rhodococcus sp. isolated from mangrove sediment. *Marine Pollution Bulletin*, *127*, pp. 15–21.
32. Yuan, J., Cao, J., Yu, F. and Ma, J., 2022. Microbial degradation of polystyrene microplastics by a novel isolated bacterium in aquatic ecosystem. *Sustainable Chemistry and Pharmacy*, *30*, p. 100873.
33. Auta, H.S., Emenike, C.U. and Fauziah, S.H., 2017. Screening of Bacillus strains isolated from mangrove ecosystems in Peninsular Malaysia for microplastic degradation. *Environmental Pollution*, *231*, pp. 1552–1559.
34. Paço, A., Duarte, K., da Costa, J.P., Santos, P.S., Pereira, R., Pereira, M.E., Freitas, A.C., Duarte, A.C. and Rocha-Santos, T.A., 2017. Biodegradation of polyethylene microplastics by the marine fungus Zalerion maritimum. *Science of the Total Environment*, *586*, pp. 10–15.
35. Zhang, J., Gao, D., Li, Q., Zhao, Y., Li, L., Lin, H., Bi, Q. and Zhao, Y., 2020. Biodegradation of polyethylene microplastic particles by the fungus Aspergillus flavus from the guts of wax moth Galleria mellonella. *Science of the Total Environment*, *704*, p. 135931.
36. Malla, M.A., Dubey, A., Kumar, A. and Yadav, S., 2023. Unlocking the biotechnological and environmental perspectives of microplastic degradation in soil-ecosystems using metagenomics. *Process Safety and Environmental Protection*, *170*, pp. 372–379.
37. Giangeri, G., Morlino, M.S., De Bernardini, N., Ji, M., Bosaro, M., Pirillo, V., Antoniali, P., Molla, G., Raga, R., Treu, L. and Campanaro, S., 2022. Preliminary investigation of microorganisms potentially involved in microplastics degradation using an integrated metagenomic and biochemical approach. *Science of The Total Environment*, *843*, p. 157017.

38. Gaur, V.K., Gupta, S., Sharma, P., Gupta, P., Varjani, S., Srivastava, J.K., Chang, J.S. and Bui, X.T., 2022. Metabolic cascade for remediation of plastic waste: A case study on microplastic degradation. *Current Pollution Reports*, *8*(1), pp. 30–50.
39. Zhou, H., Mayorga-Martinez, C.C. and Pumera, M., 2021. Microplastic removal and degradation by mussel-inspired adhesive magnetic/enzymatic microrobots. *Small Methods*, *5*(9), p. 2100230.
40. Othman, A.R., Hasan, H.A., Muhamad, M.H., Ismail, N.I. and Abdullah, S.R.S., 2021. Microbial degradation of microplastics by enzymatic processes: A review. *Environmental Chemistry Letters*, *19*, pp. 3057–3073.
41. Dos Santos, N.D.O., Busquets, R. and Campos, L.C., 2023. Insights into the removal of microplastics and microfibres by advanced oxidation processes. *Science of the Total Environment*, *861*, p. 160665.
42. Hidayaturrahman, H. and Lee, T.G., 2019. A study on characteristics of microplastic in wastewater of South Korea: Identification, quantification, and fate of microplastics during treatment process. *Marine Pollution Bulletin*, *146*, pp. 696–702.
43. Zafar, R., Park, S.Y. and Kim, C.G., 2021. Surface modification of polyethylene microplastic particles during the aqueous-phase ozonation process. *Environmental Engineering Research*, *26*(5).
44. Tang, Y., Zhang, S., Su, Y., Wu, D., Zhao, Y. and Xie, B., 2021. Removal of microplastics from aqueous solutions by magnetic carbon nanotubes. *Chemical Engineering Journal*, *406*, p. 126804.
45. Suhrhoff, T.J. and Scholz-Böttcher, B.M., 2016. Qualitative impact of salinity, UV radiation and turbulence on leaching of organic plastic additives from four common plastics—a lab experiment. *Marine Pollution Bulletin*, *102*(1), pp. 84–94.
46. Zhu, K., Jia, H., Sun, Y., Dai, Y., Zhang, C., Guo, X., Wang, T. and Zhu, L., 2020. Long-term phototransformation of microplastics under simulated sunlight irradiation in aquatic environments: Roles of reactive oxygen species. *Water Research*, *173*, p. 115564.
47. Nabi, I., Li, K., Cheng, H., Wang, T., Liu, Y., Ajmal, S., Yang, Y., Feng, Y. and Zhang, L., 2020. Complete photocatalytic mineralization of microplastic on TiO2 nanoparticle film. *Iscience*, *23*(7).
48. Liu, G., Liao, S., Zhu, D., Hua, Y. and Zhou, W., 2012. Innovative photocatalytic degradation of polyethylene film with boron-doped cryptomelane under UV and visible light irradiation. *Chemical Engineering Journal*, *213*, pp. 286–294.
49. Attri, P., Garg, P., Chauhan, M., Singh, R., Sharma, R.K., Kumar, S., Lim, D.K. and Chaudhary, G.R., 2023. Metal doped BiOCl nano-architectures (M-BiOCl, M= Ni, Mo, Cd, Co) for efficient visible light photocatalytic and antibacterial behaviour. *Journal of Environmental Chemical Engineering*, *11*(2), p. 109498.
50. Uheida, A., Mejía, H.G., Abdel-Rehim, M., Hamd, W. and Dutta, J., 2021. Visible light photocatalytic degradation of polypropylene microplastics in a continuous water flow system. *Journal of Hazardous Materials*, *406*, p. 124299.
51. Ariza-Tarazona, M.C., Villarreal-Chiu, J.F., Hernández-López, J.M., De la Rosa, J.R., Barbieri, V., Siligardi, C. and Cedillo-González, E.I., 2020. Microplastic pollution reduction by a carbon and nitrogen-doped TiO2: Effect of pH and temperature in the photocatalytic degradation process. *Journal of Hazardous Materials*, *395*, p. 122632.
52. Jiang, R., Lu, G., Yan, Z., Liu, J., Wu, D. and Wang, Y., 2021. Microplastic degradation by hydroxy-rich bismuth oxychloride. *Journal of Hazardous Materials*, *405*, p. 124247.

11 Emerging Frontiers in Environmental Biotechnology

Bridging Gaps to Confront the Microplastics Menace

Anjali Vijeata, Ganga Ram Chaudhary, and Savita Chaudhary

11.1 INTRODUCTION

Environmental biotechnology is the combination of scientific and engineering practices that are adopted to rebuild and rectify the deterioration in the environment [1]. Microorganisms and biological processes were investigated for ameliorating human lives. In 1980, environmental biotechnology acquired immense scope and importance for the execution of legislation and for the preservation of the surrounding environment.

Environmental biotechnology involves various practices that have been in use for generations like wastewater treatment, tissue culture, and resource conservation [2, 3]. Different fields like molecular biology, environmental microbiology, industrial biotechnology, genetic engineering, and bioinformatics originated after advancements in biotechnology (Figure 11.1). The environment is absorbing the impacts of rapid industrialization and rising consumer demand for products in the form of polluting gases in the air, discharged industrial waste in rivers, and nonbiodegradable plastic dumped into the soil [4, 5]. The direct disposal of waste into the environment accumulates, and the extended exposure and improper disposal of plastic waste have resulted in its pervasive presence in the environment, posing a significant threat to the environment and human life. Plastics smaller less than 5 mm enter the atmosphere, water bodies, soil, and other biological routes, leading to bioaccumulation inside microorganisms and marine aquaculture [6].

Because of their smaller sizes, MPs can enter the food web and cause serious physical and chemical disorders [7]. It has been estimated that plastic present in various types of packaging, bags, electronics, automobiles, and household appliances constitutes up to 54% of total waste in the environment. Jambeck et al. determined that 4.8 to 12.7 million tons of this plastic waste is disposed of in oceans every year,

DOI: 10.1201/9781003468141-11

FIGURE 11.1 Various fields in environmental biotechnology.

which in turn affects various activities like fishing and aquaculture and threatens to aquatic flora and fauna [8]. Paes et al. reported that the average municipal waste will reach up to 3.4 billion tons by 2050, up from 2.01 billion tons today [9].

These toxicants are harmful to human beings as the particles can easily cross the cell membrane and disrupt the cell cycle. Considering the adverse effects caused by these pollutants, various strategies have been put forward to mitigate their effects. Microorganisms like fungi, bacteria, and other worms play a significant role in the degradation of pollutants [10], and scientists and engineers are now focusing on producing bioplastics to completely eradicate plastic from the environment. Recent investigations have shown that the annual growth rate of bioplastic is expected to increase by 30% until 2025 [11].

Environmental biotechnology is an emerging tool in the fields of agriculture, environmental waste management, and resource conservation [12] in response to the multitude of environmental problems that demand attention to ensure a safe and healthy future. In this context, transitioning from the reliance on polymers and fossil fuels to sustainable energy sources will not only efficiently meet human food demand but also supplies raw materials to industries, in a more sustainable and environmentally friendly way. This would ensure a broad variety of end products like food, fiber, bio-based plastic products, textiles, industrial products, and fuels [13]. Environmental biotechnology is concerned with all the physical and biological processes and their interaction with the surrounding environment (Figure 11.1).

In this chapter, we describe how integrating biotechnology into various domains such as industrial development, agriculture, and water management serves as a viable alternative to chemical processing. We discuss recent advancements in environmental biotechnology along with promising aspects for a sustainable environment.

11.2 NAVIGATING THE ENVIRONMENTAL BIOTECHNOLOGICAL LANDSCAPE: ADDRESSING KEY CHALLENGES IN MP MANAGEMENT

11.2.1 Sources of MP Pollutants in the Environment

11.2.1.1 Terrestrial Ecosystems

Sources and occurrences of pollution are diverse, and the exact pathways of its emergence are not fully known. Researchers are now focusing on determining the routes and types of pollutants present in the environment. Nizzetto et al. found significantly greater microfibril content in soil treated with slurry than in untreated soil [14]. Studies and research have demonstrated that 90% of these pollutants arise from agricultural waste, textile industries, laboratories, and household products (Figure 11.2).

Apart from microfibrils, PS foam also yields polymer particles in the range of 22–220 nm, which increases the NP content in soil. NP particles below 5 μm are known to be present in skin care products like body scrubs and cleansers [15], and frequent pesticide and fertilizer use in agricultural practices exacerbates MP pollution in soil [16]. The commonly used insecticides, such as organochlorines and organophosphates, are highly toxic and can potentially harm mammalian liver, kidney, and brain tissues if they enter the human body.

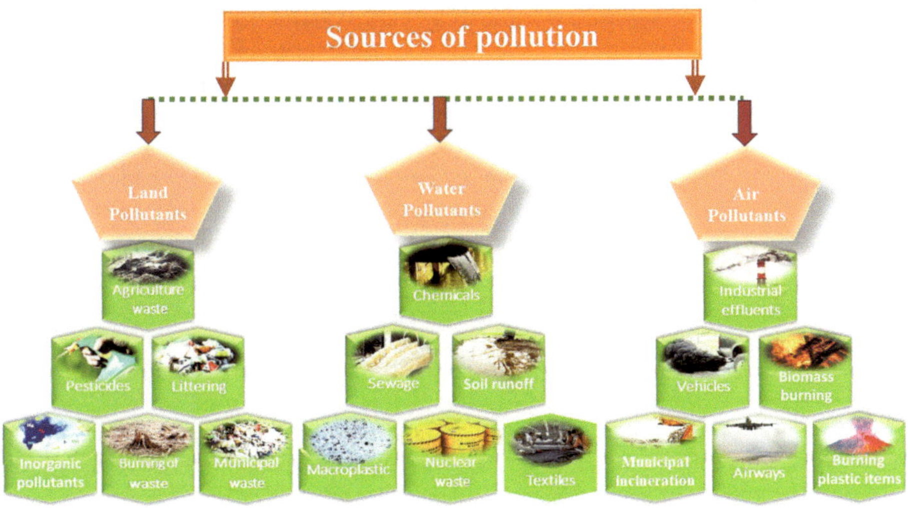

FIGURE 11.2 Major sources of microplastic pollutants

11.2.1.2 Aquatic Ecosystems

Water pollution occurs when natural water sources and other bodies of water become contaminated due to the improper release of chemicals and pollutants from human activities. The most significant source of primary pollutants is the discharge into freshwater bodies of sewage sludge containing personal and household products such as cosmetics, microbeads, and synthetic textiles made of polymers or nylon, electronics, adhesives, and paints (Figure 11.2) [17, 18]. Sewage treatment removes 99.9% of toxicants from wastewater, but a significant number of MP particles persist in the slurry; because they are so small, 1.5 mm, they easily pass through the existing debris screens.

In many countries in Europe, sewage waste is pasteurized and most often utilized as a fertilizer in agricultural land, and the slurry produced during wastewater treatment is disposed of on the land. In this way, the waste present in the wastewater enters the soil and exceeds 400,000 million tons annually [19]. Lebreton et al. estimated that 1.15 to 2.41 million tons of slurry is passed from rivers to oceans every year [20]. In a case study in California, the authors detected microbeads from cosmetic products in the seven major wastewater reclamation plants.

11.2.1.3 Atmospheric Ecosystems

The sources of air pollution are diverse and involve anthropogenic and natural emissions. Anthropogenic emissions comprise the pollutants arising from burning of fossil fuels: from motor vehicles, trains, aircraft, waste disposal, paints, and aerosol sprays, and other solvents. Yoo et al. analyze dthe inhalable airborne pollutants found in PM_{10} (particulate matter) aerosols, and found that nearly 0.008% of the particles collected through PM_{10} sampling showed high MP production, translating to approximately 800 particles per cubic meter [21]. Among the stained plastic particles smaller than 10 μm, 27% were classified under plastic category, and the remaining 73% arose from road wear or tire.

Natural emissions are most commonly related to gases like methane from waste degradation, radon from radioactive decay, and volatile organic compounds from natural vegetation. The analytes released into the atmosphere could be solid particles, gases, or liquid droplets depending on the emission source [22, 23]. Air pollutants are generally classified as primary or secondary, where primary pollutants are generated directly from ashes of volcanic eruption, vehicle soot, chimneys, etc., while secondary pollutants come from interactions or reactions of primary pollutants in the air. The most commonly emitted gases include carbon dioxide, sulfur dioxide, carbon monoxide, nitrogen oxide, and other radioactive pollutants (Figure 11.2).

11.2.2 Biotechnology in MP Remediation and Environmental Sustainability

Biotechnology is the science of using biological agents to mitigate air, water and soil pollution through physical, chemical, and biological methods. The significant use of biotechnology for economic, social, and environmental interests derives from research on older fermentation techniques.

Biotechnology is technically and commercially cost-effective for the degradation of various kinds of plastics during waste treatment; advancements in modern biotechnology have spurred the creation of novel methods for environmental monitoring, producing less toxic goods, and generating feedstock for fuel production [24]. Recognizing its strategic importance, many countries are now implementing biotechnology in industries, waste treatment, bioenergy, MP removal, etc., mostly in the forms of genetic, protein, metabolic, and biotechnological engineering and biochemical methods. Biotechnology is versatile and inherently attractive because it requires basic low-energy processes to generate a wider range of molecules using light, water, and carbon dioxide. In addition to generating cost-effective methods of producing well-established products, biotechnology also emphasizes the production of completely bio-based products. This shift from older to more modern approaches has led to more emphasis on bioplastic products that are anticipated to improve human lives.

11.2.3 Biotechnology for Environmental MP Detection and Degradation

Reducing pollution in the environment is more likely to be achieved through the implementation of the UN Sustainable Development Goals: The UN's Sustainable Development Plan involves 17 goals that deal with the prevention and reduction of marine pollution, comprising collective measures that reflect different opinions and points of interest [25, 26]. for precise and accurate monitoring at spatial and temporal levels, but the lack of information on analytical methods of detecting plastic contaminants in the environment is a limitation to the ability to conduct this monitoring.

Researchers have investigated detecting plastic pollutants with instruments that purely detect nano- and microscale materials. Some of the commonly used instrumentation techniques are UV-Vis. spectrophotometry, fluorescence spectroscopy, transmission electron microscopy (TEM), field emission scanning electron microscopy (FE-SEM), dynamic light scattering (DLS), Raman spectroscopy, etc. In 2012, Wegner et al. examined the effects of 30 nanometer polystyrene (PS) on the feeding behavior of the blue mussel (*Mytilus edulis*) by subjecting the organism to various concentrations of PS NP and different algae (*Pavlova lutheri*) concentrations. Additionally, they employed DLS to evaluate the hydrodynamic radii of the PS nanoparticles [27].

Rochman et al. analyzed the presence and behavior of polycyclic aromatic hydrocarbons (PAHs) on virgin PS and PS marine debris in San Diego Bay. Using advanced analytical techniques, the authors monitored PAH sorption of PS pellets over a year. They found that while some PAH concentrations decreased after deployment, others increased, indicating that PS debris acts both as a source and sink for PAHs in the marine environment. Comparing PS with other common polymers, PS showed higher sorption of PAHs, suggesting a potentially greater risk of PAH exposure upon ingestion [28].

Given the challenges associated with detecting, separating, and identifying nanosized pollutants in the environment using traditional techniques, there is a pressing need for analytical methods that are sensitive, cost-effective, rapid, and specific. Scientists have developed various biosensors as promising approaches for high-precision monitoring of micropollutants. The biosensors are categorized on the basis of their properties, listed in Table 11.1.

TABLE 11.1
The Role of Environmental Biotechnology in Detecting and Degrading Environmental Pollutants

S. No.	Name of species	Detection	Degradation	Type of material	Ref.
1.	*Gloecapsa gelationosa*	Determines low MP concentrations	—	Cyanobacterial extracellular polymeric substances	[59]
2.	*Comamonas thiooxidans*	Detects enzymatic PET degradation	—	Microbe-based biosensor	[60]
3.	Chamber batch-mode cube microbial cell	Detects chromium, iron, nitrate, and sodium acetate in municipal wastewater	—	Microbial fuel cell-based biosensor	[61]
4.	Laccase molecules immobilized on electrode surface	Detects pharmaceutical residues in wastewater	—	Enzyme-based electrochemical biosensor	[62]
5.	*S. maltophilia and B. velezennsis*	—	Degradation of MPs	Bacteria-based degradation	[31]
6.	Algal cells	Determines heavy (copper, lead, cadmium) and light (sodium, aluminum, lithium) metals	—	Microalgae- based biosensor	[63]
7.	*Bacillus cereus* CH_6	—	Degradation of MPs	Microbe-based degradation	[64]
8.	*Lysinibacillus, Bacillus, Candidatus Caldatribacterium*	—	Degradation of plastic in sanitary and nonsanitary landfill	Microbe-based degradation	[65]
9.	Haloalkane dehalogenase enzyme	Determines halocarbon concentrations in the environment	—	Enzyme-based biosensor	[66]
10.	Microbial fuel cells	Detects heavy metals, tetracyclines, and avermectins	—	Microbe-based biosensor	[67]
11.	*Chlorella vulgaris*	Determines herbicides	—	Optical biosensor	[68]

11.3 MP REMEDIATION

Biotechnology is the combination of biological science and engineering and has influenced various fields like agriculture, medicine, genetic engineering, food processing, crop production, etc. This technology has triggered improvements in various sectors and given rise to advanced products. Depending upon various fields, biotechnology is differentiated into four subfields.

11.3.1 Green Biotechnology

This branch of biotechnology administers the management of crops and use of fertilizers and pesticides in agricultural fields. It focuses on using agricultural crops and plants to eradicate MPs from soil involving distinct processes like phytoremediation, enzymatic degradation, and microbial degradation and is is most commonly engaged for collecting and domesticating plants via micropropagation. Green biotech is resulting in healthier and stress (biotic and abiotic) resistant plants. Further, this technology involves the use of plant cells and tissues to develop products and technologies that have applications in various fields, including environmental remediation.

The direct administration of green biotech to eradicate MPs is not as common as in other environmental applications; as a new and ongoing approach, it is still in the early stages of practical application for MP removal. For instance, Schwaminger et al. used the PET-degrading enzyme PETase and immobilized it over superparamagnetic iron oxide nanoparticles to enhance its stability and recovery via magnetic separation [29]. The high enzymatic activity of synthesized nano-biocatalyst was further utilized to degrade PET in water.

11.3.2 Blue Biotechnology

This field of biotechnology scrutinizes aquatic and marine environments and marine resources and their global contributions to meeting human demands. Marine water bodies are advanced source of biorefineries and other value-added products as well as unexplored resources to combat MPs that the environment [30]. Blue biotech uses marine microorganisms, bacteria, algae, etc. to break down and completely eradicate MPs from water bodies. For instance, Xiang et al. investigated the combined role of three microorganisms: *Stenotrophomonas maltophilia, Bacillus velezensis*, and *Acinetobacter radioresistens* on the decomposition of PS MPs [31]. The study offers a distinctive strategy by investigating the use of mixed bacterial strains for eradicating MPs from the environment.

11.3.3 White Biotechnology

White biotech is also known as grey, and it refers to generating different organisms that manipulate and manufacture products in distinct ways for industrial use [32]. White biotechnology anticipates assistance in various sectors like chemical, agriculture, and food industries in manufacturing antibiotics, paper, pulp, fibers, etc. [33].

This field also explores the use of microorganisms, enzymes, biofilms, and genetically engineered microbes for the degradation of MPs; genetic engineering biotechnology deals with the genetic makeup of bacteria for in situ bioremediation of MP waste. Cao et al. reported on bioengineered *Comamonas testosteroni* CNB-1 as an effective biocatalyst for MP degradation from sewage sludge [34]. The biocatalyst achieved high degradation efficiency of 9% within seven days of treatment. Further studies showed that the catalyst effectively mineralized the intermediates obtained during MP degradation.

11.3.4 Red Biotechnology

Red biotech uses chemical compounds to produce medicines that treats diseases such as AIDS and cancer [35]. It is applied in many different fields such as pharmaceuticals, genetic engineering, protein engineering, metabolic engineering, gene therapy, and biotechnology engineering. While this field of biotechnology may not be directly associated with MP degradation, it does play a role in reducing the use of plastics in pharmaceuticals. Nonetheless, advancements in biotechnology could potentially lead to specific applications of red biotech aimed at addressing the issue of MP usage.

11.4 RECENT ADVANCES AND FUTURE FRONTIERS IN ENVIRONMENTAL BIOTECHNOLOGY FOR MP REMEDIATION

Biotechnology involves the controlled application of biological agents for the commercial manufacture of products and techniques for the remediation of polluted environments [36]. In addition, environmental biotechnology not only prevents pollution and minimizes plastic waste but also enriches the soil with various nutrients. Recognizing its influential effect on society and the environment, several countries are now designing strategies for implementing biotechnology in different sectors. Biotechnology is versatile, and the purpose of using it for environmental remediation is often obvious. In fact it plays a role in the following MP remediations.

11.4.1 In Agriculture

The global biotechnology achievements in agriculture have immensely contributed to replacing agricultural chemicals, especially pesticides and fertilizers, and in eradicating MPs from soil [37]. In the past few years, excessive use of pesticides has enormously impacted human lives and drastically increased pollution levels, and the presence of MPs in pesticides has raised concerns as these tiny particles can persist in the environment for extended periods and have the potential to cause ecological harm. The presence of MPs in pesticides is a subject of ongoing research and scrutiny, and biotechnology has replaced agricultural chemicals with biopesticides, which are minute biological agents derived from microorganisms [38].

Biopesticides are highly specific and less toxic and produce less impact on the ecosystem. Similarly, fertilizers play a crucial role in delivering nutrients like nitrogen, potassium, and phosphorous for plant growth, but a study conducted in China revealed the presence of MPs in even organic fertilizers. The authors investigated the occurrence and distribution of MPs in 102 organic fertilizers collected from

22 provincial administrative regions of China and used sequential density floatation to separate MPs in the range of 0.1 to 5 mm. The authors found the major MP abundance in Beijing city with an average of 386 items/kg in compound organic fertilizers [39].

Biological fertilizers are considered safer and greener than commercially available chemical fertilizers [40], and agricultural biotech has been critical in offering a natural alternative to the commercially prepared organic fertilizers, specifically nitrogen-fixing rhizobia found in the root nodules of legumes. These bacteria fix nitrogen from the atmosphere to supply plants, but cultivating leguminous and nonleguminous plants together facilitates the transfer of nitrogen content, eliminating the need for external fertilizers.

Anwar et al. utilized *Trichoderma harzianum* in conjunction with copper and boron to examine its impact on the yield, grain quality, and nutrient content of wheat [41]. The incorporated yields show significant enhancement in growth traits like height, length, and chlorophyll content, and the crops were enriched with micronutrients (Figure 11.3).

Similarly, Li et al. reported the significance of molybdenum disulfide nanoparticles in enhancing the biological nitrogen fixation (BNF) capacity of Soyabean plant (Figure 11.4) [42]. Only one application increase BNF and increased grain yield by a marked 30% over the commercial fertilizer.

Further, the authors observed additional advantages like delayed nodule aging, protection against reactive oxygen species (ROS) damage, and enhanced tolerance to abiotic stress, demonstrating nanobiotechnology as an effective approach for increasing BNF capacity and plant yield in soyabean plants. More broadly, biofortification

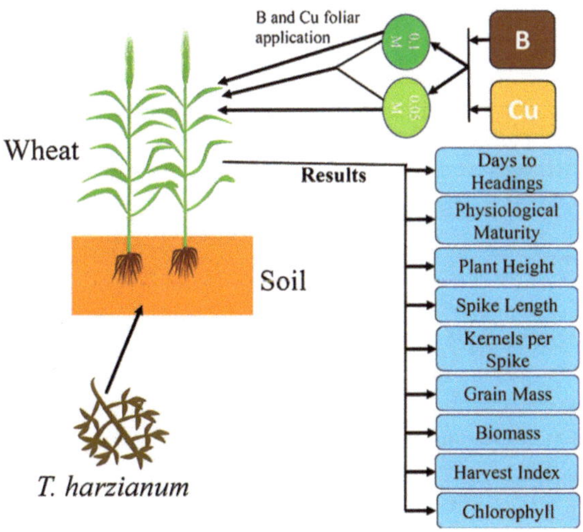

FIGURE 11.3 Visualizing the combined effects of *Trichoderma harzianum*, copper, and boron on wheat yield.

Source: Adapted from reference [41] with permission from the copyright holder, American Chemical Society (2023), Washington, DC, USA.

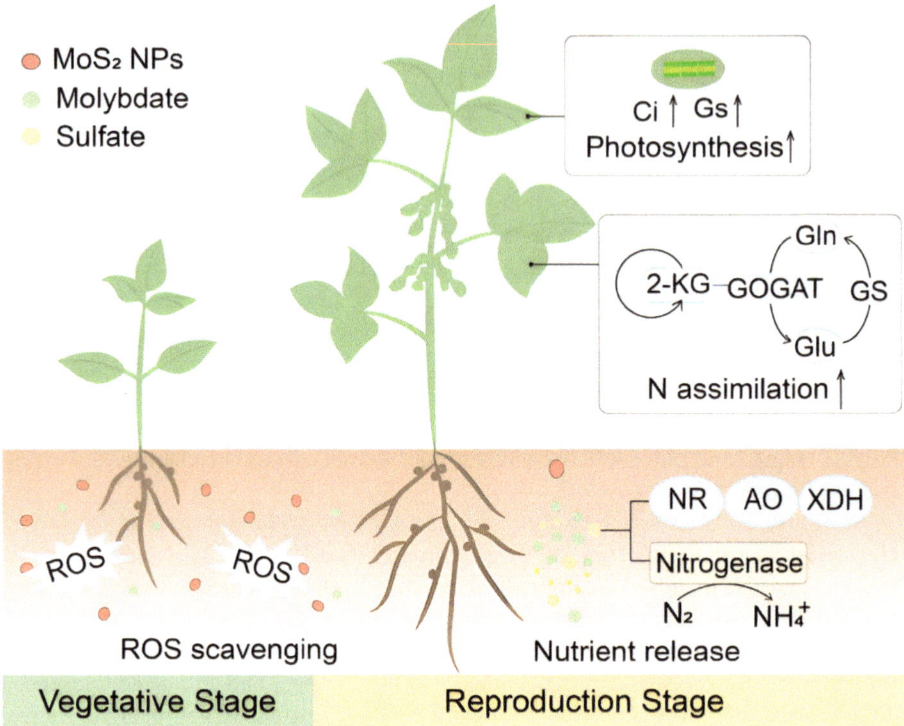

FIGURE 11.4 Illustrating the multifunctionality of molybdenum nanofertilizer.

Source: Adapted from reference [42] with permission from the copyright holder, American Chemical Society (2023), Washington, DC, USA.

and modern biotechnological approaches are emerging as crucial strategies for eliminating chemical and organic fertilizers and addressing malnutrition and the nutritional quality of crops

11.4.2 In Crop Processing

Agronomic practices embrace modern and current methods of biotechnologically and economically interrelated innovations and influence an entire phase of economic development. The MP contamination in the soil available for agricultural practices is increasing rapidly due to industrialization and modernization. MPs directly and indirectly enter soil and alter the soil's nutritional composition, properties, fauna, and microbial communities, although some have shown that MPs enter plants through their rootlets and affect plant growth through distinct pathways.

Increased food demand, malnutrition, and global increases in MP contamination have led to the rapid deployment of biotechnology in crop production. In addition to producing genetically engineered crops, biotechnology plays a key role in alleviating the synergistic MP toxicity using microorganisms, plants, fungi, etc. Liu et al.

examined the plant growth-promoting bacterium *Bacillus* (SL-413) species in sorghum plants [43]. They conducted a hydroponic study to investigate the stress induced by cadmium and MPs, which decreased sorghum plant length as well as fresh and dry plant weight. The authors studied the effects of bacterial inoculation on sorghum biomass, and the results showed that bacterial strain alleviated the stress induced by the combined action of cadmium and MPs. Plant physiological measurements revealed that inoculation decreased ROS content in sorghum from 10.5% to 27.2%. This outcome indicated the bacterium's ability to mitigate oxidative stress in the sorghum plants.

11.4.3 In Energy Production

Biotechnology also deals with the industrial effluents that result from the treatment of waste by-products, MPs, food stocks, etc. [44]. There are several examples today of enzymatic production of biofuels from renewable sources. For instance, enzymatic biocatalysis of renewable vegetable oil leads to the formation of biodiesel. Furthermore, the enzyme used in these processes is a renewable catalyst and could be recycled depending on its stability, and its by-products can be used as starting feedstock to generate many fuels [45].

Attractive substrates like residue grass, municipal organic waste, MPs and NPs, and sludge waste have been used to successfully generate biogas, yielding higher profits, lowering production costs, and reducing substrate prices. Recent researchers used microbial fuel cells (MFC) to treat industrial wastewater and recover bioenergy. MFCs demand less energy and offer faster kinetics, faster degradation and negligible slurry production [46]. Another common application of biotechnology is to generate bulk chemicals (amines, sugars, etc.) through the enzymatic treatment of waste biomass.

MPs demonstrated impacts on microalgae and potential application in biofuel generation. MPs induced oxidative stress in microalgae, leading to lipid accumulation and increased macromolecules production [47]. In 2024, Jadhav et al. generated carbon dioxide by the action of *Pseudomonas mendocina* ABF786 on plastic-containing waste residue [48]; *P. mendocina* ABF786 is an efficient plastic-degrading bacteria that readily converts the plastic waste into microalgal biomass. This microalgal biomass contains elevated amounts of oil and lipids which act as a primary source of biodiesel (Figure 11.5). The authors anticipate future applications, emphasizing enhanced efficiency through the combined action of two microorganisms.

11.4.4 In Aquaculture

In the 21st century, biotechnology is an emerging technology in the context of environmental protection, and it imparts powerful techniques for boosting production in the aquaculture sector. Due to the increasing demand for freshwater species, annual fish production has shown a tremendous increase in recent years by approximately 120 million tons. Biotechnology has supported increasing quality and quantity of fish reared in aquaculture through various techniques like genetic monitoring, chromosome engineering, and transgenesis.

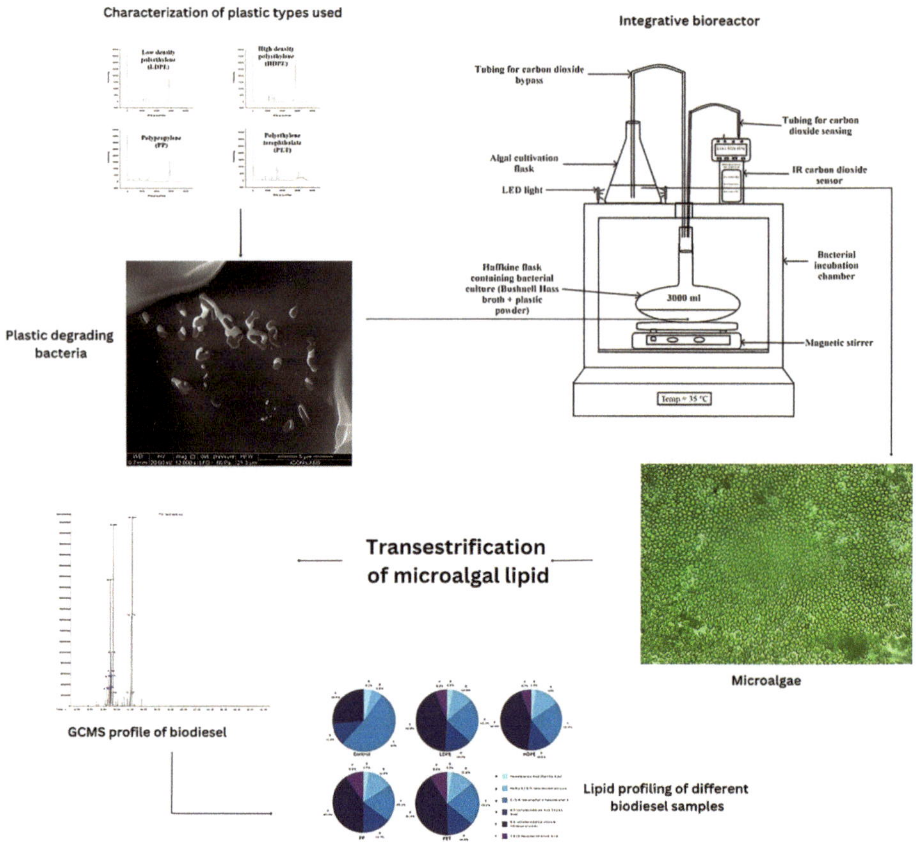

FIGURE 11.5 Plastic bio-mitigation by *P. mendocina* ABF786 and the simultaneous conversion of its byproduct to biodiesel.

Source: Adapted from reference [48] with permission from the copyright holder, Elsevier (2024).

Researchers are now seeking to ameliorate the genetic traits of mollusks, fish, and crustaceans by employing different transgenic techniques. These techniques offer enhanced growth rate, nutritional value, disease resistance, and tolerance to the low oxygen level of aquatic animals, along with greater size and weight. Karunasagar et al. introduced molecular diagnostic methods for providing resistance against pathogens in fish and other aquatic animals [49].

Additionally, there has recently been extensive research on and exploration of microbes due to their ability to degrade MPs with potential enzymes present in them. Bacteria, fungi, and actinomycetes have emerged as organisms capable of producing a diverse range of depolymerase enzymes, and these enzymes exhibit the capacity to break the bonds within MPs, resulting in the formation of monomeric residues. A report in 2022 by Hou et al. showed the biodegradation potential of *P. aeruginosa* and *P. knackmussi* bacteria for PE mulching film [50]. The degraded product was further analyzed with Fourier transform-infrared spectroscopy (FT-IR), SEM, and AFM. Over an eight-week period, the PE mulch experienced mass losses of 5.95% ± 0.03% when incubated with *P. knackmussii* N1–2 and 3.62% ± 0.32% when incubated with *P. aeruginosa* RD1–3.

However, biotechnology is a significant area for revolutionizing aquaculture as well as for environmental conservation. In a recent study, Chan et al. generated aggregators designed to form MP biofilms to facilitate their faster removal from wastewater (Figure 11.6) [51]. The researchers conducted this investigation utilizing the environmentally prevalent bacterium *P. aeruginosa* by upregulating the cyclic-di-GMP signaling inherent in this bacterial species. Additionally, a protease enzyme was utilized to break down and disrupt the biofilm matrix.

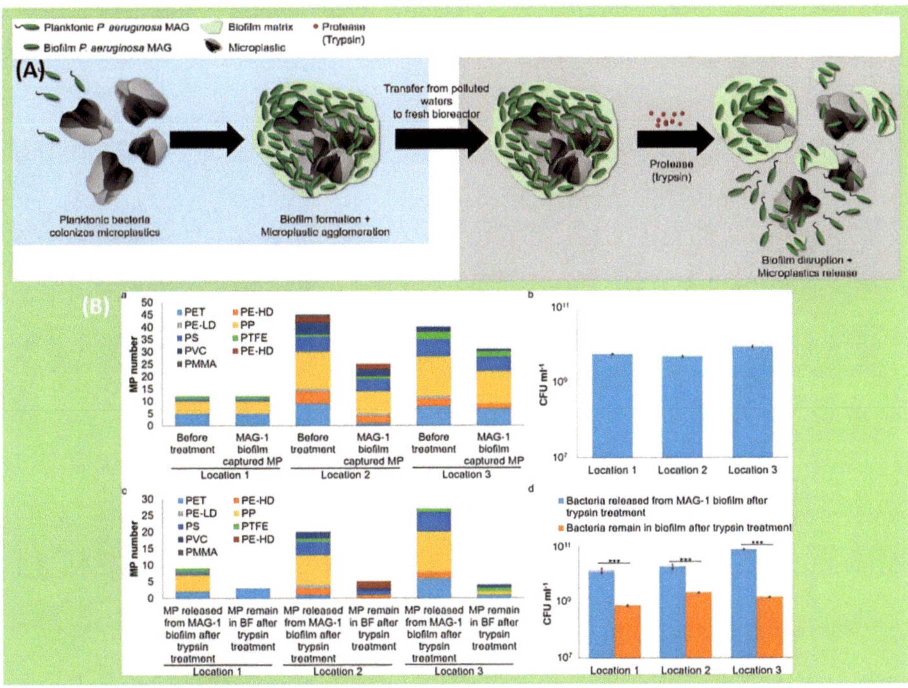

FIGURE 11.6 (A) Microbial and enzymatic entrapment and release of MPs via biofilm formation. Biofilms have the capacity to bioaccumulate microplastics from marine samples. (B) Graphical representation of (a) MP count collected before and after MAG3 aggregation; (b) bacterial colony-forming units (CFU) of MAG3 biofilms after microplastic capture. (c) The number of MPs retained within the biofilm and (d) bacterial quantities in the biofilm before and after trypsin treatment.

Source: Adapted from reference [51] with permission from the copyright holder, American Chemical Society (2022), Washington, DC, USA.

11.4.5 In Industrial Development

The potential role of biotechnology in industrial development has been investigated by various scientists; biotech has played a crucial role in transforming developing countries into industrialized ones. The bio-based industries that derive their raw material from natural biomass or through biological processes contribute to integrated bioeconomies, in contrast to those that work on synthetic raw materials [52].

Synthetic textile materials, including polyester, rayon, and acrylic, are extensively utilized in the clothing industry, accounting for approximately 70% of global clothing raw materials. The issue arises as synthetic MPs derived from these materials are released during washing, constituting 30%–35% of primary MPs that contaminate our environment and oceans. A startling statistic is that a singular laundry cycle discharges roughly 8 million microfibers into wastewater treatment plants, where they remain unrecoverable and eventually find their way into the marine environment [53]. Conclusively, the use of bio-based industries has reduced the negative impacts of industry on environmental, economic, and social life. Biotech-based paper and pulp and textile processing requires enzymes for their processing, which in turn

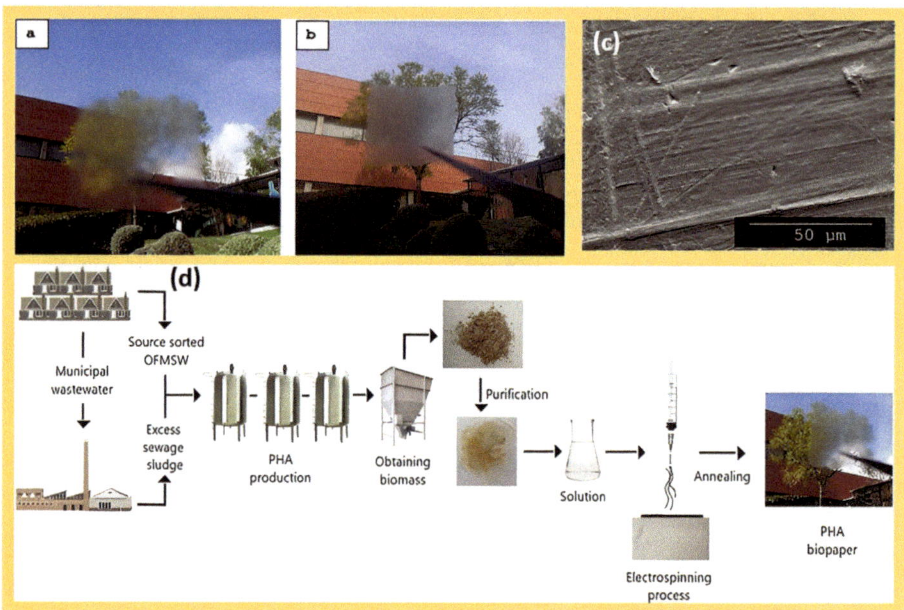

FIGURE 11.7 Electrospun PHBV biopapers from (a) MBW and (b) commercial grade sources; (c) SEM images of PHBV biopaper derived from MBW; (d) forming biopackaging material utilizing mixed bacteria culture.

Note: MBW = municipal biowaste; OFMSW = organic fraction of municipal solid waste. Source: Adapted from reference [54] with permission from the copyright holder, American Chemical Society (2020), Washington, DC, USA.

reduces the use of chemicals and energy and generates products with improved quality. Oxidative enzymes like laccases further reduce the production cost in the pulp mills. Melendez-Rodriguez et al. produced poly(3-hydroxybutyrate-co-3-hydroxyvalerate)) (PHBV) from municipal biowaste using bacterial culture colonies. The produced PHBV was further processed by ultra-spinning to yield an ultrathin fiber. The resultant fiber mats were post-treated in the temperature range of 80 °C to 150 °C for 5 seconds without any external pressure to obtain a fiber-based continuous film known as biopaper; Figure 11.7 graphically outlines the process [54].

Bio-based industries substantially contribute to higher product quality, renewable precursors, global food security, and a healthy environment. Modern biotechnology has enabled enzymatic processing in textile industries in an ecofriendly way as an alternative to traditional chemical processing.

11.4.6 In Biosensors

In biotechnology, biosensors act as an important tool for biomonitoring toxic analytes with high specificity. Biosensors are the most prominent tool in the biological and nonbiological systems in the environment based on the properties of bioreceptor, biosensors can be classified into different types (Figure 11.8) [55–57].

The most popular approach involves using nucleic acid, hormones and enzymes, and biofilms to produce physical, chemical, and electrical signals. Biosensors are most commonly employed in animal biotechnology, using molecular biology for

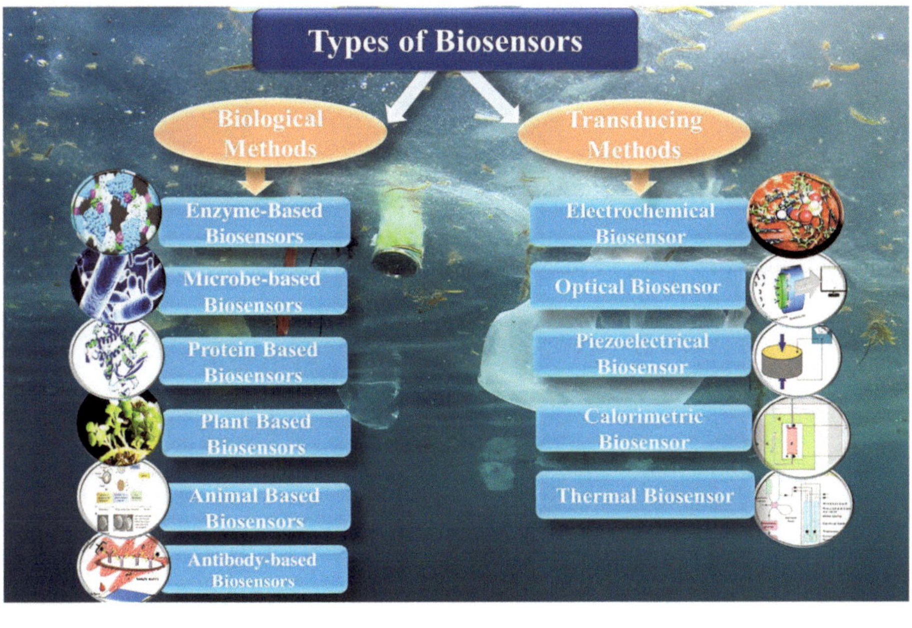

FIGURE 11.8 Different types of biosensors.

gene manipulation. For example, Dong et al. designed silver nanoparticles (AgNPs) as a nanoplatform to detect chloride ion and checked the interference of other anions, amino acids, metal ions, and reactive oxygen and nitrogen species (Figure 11.9) [58].

Dong et al. found that the oxidation peak potential of AgNPs shifted with increasing Cl^- concentrations due to the reaction between AgNPs and Cl^- to form AgCl. Methylene blue (MB) labelled oligonucleotide (SH-DNA-MB) was chosen as an inner reference molecule for establishing built-in correction against distinct environmental effects. To further enhance the response signals, three-dimensional nanostructures were assembled on carbon fiber microelectrode which resulted in high performance and good biocompatibility.

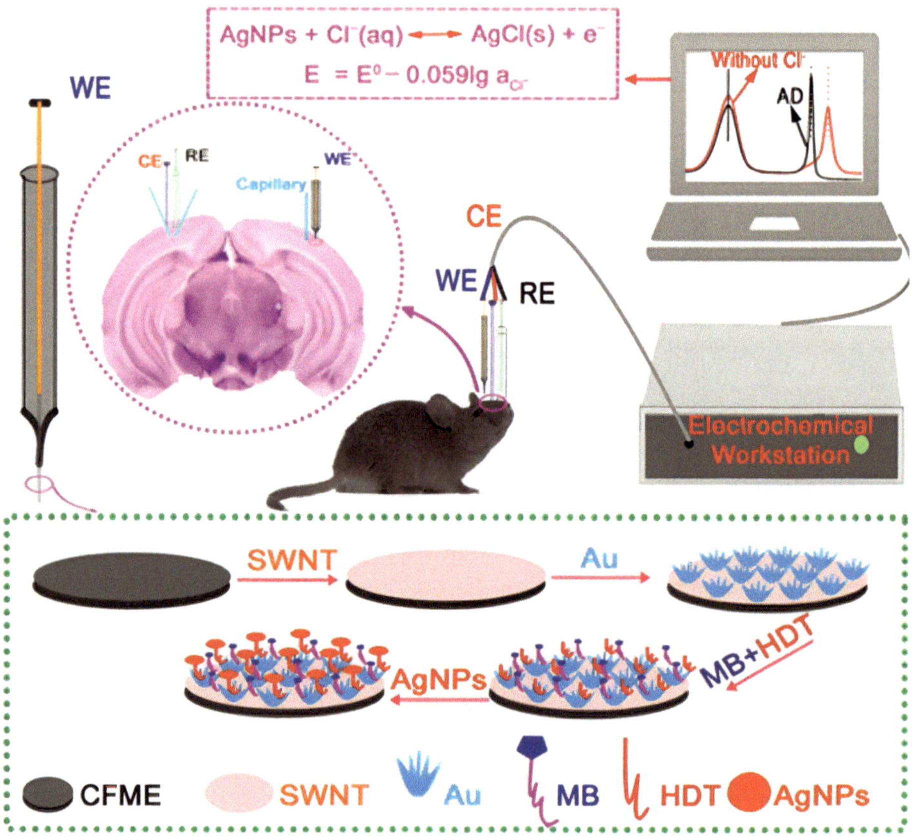

FIGURE 11.9 Constructing an electrochemical biosensor for the in vivo monitoring of chloride ion in mouse brain.

Source: Adapted from reference [58] with permission from the copyright holder, American Chemical Society (2017), Washington, DC, USA.

11.5 CONCLUSION

For this book chapter, we evaluated the key role of environmental biotechnology in meeting the environmental pollution remediation and global food demands, supporting resource recovery, and filling in knowledge gaps. Biotechnology applications across distinct areas have led to remunerative and efficient environmental benefits, including uncoupling economic drivers from the terrible environmental effects of MPs. Beyond helping with remediating air, water, and soil pollution, biotechnology is enhancing the use of functionally efficient bioproducts that are less polluting and cost effective than traditional counterparts. Biotechnology has introduced the world to wide range of advanced models like genetically engineered crops, vaccines, sensors, chemicals, textile, and bio-based papers, and these bioproducts have abated environmental issues and augmented sustainability.

ACKNOWLEDGMENTS

Anjali Vijeata is thankful to Council of Scientific and Industrial Research (09/135(0894)/2019-EMR-l) and New Delhi and Commonwealth Scholarships (INCN-2021–27) for financial support. Savita Chaudhary is thankful for the DST Inspire Faculty award [IFACH- 17] and for Haryana State Council for Science and Technology [HSCSIT/R&D/2020/476]; DST Chandigarh [S&T&RE/RP/147/e-2873/(19–20)/Sanc/12/2019/1878–1885] and [S&T&RE/RP/147/e-2873/(22–23)/Sanc/09/2022/907–915]; and DST Purse grants II for financial assistance. Ganga Ram Chaudhary would like to acknowledge the support of UGC, India, under the INDO-US 21st Century Knowledge Initiative project [F. No. 194–2/2016 (IC)]. The authors are thankful to the Sophisticated Analytical Instrumentation Facility, Panjab University, Chandigarh for technical support.

REFERENCES

1. Manzoor, M.M., 2020. Environmental biotechnology: For sustainable future. In *Bioremediation and Biotechnology, Vol 2: Degradation of Pesticides and Heavy Metals* (pp. 241–258). Cham: Springer International Publishing.
2. Gavrilescu, M., 2010. Environmental biotechnology: Achievements, opportunities and challenges. *Dynamic Biochemistry, Process Biotechnology and Molecular Biology*, *4*(1), pp. 1–36.
3. Sayler, G.S., Fox, R. and Blackburn, J. eds., 2013. *Environmental Biotechnology for Waste Treatment* (Vol. 41). New York: Springer Science & Business Media, Plenum Press.
4. Rasmussen, R.S. and Morrissey, M.T., 2007. Biotechnology in aquaculture: Transgenics and polyploidy. *Comprehensive Reviews in Food Science and Food Safety*, *6*(1), pp. 2–16.
5. Shi, J., Wang, Z., Peng, Y., Fan, Z., Zhang, Z., Wang, X., Zhu, K., Shang, J. and Wang, J., 2023. Effects of microplastics on soil carbon mineralization: The crucial role of oxygen dynamics and electron transfer. *Environmental Science & Technology*, *57*(36), pp. 13588–13600.
6. Löhr, A., Savelli, H., Beunen, R., Kalz, M., Ragas, A. and Van Belleghem, F., 2017. Solutions for global marine litter pollution. *Current Opinion in Environmental Sustainability*, *28*, pp. 90–99.

7. Setälä, O., Fleming-Lehtinen, V. and Lehtiniemi, M., 2014. Ingestion and transfer of microplastics in the planktonic food web. *Environmental Pollution*, *185*, pp. 77–83.
8. Jambeck, J.R., Geyer, R., Wilcox, C., Siegler, T.R., Perryman, M., Andrady, A., Narayan, R. and Law, K.L., 2015. Plastic waste inputs from land into the ocean. *Science*, *347*(6223), pp. 768–771.
9. Paes, L.A.B., Bezerra, B.S., Deus, R.M., Jugend, D. and Battistelle, R.A.G., 2019. Organic solid waste management in a circular economy perspective–a systematic review and SWOT analysis. *Journal of Cleaner Production*, *239*, p. 118086.
10. Liu, S.H., Zeng, G.M., Niu, Q.Y., Liu, Y., Zhou, L., Jiang, L.H., Tan, X.F., Xu, P., Zhang, C. and Cheng, M., 2017. Bioremediation mechanisms of combined pollution of PAHs and heavy metals by bacteria and fungi: A mini review. *Bioresource Technology*, *224*, pp. 25–33.
11. Coppola, G., Gaudio, M.T., Lopresto, C.G., Calabro, V., Curcio, S. and Chakraborty, S., 2021. Bioplastic from renewable biomass: A facile solution for a greener environment. *Earth Systems and Environment*, *5*, pp. 231–251.
12. Mosttafiz, S., Rahman, M. and Rahman, M., 2012. Biotechnology: Role of microbes in sustainable agriculture and environmental health. *The Internet Journal of Microbiology*, *10*(1), pp. 1–6.
13. Gómez-Salcedo, Y., Oliva-Merencio, D., Rodríguez-Díaz, J.M. and Pereda-Reyes, I., 2021. Contribution of the environmental biotechnology to the sustainability of the coffee processing industry in developing countries. In *Advances in the Domain of Environmental Biotechnology: Microbiological Developments in Industries, Wastewater Treatment and Agriculture* (pp. 565–589). Singapore: Springer.
14. Nizzetto, L., Bussi, G., Futter, M.N., Butterfield, D. and Whitehead, P.G., 2016. A theoretical assessment of microplastic transport in river catchments and their retention by soils and river sediments. *Environmental Science: Processes & Impacts*, *18*(8), pp. 1050–1059.
15. Syberg, K., Khan, F.R., Selck, H., Palmqvist, A., Banta, G.T., Daley, J., Sano, L. and Duhaime, M.B., 2015. Microplastics: Addressing ecological risk through lessons learned. *Environmental Toxicology and Chemistry*, *34*(5), pp. 945–953.
16. Molina, M., Aburto, F., Calderón, R., Cazanga, M. and Escudey, M., 2009. Trace element composition of selected fertilizers used in Chile: Phosphorus fertilizers as a source of long-term soil contamination. *Soil and Sediment Contamination*, *18*(4), pp. 497–511.
17. Neelavannan, K. and Sen, I.S., 2023. Microplastics in freshwater ecosystems of india: Current trends and future perspectives. *ACS Omega*, *8*(38), pp. 34235–34248.
18. Anagnosti, L., Varvaresou, A., Pavlou, P., Protopapa, E. and Carayanni, V., 2021. Worldwide actions against plastic pollution from microbeads and microplastics in cosmetics focusing on European policies. Has the issue been handled effectively? *Marine Pollution Bulletin*, *162*, p. 111883.
19. Awasthi, M.K., Singh, E., Binod, P., Sindhu, R., Sarsaiya, S., Kumar, A., Chen, H., Duan, Y., Pandey, A., Kumar, S. and Taherzadeh, M.J., 2022. Biotechnological strategies for bio-transforming biosolid into resources toward circular bio-economy: A review. *Renewable and Sustainable Energy Reviews*, *156*, p. 111987.
20. Lebreton, L.C., Van Der Zwet, J., Damsteeg, J.W., Slat, B., Andrady, A. and Reisser, J., 2017. River plastic emissions to the world's oceans. *Nature Communications*, *8*(1), p. 15611.
21. Yoo, H., Kim, M., Lee, Y., Park, J., Lee, H., Song, Y.C. and Ro, C.U., 2023. Novel single-particle analytical technique for inhalable airborne microplastic particles by the combined use of fluorescence microscopy, Raman Microspectrometry, and SEM/EDX. *Analytical Chemistry*, *95*, pp. 8552–8559.

22. Fischer, P.H., Marra, M., Ameling, C.B., Velders, G.J., Hoogerbrugge, R., de Vries, W., Wesseling, J., Janssen, N.A. and Houthuijs, D., 2020. Particulate air pollution from different sources and mortality in 7.5 million adults—the Dutch environmental longitudinal study (DUELS). *Science of the Total Environment*, *705*, p. 135778.
23. Deshmukh, P., Kimbrough, S., Krabbe, S., Logan, R., Isakov, V. and Baldauf, R., 2020. Identifying air pollution source impacts in urban communities using mobile monitoring. *Science of the Total Environment*, *715*, p. 136979.
24. Umamah, N. and Wathon, S., 2020, May. Identification of biotechnology urgency in the environmental knowledge course. In *IOP Conference Series: Earth and Environmental Science* (Vol. 485, No. 1, p. 012085). East Java: IOP Publishing.
25. Bourlat, S.J., Borja, A., Gilbert, J., Taylor, M.I., Davies, N., Weisberg, S.B., Griffith, J.F., Lettieri, T., Field, D., Benzie, J. and Glöckner, F.O., 2013. Genomics in marine monitoring: New opportunities for assessing marine health status. *Marine Pollution Bulletin*, *74*(1), pp. 19–31.
26. Bundy, J.G., Davey, M.P. and Viant, M.R., 2009. Environmental metabolomics: A critical review and future perspectives. *Metabolomics*, *5*, pp. 3–21.
27. Wegner, A., Besseling, E., Foekema, E.M., Kamermans, P. and Koelmans, A.A., 2012. Effects of nanopolystyrene on the feeding behavior of the blue mussel (Mytilus edulis L.). *Environmental Toxicology and Chemistry*, *31*(11), pp. 2490–2497.
28. Rochman, C.M., Manzano, C., Hentschel, B.T., Simonich, S.L.M. and Hoh, E., 2013. Polystyrene plastic: A source and sink for polycyclic aromatic hydrocarbons in the marine environment. *Environmental Science & Technology*, *47*(24), pp. 13976–13984.
29. Gongi, W., Touzi, H., Sadly, I., Ben Ouada, H., Tamarin, O. and Ben Ouada, H., 2022. A novel impedimetric sensor based on cyanobacterial extracellular polymeric substances for microplastics detection. *Journal of Polymers and the Environment*, *30*(11), pp. 4738–4748.
30. Dierkes, R.F., Wypych, A., Pérez-García, P., Danso, D., Chow, J. and Streit, W.R., 2023. An ultra-sensitive Comamonas thiooxidans biosensor for the rapid detection of enzymatic polyethylene terephthalate (PET) degradation. *Applied and Environmental Microbiology*, *89*(1), pp. e01603–22.
31. Liu, B., Lei, Y. and Li, B., 2014. A batch-mode cube microbial fuel cell based "shock" biosensor for wastewater quality monitoring. *Biosensors and Bioelectronics*, *62*, pp. 308–314.
32. Campaña, A.L., Florez, S.L., Noguera, M.J., Fuentes, O.P., Ruiz Puentes, P., Cruz, J.C. and Osma, J.F., 2019. Enzyme-based electrochemical biosensors for microfluidic platforms to detect pharmaceutical residues in wastewater. *Biosensors*, *9*(1), p. 41.
33. Xiang, P., Zhang, Y., Zhang, T., Wu, Q., Zhao, C. and Li, Q., 2023. A novel bacterial combination for efficient degradation of polystyrene microplastics. *Journal of Hazardous Materials*, p. 131856.
34. Wong, L.S., Judge, S.K., Voon, B.W.N., Tee, L.J., Tan, K.Y., Murti, M. and Chai, M.K., 2017. Bioluminescent microalgae-based biosensor for metal detection in water. *IEEE Sensors Journal*, *18*(5), pp. 2091–2096.
35. Yuan, J., Cao, J., Yu, F. and Ma, J., 2022. Microbial degradation of polystyrene microplastics by a novel isolated bacterium in aquatic ecosystem. *Sustainable Chemistry and Pharmacy*, *30*, p. 100873.
36. Li, N., Han, Z., Guo, N., Zhou, Z., Liu, Y. and Tang, Q., 2022. Microplastics spatiotemporal distribution and plastic-degrading bacteria identification in the sanitary and non-sanitary municipal solid waste landfills. *Journal of Hazardous Materials*, *438*, p. 129452.
37. Shahar, H., Tan, L.L., Ta, G.C. and Heng, L.Y., 2019. Detection of halogenated hydrocarbon pollutants using enzymatic reflectance biosensor. *Sensors and Actuators B: Chemical*, *281*, pp. 80–89.

38. Yi, Y., Xie, B., Zhao, T., Li, Z., Stom, D. and Liu, H., 2019. Effect of external resistance on the sensitivity of microbial fuel cell biosensor for detection of different types of pollutants. *Bioelectrochemistry*, *125*, pp. 71–78.
39. Védrine, C., Leclerc, J.C., Durrieu, C. and Tran-Minh, C., 2003. Optical whole-cell biosensor using Chlorella vulgaris designed for monitoring herbicides. *Biosensors and Bioelectronics*, *18*(4), pp. 457–463.
40. Schwaminger, S.P., Fehn, S., Steegmüller, T., Rauwolf, S., Löwe, H., Pflüger-Grau, K. and Berensmeier, S., 2021. Immobilization of PETase enzymes on magnetic iron oxide nanoparticles for the decomposition of microplastic PET. *Nanoscale Advances*, *3*(15), pp. 4395–4399.
41. Prabha, S.P., Nagappan, S., Rathna, R., Viveka, R. and Nakkeeran, E., 2020. Blue biotechnology: A vision for future marine biorefineries. In *Refining Biomass Residues for Sustainable Energy and Bioproducts* (pp. 463–480). Amsterdam: Elsevier.
42. Yadav, A.N., Singh, S., Mishra, S. and Gupta, A., 2019. *Recent Advancement in White Biotechnology Through Fungi* (p. 528). Cham: Springer International Publishing.
43. Heux, S., Meynial-Salles, I., O'Donohue, M.J. and Dumon, C., 2015. White biotechnology: State of the art strategies for the development of biocatalysts for biorefining. *Biotechnology Advances*, *33*(8), pp. 1653–1670.
44. Cao, Z., Xia, W., Wu, S., Ma, J., Zhou, X., Qian, X., Xu, A., Dong, W. and Jiang, M., 2023. Bioengineering Comamonas testosteroni CNB-1: A robust whole-cell biocatalyst for efficient PET microplastic degradation. *Bioresources and Bioprocessing*, *10*(1), p. 94.
45. Deniz, I., García-Vaquero, M. and Imamoglu, E., 2017. Trends in red biotechnology: Microalgae for pharmaceutical applications. In *Microalgae-Based Biofuels and Bioproducts* (pp. 429–460). Elsevier Inc. Woodhead Publishing.
46. Pessôa, M.G., Vespermann, K.A., Paulino, B.N., Barcelos, M.C., Pastore, G.M. and Molina, G., 2019. Newly isolated microorganisms with potential application in biotechnology. *Biotechnology Advances*, *37*(2), pp. 319–339.
47. Giri, B., Prasad, R., Wu, Q.S. and Varma, A. eds., 2019. *Biofertilizers for Sustainable Agriculture and Environment*. Cham: Springer International Publishing.
48. Kumar, K.K., Sridhar, J., Murali-Baskaran, R.K., Senthil-Nathan, S., Kaushal, P., Dara, S.K. and Arthurs, S., 2019. Microbial biopesticides for insect pest management in India: Current status and future prospects. *Journal of Invertebrate Pathology*, *165*, pp. 74–81.
49. Zhang, S., Li, Y., Chen, X., Jiang, X., Li, J., Yang, L., Yin, X. and Zhang, X., 2022. Occurrence and distribution of microplastics in organic fertilizers in China. *Science of the Total Environment*, *844*, p. 157061.
50. Kour, D., Rana, K.L., Yadav, A.N., Yadav, N., Kumar, M., Kumar, V., Vyas, P., Dhaliwal, H.S. and Saxena, A.K., 2020. Microbial biofertilizers: Bioresources and eco-friendly technologies for agricultural and environmental sustainability. *Biocatalysis and Agricultural Biotechnology*, *23*, p. 101487.
51. Anwar, S., Ali, A., Ullah, Z., Binjawhar, D.N., Sher, H., Ali, R., Iqbal, R., Ali, B. and Ali, I., 2023. The impact of Trichoderma harzianum together with copper and boron on wheat yield. *ACS Agricultural Science & Technology, 3*(6), p. 517.
52. Li, M., Zhang, P., Guo, Z., Cao, W., Gao, L., Li, Y., Tian, C.F., Chen, Q., Shen, Y., Ren, F. and Rui, Y., 2023. Molybdenum nanofertilizer boosts biological nitrogen fixation and yield of soybean through delaying nodule senescence and nutrition enhancement. *ACS Nano*, *17*(15), pp. 14761–14774.
53. Liu, Y.Q., Chen, Y., Ren, X.M., Li, Y.Y., Zhang, Y.J., Zhang, H., Han, H. and Chen, Z.J., 2023. Plant growth-promoting bacteria modulate gene expression and induce antioxidant tolerance to alleviate synergistic toxicity from combined microplastic and Cd pollution in sorghum. *Ecotoxicology and Environmental Safety*, *264*, p. 115439.

54. Reddy, N. and Yang, Y., 2005. Biofibers from agricultural byproducts for industrial applications. *TRENDS in Biotechnology*, *23*(1), pp. 22–27.
55. Woodley, J.M., 2020. Towards the sustainable production of bulk-chemicals using biotechnology. *New Biotechnology*, *59*, pp. 59–64.
56. Molognoni, D., Chiarolla, S., Cecconet, D., Callegari, A. and Capodaglio, A.G., 2018. Industrial wastewater treatment with a bioelectrochemical process: Assessment of depuration efficiency and energy production. *Water Science and Technology*, *77*(1), pp. 134–144.
57. Abomohra, A. and Hanelt, D., 2022. Recent advances in micro-/nanoplastic (MNPs) removal by microalgae and possible integrated routes of energy recovery. *Microorganisms*, *10*(12), p. 2400.
58. Jadhav, H.S., Fulke, A.B., Dasari, L.N., Dalai, A. and Haridevi, C.K., 2024. Plastic bio-mitigation by Pseudomonas mendocina ABF786 and simultaneous conversion of its CO_2 byproduct to microalgal biodiesel. *Bioresource Technology*, *391*, p. 129952.
59. Karunasagar, I. and Karunasagar, I., 1999. Diagnosis, treatment and prevention of microbial diseases of fish and shellfish. *Current Science*, pp. 387–399.
60. Hou, L., Xi, J., Liu, J., Wang, P., Xu, T., Liu, T., Qu, W. and Lin, Y.B., 2022. Biodegradability of polyethylene mulching film by two Pseudomonas bacteria and their potential degradation mechanism. *Chemosphere*, *286*, p. 131758.
61. Chan, S.Y., Wong, M.W.T., Kwan, B.T.C., Fang, J.K.H. and Chua, S.L., 2022. Microbial–enzymatic combinatorial approach to capture and release microplastics. *Environmental Science & Technology Letters*, *9*(11), pp. 975–982.
62. Lokko, Y., Heijde, M., Schebesta, K., Scholtès, P., Van Montagu, M. and Giacca, M., 2018. Biotechnology and the bioeconomy—towards inclusive and sustainable industrial development. *New Biotechnology*, *40*, pp. 5–10.
63. Mishra, S., Singh, R.P., Rout, P.K. and Das, A.P., 2022. Membrane bioreactor (MBR) as an advanced wastewater treatment technology for removal of synthetic microplastics. *Development in Wastewater Treatment Research and Processes*, pp. 45–60.
64. Melendez-Rodriguez, B., Torres-Giner, S., Lorini, L., Valentino, F., Sammon, C., Cabedo, L. and Lagaron, J.M., 2020. Valorization of municipal biowaste into electrospun poly (3-hydroxybutyrate-co-3-hydroxyvalerate) biopapers for food packaging applications. *ACS Applied Bio Materials*, *3*(9), pp. 6110–6123.
65. Pashaei, R., Dzingelevičienė, R., Abbasi, S., Szultka-Młyńska, M. and Buszewski, B., 2022. Determination of the pharmaceuticals–nano/microplastics in aquatic systems by analytical and instrumental methods. *Environmental Monitoring and Assessment*, *194*, pp. 1–33.
66. Bacakova, L., Pajorova, J., Tomkova, M., Matejka, R., Broz, A., Stepanovska, J., Prazak, S., Skogberg, A., Siljander, S. and Kallio, P., 2020. Applications of nanocellulose/nanocarbon composites: Focus on biotechnology and medicine. *Nanomaterials*, *10*(2), p. 196.
67. Nigam, V.K. and Shukla, P., 2015. Enzyme based biosensors for detection of environmental pollutants-a review. *Journal of Microbiology and Biotechnology*, *25*(11), pp. 1773–1781.
68. Dong, H., Zhang, L., Liu, W. and Tian, Y., 2017. Label-free electrochemical biosensor for monitoring of chloride ion in an animal model of Alzhemier's disease. *ACS Chemical Neuroscience*, *8*(2), pp. 339–346.

12 Biodecomposition of Microplastics

A Sustainable Approach for Microplastic Waste Management

Flavius Phrangsngi Nonglang, Haorangbam Arundeep Singha, Emma Magdalene Nonglang, and Divya Newar

12.1 INTRODUCTION TO MICROPLASTICS

Plastics are synthetic materials derived from hydrocarbons such as coal, petroleum, and fossils and created by chemically bonding hundreds or thousands of subunits, known as monomers, through strong covalent chemical bonds [1]. The utilization of plastic has profoundly altered our existence in numerous ways, and plastics production is continuously expanding in response to the growing demand.

Plastics possess the desirable qualities of being cost-effective, durable, bio-inert, lightweight, and resistant to corrosion and show excellent thermal and electrical insulation properties [2], which makes them suitable for various applications, spanning from food packaging to technical and medical purposes. The inherent characteristics that grant plastics their remarkable versatility in meeting human requirements, however, have concurrently led to the emergence of a global ecological peril that is extensively prevalent in freshwater, marine, and terrestrial ecosystems. Plastics make up 5%–12% of the world's waste [3]. Plastic waste production increased from approximately 200 million tons (MT) in 2008 to over 698 MT in 2020 (249% increase) [4], and about 60%–80% of plastic debris is found in marine water bodies [5]. One of the major reasons for the environmental burden of plastics is their extended degradation period, which results from their inherent properties, particularly their high molecular weight, crystalline structure, and hydrophobic nature [6].

Once plastics have reached the end of their useful life, they are inevitably disposed of in the environment. These plastics can be found in various locations and undergo physical, chemical, and mechanical degradation, forming smaller fragments called microplastics (MPs). Two categories of synthetic plastic polymers exist: C=C and C=O. Polymers with C=C backbones, including polypropylene (PP), polyethylene (PE), polyvinyl chloride (PVC), and polystyrene (PS), account for 77% of the global

 DOI: 10.1201/9781003468141-12

market and individually contribute about 21%, 36.3%, 11.8%, and 7.6%, respectively. On the other hand, C=O polymers such as polyurethane (PUR) and polyethylene terephthalate (PET) account for over 18% of the global market share and individually contribute about 8.2% and 10.2%, respectively [7].

12.2 IMPACTS OF MPs ON THE ECOSYSTEM

The increasing awareness of MPs is primarily driven by their widespread distribution and detrimental effects on human and animal health as well as the ecosystem. The prevalence of MP exposure has significantly multiplied following the outbreak of COVID-19, primarily as a result of the extensive utilization of single-use plastic products and personal protective equipment. For instance, around 396 MT of plastic waste was generated in 2018, but by 2020, the rate of waste generated escalated to 698 MT, an increased rate of 76.26% [4, 8].

12.2.1 Effects on the Ecosystem

After their usage, plastics are discarded into the environment, even in remote areas like mountaintops, Antarctica, and deep oceans. Mechanical and physical processes such as weathering, photolysis, and abrasion break them down into MPs, which cause their spread in ecosystems and pose a high risk for pollution and other irreversible environmental changes in land and oceans [9].

12.2.2 Impacts on Human and Animal Health

The effects of MPs on human and animal health are intricate and require substantial investigation due to their potential hazards to varied biota. Numerous organisms consume MPs, resulting in pathological stress, stunted growth, false appetite, and reproductive complications [10].

Human exposure to MPs is thought to occur mostly through ingestion. The estimated intake of MPs is 39,000 to 52,000 particles per person per year based on dietary consumption [11]. Particles can get past the mucociliary clearance after inhalation, through contaminated food, and into the gastrointestinal tract. Once inside, they alter the makeup and metabolism of gut microorganisms and induce an inflammatory response [12]. MPs also negatively impact marine organisms by causing digestive blockages, reduced feeding, and disrupted reproduction [13]. In a 2020 study, MPs were found in the organs of a median of 60% of fish, representing 198 species that were caught in 24 different countries [14].

In laboratory settings, nanoparticles (NPs) translocate from originally exposed tissues such as the digestive and respiratory systems to interior tissues. This phenomenon has been observed in the hemolymph found in *Mytilus edulis* mussels [15] and the hemolymph, hepatopancreas, and ovary of *Carcinus maenas* crabs [16]. A growing body of research indicates that trophic transfer of MPs occurs in aquatic systems [17]. MPs have also been detected in *Halicoerus grypus* (captive grey seals); their wild prey, *Scomber scombrus* (Atlantic mackerel); and the feces and regurgitation samples of wild Eurasian dippers, *Cinclus cinclus* [18]. Potential ingestion of these minute plastics through adulterated seafood raises concern and poses a threat to human health as they can build up and transfer through the food chain.

12.3 MP BIODEGRADATION: THE NEED OF THE HOUR

The persistence and resilience of plastic polymers contribute to the escalating environmental issue of plastic pollution. An estimated 34 billion tons (BT) of thermoplastics are projected to be produced by 2050, and at the current rate of consumption, approximately 12 BT of plastic wastes will be disposed of in landfills or marine environments [19].

Present strategies employed to reduce their impacts on the environment include incineration, recycling, and landfilling, but, these approaches have their limitations [20]. Incineration emits numerous noxious and hazardous end products into the atmosphere, including dioxins, furans, mercury, and sulfides, as well as highly toxic and carcinogenic vapors that pose serious health and environmental risks [20, 21]. Landfills require a lot of space and expensive land purchases and maintenance worldwide. Moreover, plastic byproducts can become persistent organic pollutants, contaminating soil and groundwater over time [22]. Recycling plastic is more expensive than new plastic production and thus is economically inefficient [23]. Furthermore, recycled products typically have lower value and functionality than new products [24].

In short, the existing approaches to plastic disposal are afflicted with drawbacks; consequently, there is an increasing imperative to devise more suitable and ecologically sustainable methods for plastic disposal. Among these emerging techniques, biological degradation is regarded as the most effective and environmentally sustainable strategy for the management of plastic waste [25]. A heterogeneous microbial population is employed to degrade persistent plastics into individual monomers, which can then be converted into carbon dioxide (CO_2), water (H_2O), and new biomass while simultaneously producing valuable bioproducts [26]. An overview of the biodegradation of MPs is shown in Figure 12.1.

12.4 MICROORGANISMS IN MP BIODEGRADATION

Microbes undergo adaptations in response to the detrimental conditions caused by the environmental contamination of MPs [27]. In response to such conditions, microorganisms demonstrate a range of behaviors, such as modifications in growth rate and energy metabolism, and the synthesis of additional macromolecules that improve cellular defense [28].

12.4.1 Bacterial Degradation of MPs

Bacterial species are commonly acknowledged for their capacity to break down pollutants. Many bacterial species belonging to the genera *Pseudomonas*, *Escherichia*, and *Bacillus* have the capability to degrade plastic polymers. This has been demonstrated by employing various techniques such as metagenomics, cloning, pure culturing, and computational methods [29]. These bacteria utilize various enzymes such as dehydrogenase, dioxygenase, and monooxygenase to facilitate the breakdown process.

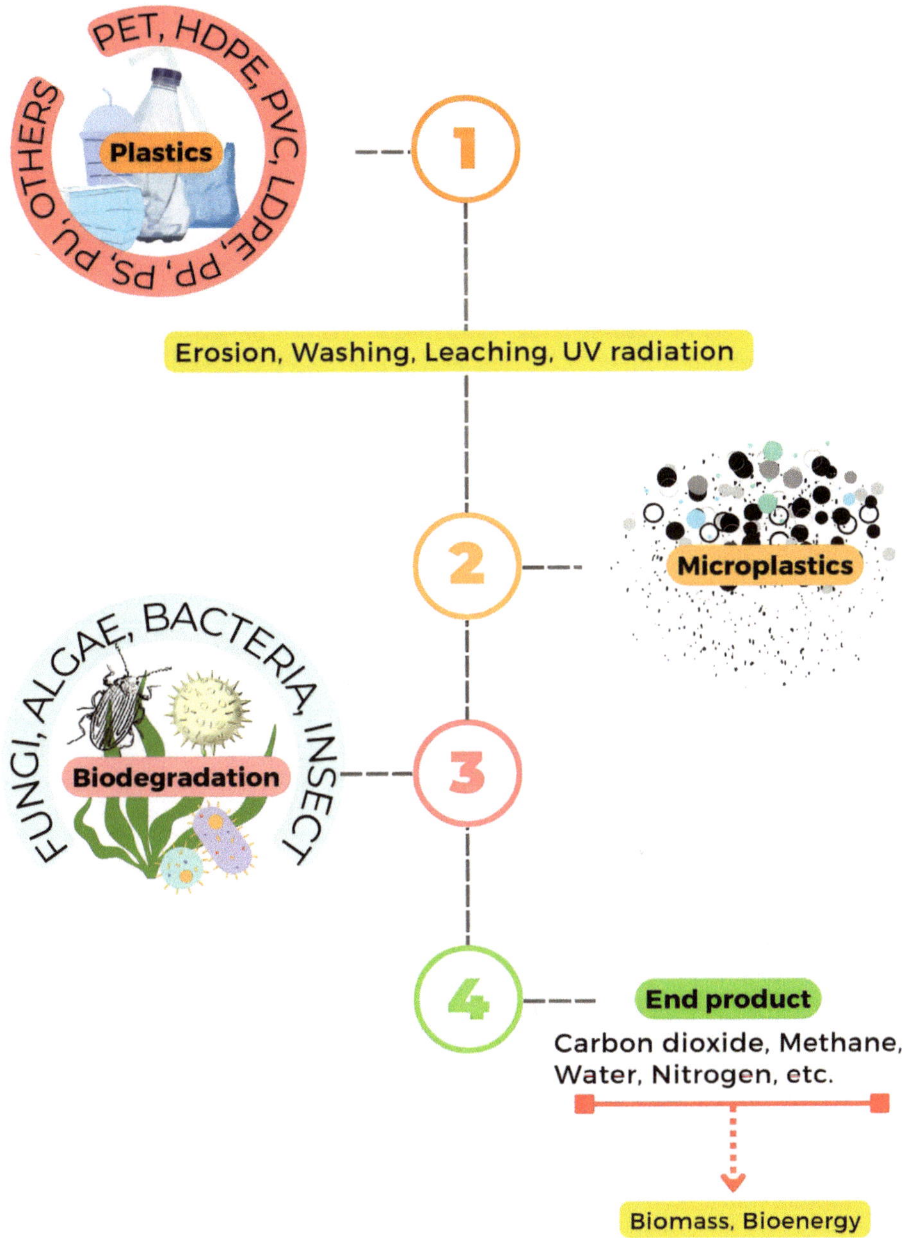

FIGURE 12.1 Biodegradation of MPs: An overview.

Laccase, for instance, is a crucial enzyme in the bioremediation of aromatic compounds from contaminated soils, industrial pollutants, and xenobiotics and is present in bacteria such as *Azospirillum lipoferum, Streptomyces griseus, Bacillus sphaericus, Marinomonas mediterranea* and *Serratia marcescens* [30]. A bacterial strain called PE3, obtained from contaminated soils, was observed to break down carboxylic acids, alkanes, and alkenes when exposed to low-density polyethylene (LDPE) sheets during incubation [31], and *B. cereus* and *B. gottheilii* extracted from mangrove silt exhibited the capability to break down plastic composed of PE and PS [32].

Thermoplastics are highly resistant to microbial biodegradation due to their homopolymeric composition. However, a remarkable discovery in plastic degradation involves a newly identified species of bacteria called *Ideonella sakaiensis.* This bacterium, isolated from dump sites, can break down PET by utilizing it as its primary energy and carbon source [33]. Studies have demonstrated that *Pseudomonas aeruginosa* [34], *Rhodococcus ruber* [35], and numerous bacterial strains can break down PS and polycarbonate (PC), which are both widely used thermoplastics [36]. Studies have demonstrated that bacteria belonging to different genera including *Bacillus, Pseudomonas, and Micrococcus* are capable of decomposing various types of thermoset plastics, with a particular emphasis on PUR [37]. Therefore, bacterial species play a vital role in MP degradation.

12.4.2 Fungal Degradation of MPs

Fungi encompass a wide range of organisms, most of which are obligate parasites, saprotrophs, or opportunistic. These organisms are highly adaptable and can thrive in various environments, both on land and in water. They are especially well-suited for living in the "plastisphere" because they can absorb nutrients in different environmental conditions.

Fungi possess a wide variety of intra- and extracellular enzymes with the capacity to catalyze a range of reactions and effectively break down fossil-derived polymers. Fungal mycelia can infiltrate the surfaces of polymeric materials, penetrating and subsequently degrading a larger amount of the substrate [38]. The fungal intracellular enzymatic system plays a vital role in their adaptation by serving as an internal mechanism for detoxification [39]. The system operates by employing cytochrome P450 family enzymes, such as transferases responsible for conjugation reactions and epoxidases involved in oxidations [40]. They carry out diverse chemical reactions, including epoxidation, sulfoxidation, desulfuration, dehalogenation, and deamination [41].

The extracellular enzymatic system has two subsystems: the hydrolytic subsystem produces hydrolases to hydrolyze complex polymers [42], while the nonspecific oxidoreductases, such as versatile peroxidases, lignin peroxidases, manganese peroxidases, and laccases, oxidize complex polymers [43]. Fungi attach to MPs by forming biofilms and attaching via their hyphae. They enhance the breakdown and conversion of intricate pollutants by generating a range of external enzymes and natural biosurfactants, such as hydrophobins, which break down complex polymers into simple monomers and supply microorganisms with electrons and carbon.

Fungi demonstrate remarkable efficiency in the biodegradation of aromatic polymers and other complex-structured materials [39]; they can utilize the chemical constituents of decomposed MPs for growth or release them for utilization by other microbes. Filamentous fungi, such as *Fusarium falciforme*, *Aspergillus* species, *Purpureocillium lilacinum*, and *F. oxysporum*, play a major role in the breakdown of micropollutants in the environment [44]. *Aspergillus niger*, *Zalerion maritimum*, and *Penicillium simplicissimum* are the main fungal species responsible for breaking down different types of polymers such as PE, PET, and PP [45–47]. They facilitate the formation of various chemical bonds (such as ester, carboxyl, and carbonyl functional groups) while also decreasing their hydrophobic nature. Various fungal strains, such as *F. solani*, *P. chrysogenum*, *A. fumigatus*, *Cladosporium pseudocladosporioides*, *Fusarium fumigatus*, and *A. tubingensis* have been found to degrade PUR [48–50]. Fungi possess a diverse array of nonspecific enzymes that can facilitate a broad spectrum of chemical reactions. As a result, they are well-suited for the process of breaking down MPs naturally and sustainably.

12.4.3 Algal Degradation of MPs

A noteworthy finding has been made concerning the capacity of filamentous algae to colonize the plastic surface [51]. Algae can attach to a plastic surface by biofilm formation and start breaking it down by releasing ligninolytic and exopolysaccharide enzymes [52]. These extracellular polymeric substances play a crucial role in transforming polymers into oligomers and monomers, which can serve as a carbon source for microbial growth that may result in the complete biomineralization, the ultimate step of biodegradation, resulting in the release of CO_2, H_2O, nitrogen (N_2), and methane (CH_4) [53].

Microalgae can be a viable alternative to bacterial and fungal systems as they do not need organic carbon sources to grow, so they can thrive on simple nutrients [54]. Furthermore, the fungal and bacterial species, namely *Pestalotiopsis microspora* and *I. sakaiensis*, that are utilized for PETase production, are not suitably acclimated to the marine habitats where most of the plastic waste is located [55]. It has been found that specific categories of algae, specifically those classified under the Bacillariophyceae, Chlorophyceae, and Cyanophyceae classes, are capable of flourishing on polythene surfaces and creating algal biofilms in various contaminated aquatic environments [56].

Specific nontoxic cyanobacteria, namely *Oscillatoria subbrevis* and *Phormidium lucidum*, exhibit rapid and effortless growth, leading to the formation of colonies on PE surfaces and their subsequent degradation without any pretreatment [57]. Another group of microalgae including *Anabaena spiroides* (blue-green alga), *Navicula pupula* (diatom), and *Scenedesmus dimorphus* (green alga) were also found to colonize and degrade the PE surface through the processes of fouling, degradation of leached components, corrosion, hydrolysis, and penetration by diffusing into the polymers [58]. The tensile strength of polymers, such as PET and PP, decreased significantly in the presence of genus *Spirulina*. This suggests that PET and PP serve as a carbon source for the *Spirulina* species [59].

Other algae species that are known to colonise plastics are *Oscillatoria princeps, Phormidium lucidum, O. tenuis, Closterium costatum, Chlorella vulgaris, Uronema africanum*, etc. [60]. Microalgae can be modified genetically to serve as a microbial cell factory, with the ability to secrete and produce enzymes that can break down polymers. For example, *Chlamydomonas reinhardtii*, a type of green microalgae, was genetically modified to produce PETase and tested for its efficacy in breaking down PET and ultimately verified through the development of perforations and cavities on the plastic surface [61].

Microalgae enhance the breakdown of MPs into valuable products like biogas, biofuels, fertilizers, and animal feed, thereby lowering water pollutants and also boosting energy production [62]. In addition to microalgae, *Fucus vesiculosus*, a macroalga, has demonstrated a remarkable capacity to bind PS of less than 20 μm at a sorption rate of 94.5%. This is made possible by the presence of phaeophycean hairs, which facilitate the absorption of nutrients, and by the release of alginate, which enhances the MPs adherence mechanism [63].

12.4.4 Insect Degradation of MPs

Because of their ability to recycle plastic, insects are good biodegradation agents. *Tenebrio molitor* beetles are capable of chewing and ingesting pieces of plastic packaging, such as PE, PP, and PVC films, and converting them to CO_2, some of which is then excreted as waste [64, 65]. Other coleopteran species, such as *Tribolium castaneum* [66], *Plesiophthalmus davidis* [67], *T. obscurus* [65], and *Zophobas atratus* [68], have been observed to degrade nonbiodegradable PE, PP, PS, and PVC polymers. Apart from beetles, lepidopteran species have also been found to break down plastic. The barn moth, *Plodia interpunctella* [67], the lesser wax moth, *Achroia grisella* [68], and the larger wax moth *Galleria mellonella* break down PS polymers independent of their gut microbioma, which may be beneficial for reducing MP pollution [69]. It was also found that *Galleria mellonella* was able to degrade 92% of plastic materials, such as PP and PE, at a rate of 1.84 mg/day/worm [70]. Plastic degradation in the gut of insect larvae of *T. molitor, T. obscurus*, and *Galleria mellonella* is also reported, facilitated by gut microbiota and digestive enzymes [71].

12.5 MICROBIAL DEGRADATION OF MPs

Microorganisms facilitate the breakdown of complex and resistant wastes into simpler and harmless forms through biochemical conversion, allowing these compounds to be cycled back into the environment. The capacity of microorganisms, including bacteria, fungi, algae, and in a few instances, insects, to generate specialized enzymes that degrade the intricate composition of polymers constitutes a vital mechanism in the microbial degradation of MPs. This process results in the conversion of MPs into CO_2 and H_2O in aerobic conditions and into CO_2 and CH_4 in anaerobic conditions. Biodegradation is distinguished by distinct indicators such as changes to the surface properties of the plastic, lower molecular weight and mechanical strength, and ultimately breakdown [72]. The microbial mechanism of MP degradation involves various biochemical processes, such as biodeterioration (colonization), biofragmentation, bioassimilation, and biomineralization [73] as shown in Figure 12.2.

12.5.1 Biodeterioration

Biodeterioration occurs when microbes or other biological agents cause plastic polymers to degrade superficially and modify their chemical, mechanical, and physical properties [74]. The first steps in the biodeterioration process begin with microbes adhering to and colonizing polymers. Since plastics are hydrophobic, microorganisms have the innate ability to change the surface structure and introduce hydrophilic functional groups, thus encouraging microorganism attachment [75]. Microorganisms such as bacteria, algae, and fungi form biofilms for better interaction and adherence to MPs [76]. Fungi produce hydrophobins that help in their adherence to polymeric MP surfaces via their hyphae [77], after which local swelling occurs and subsequently decreases the polymer's mechanical properties significantly [74]. Once microbes attach to plastic, they use polymers as their sole carbon source to grow.

12.5.2 Biofragmentation

After attachment and colonization, microorganisms excrete extracellular enzymes that bind to MPs and free radicals that fragment large molecular-weight MPs into oligomers, dimers, or monomers. Extracellular enzymes such as laccases, esterases, lignin peroxidase, lipases, and manganese peroxide increase the hydrophilicity of the plastic polymers by adding carbonyl or alcohol groups, which aids microbial attachment and further biodegradation [43]. Enzymes such as depolymerase and cutinase hydrolyze MPs glycosidic, ester, peptide, and carbon-carbon bonds on their surfaces

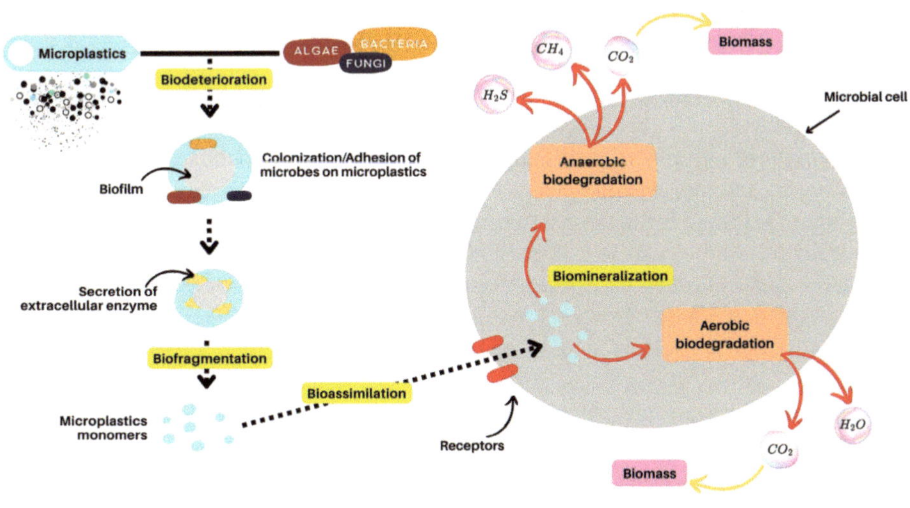

FIGURE 12.2 Mechanism of MP biodegradation.

[78]. In conclusion, extracellular enzymes secreted by microbes degrade larger MPs into smaller molecules that can pass through their semipermeable membrane and are now more susceptible to microbial degradation.

12.5.3 Bioassimilation

Fragmented monomers with decreased molecular weight can now be easily absorbed by the microbes and traverse the cellular membrane to enter the cell. Once inside, they are exposed to the action of intracellular enzymes such as monooxygenases, hydrolases, and dioxygenases. Plastic molecules are absorbed by microbial membranes through both active and passive transport mechanisms, much like hydrocarbons. Various membrane transport systems facilitate the transportation of these substances into the cytoplasm such as porins [79] for further processing.

12.5.4 Biomineralization

After being transported into the cells, the monomeric derivatives can be metabolized inside the cell by intracellular enzymes through the tricarboxylic acid cycle and β-oxidation under aerobic conditions to form into oxidized metabolites like CO_2 and H_2O or under anaerobic condition to form CO_2 and CH_4. Aerobic and anaerobic mineralization require various intracellular enzymes, including esterases, lipases, cutinase, peroxidases, and laccases [80]. They can either fully mineralize or transform into intermediates that other microorganisms can use as substrates. These intermediates can subsequently undergo additional degradation to produce simpler, nontoxic forms that can either be recycled back into the environment or used to produce energy (biomass production) [81]. In this way, microbes efficiently and effectively degrade MPs without causing any harm to the environment, making it a sustainable approach in this regard.

12.6 MICROBIAL ENZYMES AND MP DEGRADATION

Microbial enzymes play a crucial role in biodegradation as the primary environmental agents responsible for this process. Enzymes play a crucial role in both cell function and regulation, as well as in breaking down human-made pollutants, such as MPs. For instance, microorganisms possess enzymes that can selectively target the polymer structure of MPs and break them down into their individual monomers. Microorganisms then utilize these monomers as a carbon source in their energy production cycle [82]. Enzymes such as lipases, carboxylesterase, cutinase, and proteases exhibit distinct ways of breaking down MPs. One such mechanism involves modifying the surface of the enzyme through hydrolysis, which enhances the hydrophilicity of the MP surface [83].

Fungal laccase, peroxidase, and alkane hydrolase showed significant reactions toward LDPE, high-density polyethylene (HDPE), and PVC degradation [84]. *I. sakaiensis* 201-F6 bacterium has been shown to successfully express the PET-degrading enzyme, known as PETase, gathering a lot of attention related to PET degradation. The primary PETases degrade PET into intermediate components that are then acted upon by secondary mono(2-hydroxyethyl) terephthalic enzymes to produce nonhazardous end products such as terephthalic acid and ethylene glycol [85].

PS degradation involves enzyme reactions to form styrene monomer, and ultimately, the process of mineralization of the monomeric styrene occurs to form by-products that can then enter the tricarboxylic acid cycle for energy production or biofuel production. The PS enzymatic degradation involves styrene monooxygenase, isomerase, dehydrogenase, ligases, etc. Another enzyme known as serine hydrolase is proven to be involved in PS degradation. Enzymatic attack on PUR can be facilitated by hydrolases such as urease, protease, and esterase. It has been shown that proteolytic enzymes and cholesterol esterase can break down PUR in vitro [86], and cutinase, lipase, and carboxylesterase are widely used to degrade PUR [87]. Fungal laccases and peroxidases, generally used to degrade lignin, showed good results in degrading PE and PVC, while good results were also obtained on PUR by fungal protease and urease [88].

12.7 STRATEGIES TO ENHANCE MP BIODEGRADATION

Considering the inexhaustible capabilities of microorganisms and their perpetual adjustment to dynamic surroundings, it is anticipated that additional and strategic investigations in this domain to enhance the efficacy and effectiveness of biodegradation will rapidly result in commercially viable and long-term sustainable biodegradation processes.

12.7.1 Novel Microorganisms That Can Degrade MPs

Ensuring the identification and characterization of microorganisms originating from diverse ecosystems and their capacity to degrade MPs is of utmost significance. There is still a great deal of unexplored potential in the field of MP biodegradation, specifically concerning the effectiveness of various exogenous microorganisms in decomposing MPs [89]. The discovery of the novel bacterial isolate *Ideonella sakaiensis* 201-F6 that produces PET-hydrolyzing enzymes was a revelation in the effective biodegradation of PET, and continuous research on the discovery of novel microorganisms that can be more effective in degrading MPs is vital.

12.7.2 Identifying and Utilizing Microbial Consortium and Enzyme Cocktail

The preference for microbial consortia over individual strains for the biodegradation of complex natural and synthetic compounds is likely attributable to the limited metabolic capacities of individual organisms. It is reasonable to assume that microbial assemblages could degrade plastic wastes more efficiently than single microbial strains [90]. However, this field has received less attention than plastic degradation by single microbial strains.

Some studies have shown that the biodegradation efficacies of consortium-acting microorganisms can surpass those of individual strains. This is achieved through the removal of potentially toxic degradation intermediates or the direct participation of microorganisms in the biodegradation process [91]. It is imperative to conduct comprehensive research on the synergistic effects of microbial consortia and the enzymatic activities exhibited by diverse microbial networks. Simultaneously, enzyme cocktails have demonstrated promise as an alternative method for bioremediating MPs, particularly those that are resistant to change [92]. Thus, the innovation of an enzyme formulation derived from various sources to degrade a complex mixture of MPs could be effective. For this, however, the utilization of modern biotechnological interventions is necessary.

12.7.3 Pretreatment of MPs before Biodegradation

Mechanical and physical pretreatment ruptures MP chemical bonds, making polymer biodegradation easier. Thus, combining these pretreatment processes and microbial activities can effectively enhance biodegradation. Photocatalysis pretreatment of plastic wastes with biodegradation can significantly increase the efficiency of MP degradation. Photocatalytic degradation can break down a variety of plastics into smaller plastic pieces and smaller organic molecules, resulting in the release of CO_2 [93]. There are numerous photocatalysts such as ZnO, TiO_2, and Cu_2O that degrade multiple types of plastics such as PE, polyester, and PS using light sources with varying durations from 30 minutes to 816 hours [94]. Mechanical pretreatment through grinding, compounding, and pelletizing can also accelerate biodegradation [95]. Additives such as starch can activate and accelerate MP biodegradation by affecting functional groups [96], and adding surfactants like Tween 80 and sodium dodecyl sulfate can boost microbial degradation by increasing the hydrophilicity of polymer surfaces [97].

12.7.4 Controlling External Variables to Enhance Biological Degradation

Temperature, pH, oxygen content, and moisture are all variables that influence the degradation efficiency of microorganisms. Thus, controlling these variables can exponentially assist in regulating the metabolism of microorganisms, ultimately dictating the rate of MP degradation in a large-scale industrial setup [98].

12.7.5 Biotechnological Advancements

Enhanced or combined approaches to polymer biodegradation may be utilized to optimize the removal of MPs from particular environments [99]. Enzyme variants with enhanced enzymatic activity and temperature stability are generated via structural-based modeling, a biotechnological tool that is extensively employed in protein engineering. The *Ideonella sakaiensis* PETase variant IsPETase exhibited an improved affinity for substrates and thermostability and demonstrated 58-fold more activity than the wild-type IsPETase [100].

Additionally, site-directed mutagenesis has been effectively implemented in investigations pertaining to MP degradation. *Escherichia coli* strains are typically employed to transport plasmid-encoding modified genes and to encode desired enzymes. Research has shown that the hydrolysis activity of modified thermostable cutinase from *Thermobifida fusca* KW3 (TfCut2) expressed in the *E. coli* model system was 12.7 times greater than that of wild-type TfCut2 [101]. Researchers have cloned and expressed a type IIa PET recombinant enzyme, a polyester hydrolase, from *Pseudomonas aestusnigri* in *E. coli.* The enzyme could degrade PE, bis(2-Hydroxyethyl) terephthalate, and amorphous PET but not conventional PET bottles. However, the enhanced enzyme activity was achieved by site-directed mutagenesis to degrade PET bottle films [102]. Site-directed mutagenesis was performed on a *Saccharomonospora viridis* AHK190 putative cutinase whereby substitution of serine-226 and arginine-228 residue with proline and serine effectively improved the cutinase's thermostability and biodegradation activity [103].

Through molecular engineering techniques, successful cloning and expression of the PETase gene originating from the bacterial *I. sakaiensis* into host microorganisms like *Phaeodactylum tricornutum* and *E. coli* was reported in an effort to overcome the instability, recovery, and solubility issues that were encountered with the wild-type organism [51, 104]. In another instance, to augment the capacity of *Pseudomonas aeruginosa* to accumulate MPs such as PVC in its biofilm, the wspF gene was deleted during genetic engineering to promote the formation of sticky exopolymeric substances [105]. Thus, extensive research in protein engineering focusing on enzyme reusability and the economic viability of large-scale enzymatic degradation of plastics to handle the massive environmental plastic wastes is very much a future strategy. Biotechnological intervention and protein engineering have the potential to increase the catalytic efficiency and stability of enzymes involved in plastic biodegradation, as demonstrated by these studies.

12.8 CONCLUSION

Plastic pollution is a global problem caused by the plastic industry's success. Throughout ecosystems, oceans, and landforms, plastics pose environmental hazards. Exposure of these plastic waste to various processes causes their gradual fragmentation to form MPs. The ingestion of these MPs by various life forms, including humans, can cause growth and development mutations and related disorders. Thus, there is a need for effective methods to harness and degrade the MPs and, at the same time, find ways to transform these pollutants into energy.

Existing plastic disposal methods have drawbacks, so there is a growing need for more sustainable and suitable methods. Among these emerging plastic waste management methods, biological degradation is the most effective and environmentally friendly. Using a heterogeneous microbial population in combination with pretreatment methods, MPs are broken down into monomers that can be converted into CO_2, H_2O, and biomass for conversion to energy while also producing valuable bioproducts. With more research and development, biodegradation can have a huge impact on MP management, degradation, and sustainable development.

REFERENCES

1. Thompson, R.C., Moore, C.J., Vom Saal, F.S. and Swan, S.H., 2009. Plastics, the environment and human health: Current consensus and future trends. *Philosophical Transactions of the Royal Society B: Biological Sciences*, *364*(1526), pp. 2153–2166.
2. Muthukumar, A. and Veerappapillai, S., 2015. Biodegradation of plastics: A brief review. *International Journal of Pharmaceutical Science Review and Research*, *31*(2), pp. 204–209.
3. Kaza, S., Yao, L., Bhada-Tata, P. and Van Woerden, F., 2018. *What a waste 2.0: A global snapshot of solid waste management to 2050*. World Bank Publications.
4. Shams, M., Alam, I. and Mahbub, M.S., 2021. Plastic pollution during COVID-19: Plastic waste directives and its long-term impact on the environment. *Environmental Advances*, *5*, pp. 100119.
5. Peng, L., Fu, D., Qi, H., Lan, C.Q., Yu, H. and Ge, C., 2020. Micro-and nano-plastics in marine environment: Source, distribution and threats—a review. *Science of the Total Environment*, *698*, p. 134254.
6. Devi, R.S., Kannan, V.R., Natarajan, K., Nivas, D., Kannan, K., Chandru, S. and Antony, A.R., 2016. The role of microbes in plastic degradation. *Environment and Waste Management*, *341*, pp. 341–370.
7. Ali, S.S., Elsamahy, T., Al-Tohamy, R., Zhu, D., Mahmoud, Y.A.G., Koutra, E., Metwally, M.A., Kornaros, M. and Sun, J., 2021. Plastic wastes biodegradation: Mechanisms, challenges and future prospects. *Science of the Total Environment*, *780*, p. 146590.
8. Plastics Europe, E.P.R.O., 2019. Plastics-the facts, 2019. An analysis of European plastics production, demand, and waste data. *Plastics Europe*. https://plasticseurope.org/wp-content/uploads/2021/10/2019-Plastics-the-facts.pdf. Accessed on 23 December, 2023.
9. Wang, J., Zheng, L. and Li, J., 2018. A critical review on the sources and instruments of marine microplastics and prospects on the relevant management in China. *Waste Management and Research*, *36*(10), pp. 898–911.
10. Batel, A., Linti, F., Scherer, M., Erdinger, L. and Braunbeck, T., 2016. Transfer of benzo [a] pyrene from microplastics to Artemia nauplii and further to zebrafish via a trophic food web experiment: CYP1A induction and visual tracking of persistent organic pollutants. *Environmental Toxicology and Chemistry*, *35*(7), pp. 1656–1666.
11. Cox, K.D., Covernton, G.A., Davies, H.L., Dower, J.F., Juanes, F. and Dudas, S.E., 2019. Human consumption of microplastics. *Environmental Science and Technology*, *53*(12), pp. 7068–7074.
12. Salim, S.Y., Kaplan, G.G. and Madsen, K.L., 2014. Air pollution effects on the gut microbiota: A link between exposure and inflammatory disease. *Gut Microbes*, *5*(2), pp. 215–219.
13. Guo, X. and Wang, J., 2019. The chemical behaviors of microplastics in marine environment: A review. *Marine Pollution Bulletin*, *142*, pp. 1–14.
14. Sequeira, I.F., Prata, J.C., da Costa, J.P., Duarte, A.C. and Rocha-Santos, T., 2020. Worldwide contamination of fish with microplastics: A brief global overview. *Marine Pollution Bulletin*, *160*, p. 111681.
15. Browne, M.A., Dissanayake, A., Galloway, T.S., Lowe, D.M. and Thompson, R.C., 2008. Ingested microscopic plastic translocates to the circulatory system of the mussel, Mytilus edulis (L.). *Environmental Science and Technology*, *42*(13), pp. 5026–5031.
16. Farrell, P. and Nelson, K., 2013. Trophic level transfer of microplastic: Mytilus edulis (L.) to Carcinus maenas (L.). *Environmental Pollution*, *177*, pp. 1–3.
17. Welden, N.A., Abylkhani, B. and Howarth, L.M., 2018. The effects of trophic transfer and environmental factors on microplastic uptake by plaice, Pleuronectes plastessa, and spider crab, Maja squinado. *Environmental Pollution*, *239*, pp. 351–358.
18. D'Souza, J.M., Windsor, F.M., Santillo, D. and Ormerod, S.J., 2020. Food web transfer of plastics to an apex riverine predator. *Global Change Biology*, *26*(7), pp. 3846–3857.

19. Geyer, R., 2020. Production, use, and fate of synthetic polymers. In *Plastic waste and recycling* (pp. 13–32). Academic Press.
20. Crowley, D., Staines, A., Collins, C., Bracken, J., Bruen, M., Fry, J., Hrymak, V., Malone, D., Magette, B., Ryan, M. and Thunhurst, C., 2003. Health and environmental effects of landfilling and incineration of waste-a literature review. *Reports*, 3.
21. Dziok, T., Bury, M. and Burmistrz, P., 2022. Mercury release from municipal solid waste in the thermal treatment process. *Fuel*, *329*, 125528.
22. Hopewell, J., Dvorak, R. and Kosior, E., 2009. Plastics recycling: Challenges and opportunities. *Philosophical Transactions of the Royal Society B: Biological Sciences*, *364*(1526), pp. 2115–2126.
23. Gradus, R., van Koppen, R., Dijkgraaf, E. and Nillesen, P., 2016. A cost-effectiveness analysis for incineration or recycling of Dutch household plastics, *Tinbergen Institute Discussion Paper*, 16-039/VI, Tinbergen Institute, Amsterdam and Rotterdam.
24. Balu, R., Dutta, N.K. and Roy Choudhury, N., 2022. Plastic waste upcycling: A sustainable solution for waste management, product development, and circular economy. *Polymers*, *14*(22), p. 4788.
25. Sarkhel, R., Sengupta, S., Das, P. and Bhowal, A., 2020. Comparative biodegradation study of polymer from plastic bottle waste using novel isolated bacteria and fungi from marine source. *Journal of Polymer Research*, *27*, pp. 1–8.
26. Montazer, Z., Habibi Najafi, M.B. and Levin, D.B., 2020. Challenges with verifying microbial degradation of polyethylene. *Polymers*, *12*(1), p. 123.
27. Othman, A.R., Hasan, H.A., Muhamad, M.H., Ismail, N.I. and Abdullah, S.R.S., 2021. Microbial degradation of microplastics by enzymatic processes: A review. *Environmental Chemistry Letters*, *19*, pp. 3057–3073.
28. Guan, N. and Liu, L., 2020. Microbial response to acid stress: Mechanisms and applications. *Applied Microbiology and Biotechnology*, *104*(1), pp. 51–65.
29. Gan, Z. and Zhang, H., 2019. PMBD: A comprehensive plastics microbial biodegradation database. *Database*, *2019*, pp. 119.
30. Sheikhi, F., Roayaei Ardakani, M., Enayatizamir, N. and Rodriguez-Couto, S., 2012. The determination of assay for laccase of Bacillus subtilis WPI with two classes of chemical compounds as substrates. *Indian Journal of Microbiology*, *52*, pp. 701–707.
31. Kavitha, R., 2017. Evaluation of polyethylene degrading potential of isolated soil bacteria, Ph.D., Avinashilingam Institute for Home Science and Higher Education For Women University (Deemed to be University), Coimbatore, Tamil Nadu, India., pp. 1–157.
32. Auta, H.S., Emenike, C.U., Jayanthi, B. and Fauziah, S.H., 2018. Growth kinetics and biodeterioration of polypropylene microplastics by Bacillus sp. and Rhodococcus sp. isolated from mangrove sediment. *Marine Pollution Bulletin*, *127*, pp. 15–21.
33. Yoshida, S., Hiraga, K., Takehana, T., Taniguchi, I., Yamaji, H., Maeda, Y., Toyohara, K., Miyamoto, K., Kimura, Y. and Oda, K., 2016. A bacterium that degrades and assimilates poly (ethylene terephthalate). *Science*, *351*(6278), pp. 1196–1199.
34. Naima, A., Safia, A., Ali, M.I., Saadia, A., Bashir, A. and Robson, G., 2010. Isolation and identification of polystyrene biodegrading bacteria from soil. *African Journal of Microbiology Research*, *4*(14), pp. 1537–1541.
35. Mor, R. and Sivan, A., 2008. Biofilm formation and partial biodegradation of polystyrene by the actinomycete Rhodococcus ruber: Biodegradation of polystyrene. *Biodegradation*, *19*, pp. 851–858.
36. Arefian, M., Tahmourespour, A. and Zia, M., 2020. Polycarbonate biodegradation by newly isolated Bacillus strains. *Archives of Environmental Protection*, *46*(1), pp. 14–20.
37. Espinosa, M.J.C., Blanco, A.C., Schmidgall, T., Atanasoff-Kardjalieff, A.K., Kappelmeyer, U., Tischler, D., Pieper, D.H., Heipieper, H.J. and Eberlein, C., 2020. Toward biorecycling: Isolation of a soil bacterium that grows on a polyurethane oligomer and monomer. *Frontiers in Microbiology*, *11*, p. 404.

38. Sánchez, C., 2020. Fungal potential for the degradation of petroleum-based polymers: An overview of macro-and microplastics biodegradation. *Biotechnology Advances*, *40*, p. 107501.
39. Olicón-Hernández, D.R., González-López, J. and Aranda, E., 2017. Overview on the biochemical potential of filamentous fungi to degrade pharmaceutical compounds. *Frontiers in Microbiology*, *8*, p. 1792.
40. Schwartz, M., Perrot, T., Aubert, E., Dumarçay, S., Favier, F., Gérardin, P., Morel-Rouhier, M., Mulliert, G., Saiag, F., Didierjean, C. and Gelhaye, E., 2018. Molecular recognition of wood polyphenols by phase II detoxification enzymes of the white rot Trametes versicolor. *Scientific Reports*, *8*(1), p. 8472.
41. Shin, J., Kim, J.E., Lee, Y.W. and Son, H., 2018. Fungal cytochrome P450s and the P450 complement (CYPome) of Fusarium graminearum. *Toxins*, *10*(3), p. 112.
42. Sánchez, C., 2009. Lignocellulosic residues: Biodegradation and bioconversion by fungi. *Biotechnology Advances*, *27*(2), pp. 185–194.
43. Anand, U., Dey, S., Bontempi, E., Ducoli, S., Vethaak, A.D., Dey, A. and Federici, S., 2023. Biotechnological methods to remove microplastics: A review. *Environmental Chemistry Letters*, pp. 1–24.
44. Vaksmaa, A., Guerrero-Cruz, S., Ghosh, P., Zeghal, E., Hernando-Morales, V. and Niemann, H., 2023. Role of fungi in bioremediation of emerging pollutants. *Frontiers in Marine Science*, *10*, p. 1070905.
45. Paço, A., Duarte, K., da Costa, J.P., Santos, P.S., Pereira, R., Pereira, M.E., Freitas, A.C., Duarte, A.C. and Rocha-Santos, T.A., 2017. Biodegradation of polyethylene microplastics by the marine fungus Zalerion maritimum. *Science of the Total Environment*, *586*, pp. 10–15.
46. de Oliveira, T.A., Barbosa, R., Mesquita, A.B., Ferreira, J.H., de Carvalho, L.H. and Alves, T.S., 2020. Fungal degradation of reprocessed PP/PBAT/thermoplastic starch blends. *Journal of Materials Research and Technology*, *9*(2), pp. 2338–2349.
47. Devi, R.S., Kannan, V.R., Nivas, D., Kannan, K., Chandru, S. and Antony, A.R., 2015. Biodegradation of HDPE by Aspergillus spp. from marine ecosystem of Gulf of Mannar, India. *Marine Pollution Bulletin*, *96*(1–2), pp. 32–40.
48. Khan, S., Nadir, S., Shah, Z.U., Shah, A.A., Karunarathna, S.C., Xu, J., Khan, A., Munir, S. and Hasan, F., 2017. Biodegradation of polyester polyurethane by Aspergillus tubingensis. *Environmental Pollution*, *225*, pp. 469–480.
49. Álvarez-Barragán, J., Domínguez-Malfavón, L., Vargas-Suárez, M., González-Hernández, R., Aguilar-Osorio, G. and Loza-Tavera, H., 2016. Biodegradative activities of selected environmental fungi on a polyester polyurethane varnish and polyether polyurethane foams. *Applied and Environmental Microbiology*, *82*(17), pp. 5225–5235.
50. Magnin, A., Pollet, E., Phalip, V. and Avérous, L., 2020. Evaluation of biological degradation of polyurethanes. *Biotechnology Advances*, *39*, p. 107457.
51. Moog, D., Schmitt, J., Senger, J., Zarzycki, J., Rexer, K.H., Linne, U., Erb, T. and Maier, U.G., 2019. Using a marine microalga as a chassis for polyethylene terephthalate (PET) degradation. *Microbial Cell Factories*, *18*(1), pp. 1–15.
52. Sarmah, P. and Rout, J., 2018. Algal colonization on polythene carry bags in a domestic solid waste dumping site of Silchar town in Assam. *Phykos*, *48*(67), p. e77.
53. Jacquin, J., Cheng, J., Odobel, C., Pandin, C., Conan, P., Pujo-Pay, M., Barbe, V., Meistertzheim, A.L. and Ghiglione, J.F., 2019. Microbial ecotoxicology of marine plastic debris: A review on colonization and biodegradation by the "plastisphere". *Frontiers in Microbiology*, *10*, p. 865.
54. Yan, N., Fan, C., Chen, Y. and Hu, Z., 2016. The potential for microalgae as bioreactors to produce pharmaceuticals. *International Journal of Molecular Sciences*, *17*(6), p. 962.
55. Tanasupawat, S., Takehana, T., Yoshida, S., Hiraga, K. and Oda, K., 2016. Ideonella sakaiensis sp. nov., isolated from a microbial consortium that degrades poly (ethylene terephthalate). *International Journal of Systematic and Evolutionary Microbiology*, *66*(8), pp. 2813–2818.

56. Sharma, M., Dubey, A. and Pareek, A., 2014. Algal flora on degrading polythene waste. *Centre for Info Bio Technology Journal of Microbiology*, *3*, pp. 43–47.
57. Sarmah, P. and Rout, J., 2018. Efficient biodegradation of low-density polyethylene by cyanobacteria isolated from submerged polyethylene surface in domestic sewage water. *Environmental Science & Pollution Research*, *25*, pp. 33508–33520.
58. Kumar, R.V., Kanna, G.R. and Elumalai, S., 2017. Biodegradation of polyethylene by green photosynthetic microalgae. *Journal of Bioremediation and Biodegradation*, *8*(1), p. 2.
59. Khoironi, A., Anggoro, S. and Sudarno., 2019. Evaluation of the interaction among microalgae Spirulina sp, plastics polyethylene terephthalate and polypropylene in freshwater environment. *Journal of Ecological Engineering*, *20*(6), pp. 161–173.
60. Orlando, M., Molla, G., Castellani, P., Pirillo, V., Torretta, V. and Ferronato, N., 2023. Microbial enzyme biotechnology to reach plastic waste circularity: Current status, problems and perspectives. *International Journal of Molecular Sciences*, *24*(4), p. 3877.
61. Kim, J.W., Park, S.B., Tran, Q.G., Cho, D.H., Choi, D.Y., Lee, Y.J. and Kim, H.S., 2020. Functional expression of polyethylene terephthalate-degrading enzyme (PETase) in green microalgae. *Microbial Cell Factories*, *19*(1), pp. 1–9.
62. Raheem, A., Azlina, W.W., Yap, Y.T., Danquah, M.K. and Harun, R., 2015. Thermochemical conversion of microalgal biomass for biofuel production. *Renewable and Sustainable Energy Reviews*, *49*, pp. 990–999.
63. Sundbæk, K.B., Koch, I.D.W., Villaro, C.G., Rasmussen, N.S., Holdt, S.L. and Hartmann, N.B., 2018. Sorption of fluorescent polystyrene microplastic particles to edible seaweed Fucus vesiculosus. *Journal of Applied Phycology*, *30*, pp. 2923–2927.
64. Yang, Y., Yang, J., Wu, W.M., Zhao, J., Song, Y., Gao, L., Yang, R. and Jiang, L., 2015. Biodegradation and mineralization of polystyrene by plastic-eating mealworms: Part 1. Chemical and physical characterization and isotopic tests. *Environmental Science and Technology*, *49*(20), pp. 12080–12086.
65. Peng, B.Y., Su, Y., Chen, Z., Chen, J., Zhou, X., Benbow, M.E., Criddle, C.S., Wu, W.M. and Zhang, Y., 2019. Biodegradation of polystyrene by dark (Tenebrio obscurus) and yellow (Tenebrio molitor) mealworms (Coleoptera: Tenebrionidae). *Environmental Science and Technology*, *53*(9), pp. 5256–5265.
66. Wang, Z., Xin, X., Shi, X. and Zhang, Y., 2020. A polystyrene-degrading Acinetobacter bacterium isolated from the larvae of Tribolium castaneum. *Science of the Total Environment*, *726*, p. 138564.
67. Yang, J., Yang, Y., Wu, W.M., Zhao, J. and Jiang, L., 2014. Evidence of polyethylene biodegradation by bacterial strains from the guts of plastic-eating waxworms. *Environmental Science and Technology*, *48*(23), pp. 13776–13784.
68. Kundungal, H., Gangarapu, M., Sarangapani, S., Patchaiyappan, A. and Devipriya, S.P., 2019. Efficient biodegradation of polyethylene (HDPE) waste by the plastic-eating lesser waxworm (Achroia grisella). *Environmental Science and Pollution Research*, *26*, pp. 18509–18519.
69. Wang, S., Shi, W., Huang, Z., Zhou, N., Xie, Y., Tang, Y., Hu, F., Liu, G. and Zheng, H., 2022. Complete digestion/biodegradation of polystyrene microplastics by greater wax moth (Galleria mellonella) larvae: Direct in vivo evidence, gut microbiota independence, and potential metabolic pathways. *Journal of Hazardous Materials*, *423*, p. 127213.
70. Bombelli, P., Howe, C.J. and Bertocchini, F., 2017. Polyethylene bio-degradation by caterpillars of the wax moth Galleria mellonella. *Current Biology*, *27*(8), pp. R292–R293.
71. Goveas, L.C., Nayak, S., Kumar, P.S., Rangasamy, G., Vidya, S.M., Vinayagam, R., Selvaraj, R. and Dai Viet, N.V., 2023. Microplastics occurrence, detection and removal with emphasis on insect larvae gut microbiota. *Marine Pollution Bulletin*, *188*, p. 114580.
72. Ho, B.T., Roberts, T.K. and Lucas, S., 2018. An overview on biodegradation of polystyrene and modified polystyrene: The microbial approach. *Critical Reviews in Biotechnology*, *38*(2), pp. 308–320.

73. Padervand, M., Lichtfouse, E., Robert, D. and Wang, C., 2020. Removal of microplastics from the environment. A review. *Environmental Chemistry Letters*, *18*, pp. 807–828.
74. Amobonye, A., Bhagwat, P., Singh, S. and Pillai, S., 2021. Plastic biodegradation: Frontline microbes and their enzymes. *Science of the Total Environment*, *759*, p. 143536.
75. Nauendorf, A., Krause, S., Bigalke, N.K., Gorb, E.V., Gorb, S.N., Haeckel, M., Wahl, M. and Treude, T., 2016. Microbial colonization and degradation of polyethylene and biodegradable plastic bags in temperate fine-grained organic-rich marine sediments. *Marine Pollution Bulletin*, *103*(1–2), pp. 168–178.
76. Schwibbert, K., Menzel, F., Epperlein, N., Bonse, J. and Krüger, J., 2019. Bacterial adhesion on femtosecond laser-modified polyethylene. *Materials*, *12*(19), p. 3107.
77. Wösten, H.A. and Scholtmeijer, K., 2015. Applications of hydrophobins: Current state and perspectives. *Applied Microbiology and Biotechnology*, *99*, pp. 1587–1597.
78. Mohanan, N., Montazer, Z., Sharma, P.K. and Levin, D.B., 2020. Microbial and enzymatic degradation of synthetic plastics. *Frontiers in Microbiology*, *11*, p. 580709.
79. Duret, G. and Delcour, A.H., 2010. Size and dynamics of the Vibrio cholerae porins OmpU and OmpT probed by polymer exclusion. *Biophysical Journal*, *98*(9), pp. 1820–1829.
80. Alshehrei, F., 2017. Biodegradation of synthetic and natural plastic by microorganisms. *Journal of Applied & Environmental Microbiology*, *5*(1), pp. 8–19.
81. Yang, Y., Wang, J. and Xia, M., 2020. Biodegradation and mineralization of polystyrene by plastic-eating superworms Zophobas atratus. *Science of the Total Environment*, *708*, p. 135233.
82. Kawai, F., Kawabata, T. and Oda, M., 2019. Current knowledge on enzymatic PET degradation and its possible application to waste stream management and other fields. *Applied Microbiology and Biotechnology*, *103*, pp. 4253–4268.
83. Vertommen, M.A.M.E., Nierstrasz, V.A., Van Der Veer, M. and Warmoeskerken, M.M.C.G., 2005. Enzymatic surface modification of poly (ethylene terephthalate). *Journal of Biotechnology*, *120*(4), pp. 376–386.
84. Montazer, Z., Habibi Najafi, M.B. and Levin, D.B., 2020. Challenges with verifying microbial degradation of polyethylene. *Polymers*, *12*(1), p. 123.
85. Othman, A.R., Hasan, H.A., Muhamad, M.H., Ismail, N.I. and Abdullah, S.R.S., 2021. Microbial degradation of microplastics by enzymatic processes: A review. *Environmental Chemistry Letters*, *19*, pp. 3057–3073.
86. Ruiz, C., Main, T., Hilliard, N.P. and Howard, G.T., 1999. Purification and characterization of two polyurethanase enzymes from Pseudomonas chlororaphis. *International Biodeterioration and Biodegradation*, *43*(1–2), pp. 43–47.
87. Danso, D., Chow, J. and Streit, W.R., 2019. Plastics: Environmental and biotechnological perspectives on microbial degradation. *Applied and Environmental Microbiology*, *85*(19), pp. e01095–19.
88. Temporiti, M.E.E., Nicola, L., Nielsen, E. and Tosi, S., 2022. Fungal enzymes involved in plastics biodegradation. *Microorganisms*, *10*(6), p. 1180.
89. Lear, G., Kingsbury, J.M., Franchini, S., Gambarini, V., Maday, S.D.M., Wallbank, J.A., Weaver, L. and Pantos, O., 2021. Plastics and the microbiome: Impacts and solutions. *Environmental Microbiome*, *16*(1), pp. 1–19.
90. Cifuentes, I.E.M. and Öztürk, B., 2021. Exploring microbial consortia from various environments for plastic degradation. In *Methods in enzymology* (Vol. 648, pp. 47–69). Academic Press.
91. Yu, K., Yi, S., Li, B., Guo, F., Peng, X., Wang, Z., Wu, Y., Alvarez-Cohen, L. and Zhang, T., 2019. An integrated meta-omics approach reveals substrates involved in synergistic interactions in a bisphenol A (BPA)-degrading microbial community. *Microbiome*, *7*(1), pp. 1–13.

92. Contreras, F., Pramanik, S., Rozhkova, A.M., Zorov, I.N., Korotkova, O., Sinitsyn, A.P., Schwaneberg, U. and Davari, M.D., 2020. Engineering robust cellulases for tailored lignocellulosic degradation cocktails. *International Journal of Molecular Sciences*, *21*(5), p. 1589.
93. Chu, S., Zhang, B., Zhao, X., Soo, H.S., Wang, F., Xiao, R. and Zhang, H., 2022. Photocatalytic conversion of plastic waste: From photodegradation to photosynthesis. *Advanced Energy Materials*, *12*(22), p. 2200435.
94. He, Y., Rehman, A.U., Xu, M., Not, C.A., Ng, A.M. and Djurišić, A.B., 2023. Photocatalytic degradation of different types of microplastics by TiOx/ZnO tetrapod photocatalysts. *Heliyon*, *9*(11).
95. Ragaert, K., Delva, L. and Van Geem, K., 2017. Mechanical and chemical recycling of solid plastic waste. *Waste Management*, *69*, pp. 24–58.
96. Mehmood, C.T., Qazi, I.A., Hashmi, I., Bhargava, S. and Deepa, S., 2016. Biodegradation of low density polyethylene (LDPE) modified with dye sensitized titania and starch blend using Stenotrophomonas pavanii. *International Biodeterioration and Biodegradation*, *113*, pp. 276–286.
97. Ghatge, S., Yang, Y., Ahn, J.H. and Hur, H.G., 2020. Biodegradation of polyethylene: A brief review. *Applied Biological Chemistry*, *63*(1), pp. 1–14.
98. Lin, C., Cheruiyot, N.K., Bui, X.T. and Ngo, H.H., 2022. Composting and its application in bioremediation of organic contaminants. *Bioengineered*, *13*(1), pp. 1073–1089.
99. Miri, S., Saini, R., Davoodi, S.M., Pulicharla, R., Brar, S.K. and Magdouli, S., 2022. Biodegradation of microplastics: Better late than never. *Chemosphere*, *286*, p. 131670.
100. Son, H.F., Joo, S., Seo, H., Sagong, H.Y., Lee, S.H., Hong, H. and Kim, K.J., 2020. Structural bioinformatics-based protein engineering of thermo-stable PETase from Ideonella sakaiensis. *Enzyme and Microbial Technology*, *141*, p. 109656.
101. Furukawa, M., Kawakami, N., Tomizawa, A. and Miyamoto, K., 2019. Efficient degradation of poly (ethylene terephthalate) with Thermobifida fusca cutinase exhibiting improved catalytic activity generated using mutagenesis and additive-based approaches. *Scientific Reports*, *9*(1), p. 16038.
102. Bollinger, A., Thies, S., Knieps-Grünhagen, E., Gertzen, C., Kobus, S., Höppner, A., Ferrer, M., Gohlke, H., Smits, S.H. and Jaeger, K.E., 2020. A novel polyester hydrolase from the marine bacterium Pseudomonas aestusnigri–structural and functional insights. *Frontiers in Microbiology*, *11*, p. 114.
103. Kawai, F., Oda, M., Tamashiro, T., Waku, T., Tanaka, N., Yamamoto, M., Mizushima, H., Miyakawa, T. and Tanokura, M., 2014. A novel Ca 2+-activated, thermostabilized polyesterase capable of hydrolyzing polyethylene terephthalate from Saccharomonospora viridis AHK190. *Applied Microbiology and Biotechnology*, *98*, pp. 10053–10064.
104. Joo, S., Cho, I.J., Seo, H., Son, H.F., Sagong, H.Y., Shin, T.J., Choi, S.Y., Lee, S.Y. and Kim, K.J., 2018. Structural insight into molecular mechanism of poly (ethylene terephthalate) degradation. *Nature Communications*, *9*(1), p. 382.
105. Liu, S.Y., Leung, M.M.L., Fang, J.K.H. and Chua, S.L., 2021. Engineering a microbial 'trap and release' mechanism for microplastics removal. *Chemical Engineering Journal*, *404*, p. 127079.

13 Plastic Recycling
Current Trends, Challenges, and Opportunities

Jaskirat Suri, Bunty Sharma, Moondeep Chauhan, and Ganga Ram Chaudhary

13.1 INTRODUCTION

Plastic recycling is the process of collecting waste or scrap plastics and transforming them into items of different usable materials. Reusing plastic involves utilizing materials that have previously been processed. This process further conserves natural resources by diminishing the necessity to extract raw materials from the earth [1], which could reduce the quantity of gases that trap heat in the air. It likewise prevents extra waste from being unloaded in landfills.

Plastic waste exhibits chemical stability and can persist in dump yards and landfill sites for hundreds if not thousands of years after its disposal [2]. The process of managing waste curated in today's world and properly sorting the plastics that can be recycled is dependent on several factors including (but not limited to) the amount of space available for facilities, the economic capacity to manage and operate said facilities, and human willingness to actively participate in the collection process. Recycling plastic is beneficial to the environment but has some economic drawbacks. Nevertheless, there are various methods of ensuring constructive plastic recycling. Figure 13.1 shows the global plastic waste treatment and disposal methods [3].

13.2 WASTE MANAGEMENT: AN OVERVIEW

Waste management is completed in three steps: collection, treatment, and disposal. It encompasses various methods such as recycling, composting, landfill management, and waste-to-energy conversion. The goal of waste management is to minimize the environmental impact of waste and promote resource conservation. Effective waste management practices are crucial for the preservation of the environment and public health.

13.2.1 Landfill

Landfills are regulated systems of garbage and trash disposal in which the waste is buried between layers of the earth. The major components of landfills are liners,

DOI: 10.1201/9781003468141-13

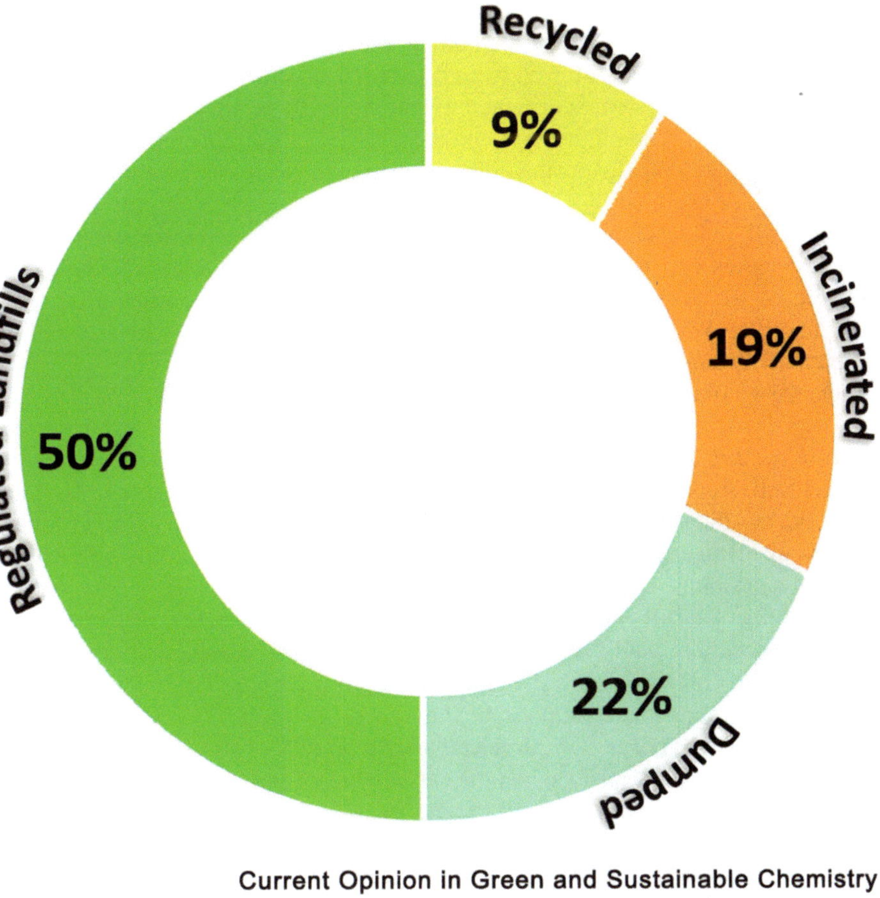

FIGURE 13.1 Global plastic waste treatment and disposal shares by method [3].

leachate collection systems, gas control systems, and landfill caps. Landfills typically use geosynthetic clay liners to help construct a barrier system that keeps the pollutants in the landfill from contaminating the soil and groundwater. These liners are highly favored due to their low permeability rates. Sodium bentonite is a naturally occurring compound in geosynthetic liners that is derived from volcanic ash [4, 5].

The greatest threat to groundwater from landfills is leachates. Leachates are contaminated liquids that have come into contact with waste. They contain high concentrations of heavy metals, inorganic chemicals, and dissolved suspended organic matter that are dangerous to the environment as well as human health. A collection system is intended with a minimum 2% slope to help channel the runoff to the collection sumps. A collection sump typically consists of a submersible pump and an HDPE (high-density polyethylene) riser [6].

The leachate drains to the collection system, where it is then transferred to a central pump and sent to the onsite holding pond. If the onsite area reaches capacity, the leachate will be transferred to a sanctioned offsite wastewater treatment plant for disposal. This system is designed to automatically pump the leachates out of the landfills to ensure compliance with state and federal requirements. Independent technicians complete monthly evaluations and testing to ensure the collection systems are operating as designed.

Landfill gas forms naturally as organic material decomposes within landfills. Landfill gases can be collected by either passive or active systems. The collection system comprises a succession of gas collection wells that are scattered throughout the landfills. Waste volume, density, depth, and the area of the specific landfills are used to determine the number and spacing of the wells [7].

As the gas develops in the landfill, the assortment wells offer favored pathways for gas movement. The majority of collection systems have redundant features to ensure that in the event of well failure, there are extra gas collection wells to replace the failed well. Passive gas collection systems exploit existing variances in landfill tension and gas focuses to vent landfill gases to the environment or a control framework. Latent assortment frameworks can be introduced while the landfill is in activity or after conclusion.

To collect landfill gas, passive systems make use of collection wells or extraction wells. Assortment wells are regularly made of opened plastic and are fitted in an upward direction all through the landfill to a profundity of half to 90% of the waste [8]. When groundwater happens in the waste, the well stops at the water table. Vertical wells are normally introduced after a landfill or part of a landfill has been shut, and horizontal wells beneath the earth's surface that serve as conduits for moving gas within the landfill are part of passive collection systems. Flat wells might be fitting for landfills where gas should be recuperated rapidly (for example, landfills with underground gas relocation issues), profound landfills, or dynamic landfills. Collection wells may occasionally discharge directly into the atmosphere. Collection wells often transport gas to a processing or control system (such as a flare).

The efficiency of a passive collection system is determined in part by how long the gas remains in the landfill [6]. Gas containment can be controlled and adjusted through the landfill assortment framework. Gases can be contained by utilizing liners at the top, sides, and lower part of the landfill. Impermeable liners (like mud or geosynthetic layers) can be utilized to catch landfill gas and make special gas movement pathways. According to the US Environmental Protection Agency (EPA), dynamic gathering frameworks are the most proficient method for catching landfill gas. Active gas collection systems use both vertical and horizontal gas collecting wells, just like passive collection systems. Be that as it may, wells in a functioning framework ought to incorporate valves to control the gas stream and capability as an examining port, as opposed to gas assortment wells in an uninvolved framework [9].

By sampling, the system administrator can measure gas production, composition, and pressure. Siphons or vacuums are utilized in dynamic gas assortment frameworks to eliminate gas from landfills, and a line is utilized to connect the vacuum to the gathering wells. Low tension is made inside the gas-gathering wells by vacuums or siphons, which concentrate gas from the waste. The type, size, and quantity of vacuums required in an active system to extract the gas from the landfill are determined by the quantity of gas produced [6].

Landfill caps are impermeable layers that are resistant to degradation and serve to reduce the amount of water that saturates into the final cover by maximizing evapotranspiration. Landfill final covers promote the drainage of precipitation, isolating the waste from the environment, and promoting the drainage of precipitation. For the final covers to be able to attain these goals, they must be resistant to the following: wind and water erosion, cracking, cold weather freeze/thaw cycles, and disruption by animals and plants. The base plan guidelines for landfill covers require a subgrade layer, a gas venting layer (with a base pressure driven conductivity of 1×10^{-3} cm/sec), A low porousness layer with a thickness of in any event—18 inches and a most extreme water powered conductivity of 1×10^{-7} cm/sec, a seepage layer comprising of 6 creeps of soil with a water driven conductivity of 1×10^{-3} cm/sec, and a vegetative help/security layer with a general thickness of no less than 18 inches, of which no less than 12 inches should be equipped for supporting vegetation [4, 10].

The subgrade layer must have low permeability to ensure that there are no substances present that could wear or penetrate the cap. There must be sufficient thickness and structural strength to support construction operations and ensure the long-term integrity of the final cover system. The thickness of this layer must also be at least 12 inches; this may also include a 6 inch soil–gas ventilation layer. The gas venting layer must be self-filtering. If it is not self-filtering, appropriate filters must be installed both above and below the gas filtering layer [4]. In addition, gas vent pipe penetrations must be designed so that the geomembrane is not damaged in the event of differential settlements between the pipes and the geomembrane. If necessary, horizontal headers embedded in the layer should be considered. It can also be a component of an active gas collection system if required.

According to the US Resource Conservation and Recovery Act, waste is separated into two major categories: solid and hazardous. A municipal solid waste (MSW) landfill is a discrete area of land where trash or garbage consisting of everyday items are disposed of. Total annual MSW production in the United States increased by 93% since 1980 to 292 million tons per year in 2018. MSW production per person also grew by 34% from 3.7 to 4.9 pounds per person over the same period. For comparison, the household waste production rates (per capita) are 2.5 in Sweden, 2.8 in the United Kingdom, and 3.9 in Germany. In 2018, plastics found in containers and packaging constituted over 14.5 million tons. Overall, about 36 million tons of plastic waste was generated, which is about 12.2% of the total MSW generated, and the landfills received most of it. As shown in Table 13.1, only about 9% of the total waste generated was recycled in 2018 [11].

The rate of waste generation differs based on the classification of cities. Projections suggest that by 2051, India's population will reach approximately 1,823 million, generating nearly 300 million tons of MSW annually, necessitating 1,450 km^2 of land for organized disposal. Although alternatives for plastic waste, such as recycling and combustion with energy recovery, are on the rise, most plastic waste is going directly into landfills. Once plastics are processed through landfills, they go through aerobic biodegradation and then conditions transition to anaerobic [2, 8].

TABLE 13.1
Plastics in Municipal Solid Waste in Thousands of US Tons [7]

Management Pathway	1960	1980	1990	2000	2005	2010	2015	2017	2018
Generation	390	6,830	17,130	25,550	29,380	31,400	34,480	35,410	35,680
Recycled	—	20	370	1,480	1,780	2,500	3,120	3,000	3,090
Combustion with Energy Recovery	—	140	2,980	4,120	4,330	4,530	5,330	5,590	5,620
Landfilled	390	6,670	13,780	19,950	23,270	24,370	26,030	26,820	26,970

13.2.2 Incineration

Incineration requires two key factors to be operational: fuel and oxygen. In this process, waste serves as fuel, and air serves as an oxygen source. Most incinerators operate at an average temperature of 1100 ºC to ensure that most organic and inorganic molecules are broken down [12]. To ensure adequate combustion while also reducing emissions, these facilities are required to maintain a stable environment with little to no fluctuations in waste/feed supply and combustion/airflow; counterproductive combustion results incomplete combustion. In addition, the pace at which a facility resumes and suspends operations may also affect the adequacy of the combustion and may lead to increased emissions.

For maximum efficiency and operation of a furnace, incineration temperatures and the turbulence of the gas mixture being combusted must be closely inspected at all times [10, 13]. The gas stream must reach high enough temperatures for an acceptable amount of time and be exposed to adequate fuel and oxygen to ensure effective combustion. The high temperatures of the incinerators are reached through oxidation [11, 14].

Temperatures must be high enough for combustion to be completed in its entirety while ensuring that the equipment is not being damaged or generating an extreme amount of nitrogen oxides. A common practice for controlling the temperatures is limiting the amount of waste in the furnace to ensure that the heat-release rate is as desired. Upon entering the desired range, the conditions can be adjusted by changing the amount of excess air. Turbulence provides a connection between oxygen and hydrocarbons to allow for proper airflow. Sufficient oxygen levels in conjunction with properly mixed gases support proper operation of the furnace [15].

The durability and lightweight nature of plastic are both valuable to humans while also being detrimental to the environment. Burning plastic releases a variety of gases that are toxic to the environment including mercury, dioxins, and polychlorinated biphenyls. Most plastic contains compounds such as polyethylene and polystyrene. The incomplete combustion of these compounds during the incineration process can lead to dangerously high concentrations of carbon monoxide [13, 14, 16]. In Mumbai, about 10,000 grams of dioxins and furans emanate into the atmosphere as a result of open burning of MSW.

Burning polystyrene can also affect the central nervous system (CNS) [15, 16] which is responsible for receiving, processing, and responding to sensory information.

The brain and the spinal cord comprise the CNS. The brain is responsible for movement, emotions, and communication among several other key functions, while the spinal cord serves as a message board that conveys messages from the brain to the rest of the body. Damage to either or both of these structures would be detrimental to a person's state of being [17].

Incinerating plastic waste also leads to more microplastics (MPs) in the environment. Plastic in general has a large area-to-volume ratio, which makes it very easy to absorb other harmful substances such as toxic heavy metals and hydrophobic pollutants [17]. The repercussions of MPs remain uncertain in terms of how they are manifested through the body. Some studies have shown that if inhaled, MPs can affect the vagus nerve, which is the messenger between the brain and the gut, and play a significant role in neurodegeneration. The vagus nerve is also linked to several motor and perceptive conditions such as Alzheimer's disease and Parkinson's disease [12, 18, 19].

13.2.3 Waste to Energy

Waste to energy is essentially transforming MSW into reusable energy. Waste-to-energy systems typically work in conjunction with incineration techniques since one of the key components of the process is burning the waste to create an energy source. The overall process is quite simple in comparison with other disposal methods.

First, the waste is gathered and placed in a large collection area. The waste is then separated into smaller segments and placed into the combustion chamber. The smaller waste portions ensure thorough combustion. The waste that is being burned lets off a significant amount of heat. This heat is used to transform water into steam with the help of a boiler. The high concentrations of steam assist in turning the blades of a turbine generator that is used to produce electricity. An air-pollutant control system then collects the remaining ash from the combustion [20].

As of 2015, there are only about 75 facilities that are capable of waste-to-energy recovery in the United States. One of the reasons why this method of waste removal is not as widely practiced is because of public opposition. These facilities have not had the best air emissions control equipment, which often meant that the facilities were letting off large amounts of pollution into the atmosphere. These facilities are also very expensive to operate. The initial cost of a waste-to-energy plant is about USD 100 million, but depending on the size of the plant, some facilities could cost up to three times that initial amount [21].

13.3 RECYCLING

Recycling is the process of turning waste into usable products. It provides the opportunity to reduce carbon emissions in the environment. It is a common misunderstanding that all plastic is recyclable; rather, most bottles and containers have a triangle with a number inside to indicate the type of plastic called the resin identification code. These codes typically range from one to six, and there can be letters under the triangle (Figure 13.2) [5, 22].

#1 or PET/PETE: The product is made of polyethylene terephthalate (PET). After recycling, this type of plastic is used to make bottles for cleaning products, other nonfood containers, or fiber products such as t-shirts or carpets.

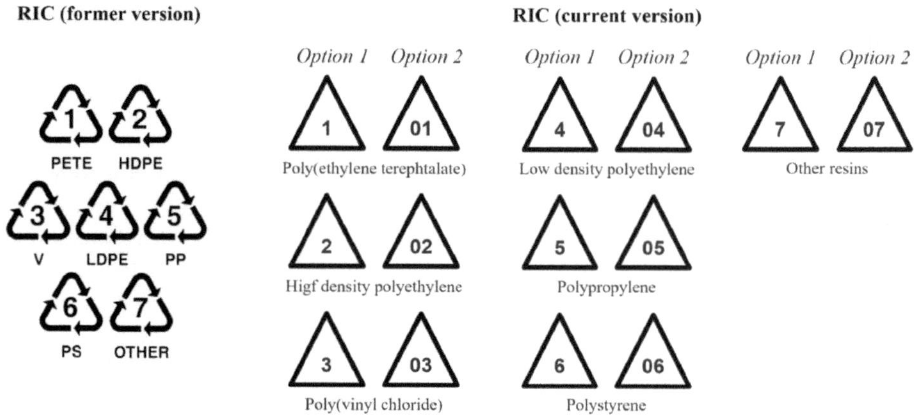

FIGURE 13.2 Plastic recycling symbols [5].

#2 or HDPE: This indicates that the plastic is made of HDPE. This type of plastic is most commonly found in household items and is used in plastic toys, traffic barrier cones, and grocery bags after being recycled.

#3 or PVC: This indicates that the object is made of polyvinyl chloride (PVC). This type of plastic is typically used for plumbing, house siding, and garden hoses and is generally harder to recycle because it contains unusually large amounts of plasticizers. High quantities of plasticizers contaminate the recycling stream. Because of this, this type of plastic is reused if possible but rarely recycled.

#4 or LDPE: This stands for low-density polyethylene (LDPE). Most plastic and disposable liners are made of this type of plastic, which is also harder to recycle due to its lower melting temperature.

#5 or PP: The product is made of polypropylene. Most car battery casings are made of this material. Even though this type of plastic is not typically recycled, it can be reused in other automotive parts and industrial fibers.

#6 or PS: This refers to polystyrene (PS) or polystyrene foam. This type of plastic is typically found in coffee cups, plastic cutlery, and most carry-out food packages. The interesting thing about recycled PS is that it is often used to create other PS products such as plastic lumber and hard plastic pens.

#7 or other: This refers to other types of plastic that do not easily fall into any of the other categories, like melamine is often used in plastic plates and cups.

Typically, plastic recycling involves six main stages: collection, sorting, cleansing, shredding, categorization, and compounding.

13.3.1 Collection

The collection phase involves collecting recyclable material from homes and businesses. Curbside collection, drop-off centers, and deposit centers are just a few of the collection options; of the three, curbside collection is the most common. For this method, the general population will place their recyclable goods into curbside collection boxes. Some regions separate fiber products (such as paper and cardboard) and containers (such as metals and plastics) into separate bins in dual streams. Vehicles will then come out to collect these bins and transport them to a processing facility. Dual-stream systems produce higher revenue per ton because the products are cleaner than they are from single-stream collection, where contamination is more likely [23].

13.3.2 Sorting

Sorting is typically completed using sophisticated machinery to increase efficiency and scale. Most plastic waste is sorted by color and thickness. These characteristics usually help determine the type of plastic and its route of disposal. Sorting is crucial in recycling because it allows for the correct handling of the materials. Manual sorting and magnetic separation are used to further separate the plastic for processing, and eddy current separators are used to extract nonferrous metals that may be located within the plastic waste. These separators operate with the help of giant magnets that generate currents that separate the metals and nonmetals [23]. Once the metals are sorted, optical sorting machines are used to determine the type of plastic in each substance. These machines operate using infrared measurements and sensors that measure absorbances to help accurately identify the type of plastic [24].

13.3.3 Washing

Washing removes impurities such as dirt and organic residue to avoid obstructions during the recycling process. The plastic should be rinsed, dried, and then sent to a facility where several washing methods are utilized to eliminate all contaminants. The most common method uses friction; this method is cost-effective and requires minimal labor. This washer decontaminates the plastic by using heat or kinetic energy to remove contaminates [25].

13.3.4 Shredding

Shredders are employed to minimize the size of the plastic. Shredding is another vital part of the recycling process because it allows for impurities to surface and be removed. Different shredders are used depending on the type of plastic. Although these industrial shredders come in a variety of different designs and sizes, the shape of the blade does not change: The blades are kept at less than five millimeters in thickness. Due to the thinness of the blades, the shredded plastic is also less than or equal to five millimeters in thickness [24, 26].

13.3.5 Identification

Identification takes place in two main stages, determining density and determining thickness. In the first stage, the plastic is tested and separated into different classes depending on its density: Higher-density items sink, while less dense plastics will float to the top. In the second stage, an air classification test is conducted to determine thickness; in this test, small plastic shreds are placed in a wind tunnel, and thicker pieces fly higher in the tunnel than the thinner pieces. If the type of plastic is still undetermined after the previous tests, further testing is used to establish color and melting points [25, 27].

13.3.6 Compounding

Compounding is the process of incorporating compounds and additives to create something new, via cold or hot mixing; cold mixing takes place at room temperature, whereas hot mixing takes place under heat. In hot mixing, the polymers are kept solid while the additives are liquefied for better mixing [28]. Figure 13.3 illustrates the sorting, reprocessing, and recycling of common plastics through either mechanical or chemical methods [3].

13.4 IMPORTANCE OF RECYCLING

Recycling is a global initiative that has increased the overall consciousness of our environment. It is taught to children at early ages to help increase overall awareness of the health of the planet. Recycling programs at schools teach kids that everyday items such as water bottles and plastic containers can and should be reused to reduce our overall carbon footprints on Earth. Although recycling sounds like the solution to global warming, it has advantages *and* disadvantages.

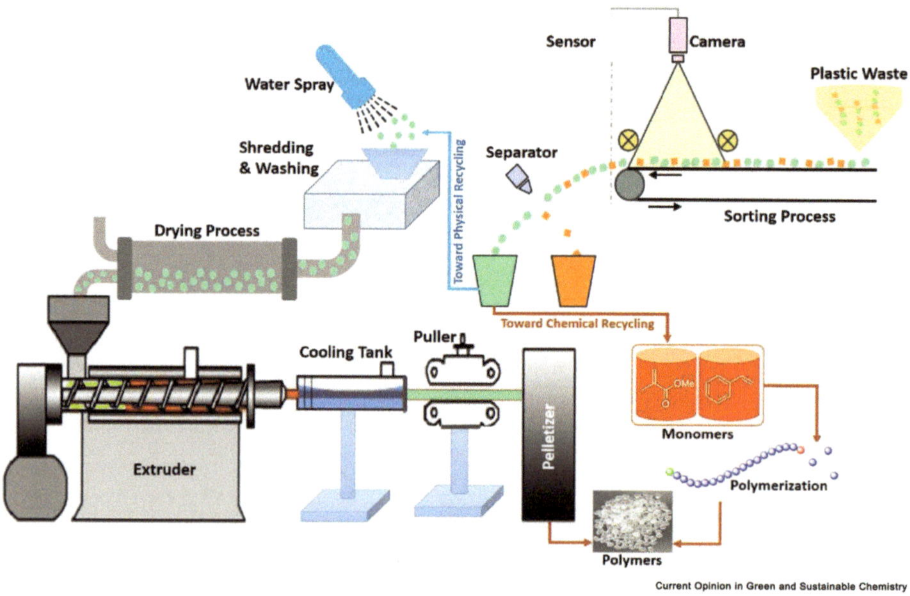

FIGURE 13.3 A visual representation outlining the recycling of plastic waste through mechanical and chemical methods [3].

13.4.1 Advantages

Reduce, reuse, and recycle. These are the words that are most commonly associated with the recycling process. Since plastic is malleable, it is easy to shape and reshape into other items. Plastic also does not easily corrode or decompose, so it can easily be altered by adding layers and additives to fit specific needs [26]. Although reducing the use of or reusing plastic goods are more environmentally beneficial, recycling is the most common practice.

Plastic recycling helps decrease the dependence on fossil fuels by diverting reusable waste from landfills and incinerators. This method of plastic disposal is another way to increase job opportunities within the community. It provides a good source of income for those without intensive education since most of the operational knowledge is gained through hands-on experience and training. The United States alone provided an estimated 680,000 so-called green jobs, about USD 38 million in wages, and generated approximately USD 5.5 billion in tax revenues [27, 28].

These jobs are on the rise due to the increased awareness of the declining health of the planet and will drive a strong economy in the 21st century. Recycling is the first step toward improving the health of the planet. Learning about the benefits of recycling and knowing how the small things can make a difference can possibly spark interest in other environment-friendly practices such as composting and expanding solar power [29].

13.4.2 Disadvantages

One of the biggest disadvantages of recycling is the pollution associated with the process. Incineration releases chemicals into the air that can be fatal to those with compromised immune systems [29]. In addition to the release of toxins from plastic incinerators, electronic waste is also responsible for releasing multiple known and suspected neurotoxins such as lead and mercury [30, 31]. Many less-developed countries are tasked with recycling these electronics due to the amount of manpower required to manually sort the devices [31]. However, this also increases their risk of health issues. Recycling is also not the most cost-effective process. Many towns and cities are unable to afford to operate sustainable recycling programs or facilities. Landfills are cheaper than recycling facilities, so many cities opt for the cheaper option regardless of the environmental hazards [32].

13.5 CURRENT TRENDS IN RECYCLING

Worldwide, plastic consumption rose from 279 million tons (MT) in 2011 to 322 MT in 2015 [32]. It is anticipated that plastic production will double by 2040 and is projected to reach approximately 1,600 MT by 2050 [33]. In 2019, plastic waste generation in India was approximately 9.4 MT per year, whereas the global generation rate exceeded 380 MT per year. Consequently, India's contribution accounted for roughly 3.1% of the global plastic waste generation.

As of November 2023, the current recycling rate in the United States is at 32%, which is a significant improvement from less than seven percent in 1960. According to research, approximately 160 MT of plastic waste will be generated globally in

2023. About 60% of the world's population will not enough have resources to manage the plastic MSW by the end of 2023 [33]. According to a 2018 EPA report, nearly 50% of total MSW and 77% of plastic waste were landfilled in the USA, with the recycling rate declining from 8.7% in 2018 to 5%–6% in 2021 [34].

13.6 CHALLENGES AND OPPORTUNITIES FOR IMPROVING RECYCLING

The recycling process is very sophisticated in terms of machinery, practice, and procedure. Each type of plastic has its own set of rules and regulations in terms of handling and disposal. Many plastic objects are made of a combination of ingredients, which makes it difficult to know how to best process those items. These combined items often do not reach total combustion, and that leads to the release of unnecessary contaminants in the atmosphere or increased waste for the landfills [34, 35]. Additionally, it is likely that the different polymers are contaminated with food particles, metal, and paper among a multitude of other variables, which only further complicates the recycling process [24, 36]. These challenges are only made worse by the shortage of collecting and sorting facilities, currently ineffective sorting, and the high cost of managing and processing the waste. High costs are one of the biggest obstacles in the creation of new, intricate recycling facilities [36].

Increased policies surrounding the use of environmentally safe design principles by industries could help promote cleaner packaging that can be more easily collected, recycled, and diverted from landfills [34]. The increased use of rigid plastic versus more problematic flexible packaging could also decrease the number of issues during the collection and sorting stages [36]. Many facilities have difficulties with flexible packaging due to its varying handling characteristics. A separate collection system or increased revenue for sorting and processing facilities can help increase flexible plastic recycling and stimulate a positive shift in the handling of mixed plastic wastes. These wastes require specialized sorting of the input materials to ensure the highest levels of purity per item [37].

13.7 CONCLUSION

Each community needs to work together to find more cost-effective ways to better the environment through these methods. Though there are some challenges such as pollution and cost, there are also various solutions that can make the planet green. The current trends of using technology and social behavior will both promote the use of recyclable wastes. The recycling of plastic waste will not only make the environment more sustainable to live in but will also improve the overall quality of life.

REFERENCES

1. Zheng, Y., Yanful, E.K. and Bassi, A.S., 2005. A review of plastic waste biodegradation. *Critical Reviews in Biotechnology*, 25(4), pp. 243–250.
2. Barnes, D.K., Galgani, F., Thompson, R.C. and Barlaz, M., 2009. Accumulation and fragmentation of plastic debris in global environments. *Philosophical Transactions of the Royal Society B: Biological Sciences*, 364(1526), pp. 1985–1998.

3. Kalali, E.N., Lotfian, S., Shabestari, M.E., Khayatzadeh, S., Zhao, C. and Nezhad, H.Y., 2023. A critical review of the current progress of plastic waste recycling technology in structural materials. *Current Opinion in Green and Sustainable Chemistry*, p. 100763.
4. Chai, J.C. and Miura, N., 2002. Comparing the performance of landfill liner systems. *Journal of Material Cycles and Waste Management*, 4, pp. 135–142.
5. Pivato, A., 2011. Landfill liner failure: An open question for landfill risk analysis. *Journal of Environmental Protection*, 2(03), p. 287.
6. Ramke, H.G., 2018. Leachate collection systems. *Solid Waste Landfilling: Concepts, Processes, Technology*, p. 345.
7. Barlaz, M.A., Chanton, J.P. and Green, R.B., 2009. Controls on landfill gas collection efficiency: Instantaneous and lifetime performance. *Journal of the Air & Waste Management Association*, 59(12), pp. 1399–1404.
8. Shen, L. and Worrell, E., 2024. Plastic recycling. In *Handbook of recycling*. Elsevier, pp. 497–510.
9. Fredenslund, A.M., Scheutz, C. and Kjeldsen, P., 2010. Tracer method to measure landfill gas emissions from leachate collection systems. *Waste Management*, 30(11), pp. 2146–2152.
10. Cdc.gov., 2019. Landfill gas primer—an overview for environmental health professionals-chapter 5: Landfill gas control measures. [online] Available at: www.atsdr.cdc.gov/HAC/landfill/html/ch5.html.
11. EPA, 2018. Plastics: Material-specific data. *US EPA*. [online] Available at: www.epa.gov/facts-and-figures-about-materials-waste-and-recycling/plastics-material-specific-data.
12. National Research Council, 2000. Incineration processes and environmental releases. In *Waste incineration & public health*. National Academies Press.
13. Ruth, L.A., 1998. Energy from municipal solid waste: A comparison with coal combustion technology. *Progress in Energy and Combustion Science*, 24(6), pp. 545–564.
14. Hu, H., Xu, Z., Liu, H., Chen, D., Li, A., Yao, H. and Naruse, I., 2015. Mechanism of chromium oxidation by alkali and alkaline earth metals during municipal solid waste incineration. *Proceedings of the Combustion Institute*, 35(2), pp. 2397–2403.
15. Calderón-Garcidueñas, L., Reynoso-Robles, R., Pérez-Guillé, B., Mukherjee, P.S. and Gónzalez-Maciel, A., 2017. Combustion-derived nanoparticles, the neuroenteric system, cervical vagus, hyperphosphorylated alpha synuclein and tau in young Mexico City residents. *Environmental Research*, 159, pp. 186–201.
16. Maqsood, T., Dai, J., Zhang, Y., Guang, M. and Li, B., 2021. Pyrolysis of plastic species: A review of resources and products. *Journal of Analytical and Applied Pyrolysis*, 159, p. 105295.
17. Sarkingobir, Y. and Tukur, U., 2020, January. Harmful biological effects of plastics. In *Virtual international conference on advancement of nanotechnology*. ISBN: 978-81-948129-1-3, p. 63.
18. Lee, C.W., Hsu, L.F., Wu, I.L., Wang, Y.L., Chen, W.C., Liu, Y.J., Yang, L.T., Tan, C.L., Luo, Y.H., Wang, C.C. and Chiu, H.W., 2022. Exposure to polystyrene microplastics impairs hippocampus-dependent learning and memory in mice. *Journal of Hazardous Materials*, 430, p. 128431.
19. Vargas-Caballero, M., Warming, H., Walker, R., Holmes, C., Cruickshank, G. and Patel, B., 2022. Vagus nerve stimulation as a potential therapy in early Alzheimer's disease: A review. *Frontiers in Human Neuroscience*, 16, p. 866434.
20. Miandad, R., Barakat, M.A., Aburiazaiza, A.S., Rehan, M., Ismail, I.M.I. and Nizami, A.S., 2017. Effect of plastic waste types on pyrolysis liquid oil. *International Biodeterioration & Biodegradation*, 119, pp. 239–252.
21. Mukherjee, C., Denney, J., Mbonimpa, E.G., Slagley, J. and Bhowmik, R., 2020. A review on municipal solid waste-to-energy trends in the USA. *Renewable and Sustainable Energy Reviews*, 119, p. 109512.

22. Shamsuyeva, M. and Endres, H.J., 2021. Plastics in the context of the circular economy and sustainable plastics recycling: Comprehensive review on research development, standardization and market. *Composites Part C: Open Access*, 6, p. 100168.
23. Liljenström, C. and Finnveden, G., 2015. *Data for separate collection and recycling of dry recyclable materials*. KTH Royal Institute of Technology, p. 53.
24. Al-Salem, S.M., Lettieri, P. and Baeyens, J., 2009. Recycling and recovery routes of plastic solid waste (PSW): A review. *Waste Management*, 29(10), pp. 2625–2643.
25. Sethi, B., 2016. *Methods of recycling. Recycling of polymers: Methods, characterization and applications*. John Wiley & Sons, pp. 55–114.
26. Sanchez, F.A.C., Boudaoud, H., Camargo, M. and Pearce, J.M., 2020. Plastic recycling in additive manufacturing: A systematic literature review and opportunities for the circular economy. *Journal of Cleaner Production*, 264, p. 121602.
27. Tsuchimoto, I. and Kajikawa, Y., 2022. Recycling of plastic waste: A systematic review using bibliometric analysis. *Sustainability*, 14(24), p. 16340.
28. Al-Salem, S.M., Lettieri, P. and Baeyens, J., 2009. Recycling and recovery routes of plastic solid waste (PSW): A review. *Waste Management*, 29(10), pp. 2625–2643.
29. Verma, R., Vinoda, K.S., Papireddy, M. and Gowda, A.N.S., 2016. Toxic pollutants from plastic waste-a review. *Procedia Environmental Sciences*, 35, pp. 701–708.
30. Mackey, G., 1995. *A review of advanced recycling technology*, Vol. 609. ACS Symposium Series.
31. Walker, T., Gramlich, D. and Dumont-Bergeron, A., 2020. The case for a plastic tax: A review of its benefits and disadvantages within a circular economy. *Sustainability*, pp. 185–211.
32. Europe, P., 2015. An analysis of European plastics production, demand and waste data. *Plastics–the Facts*, 147.
33. Rafey, A. and Siddiqui, F.Z., 2023. A review of plastic waste management in India–challenges and opportunities. *International Journal of Environmental Analytical Chemistry*, 103(16), pp. 3971–3987.
34. Tumu, K., Vorst, K. and Curtzwiler, G., 2023. Global plastic waste recycling and extended producer responsibility laws. *Journal of Environmental Management*, 348, p. 119242.
35. Khalid, M.Y., Arif, Z.U., Ahmed, W. and Arshad, H., 2022. Recent trends in recycling and reusing techniques of different plastic polymers and their composite materials. *Sustainable Materials and Technologies*, 31, p. e00382.
36. Hopewell, J., Dvorak, R. and Kosior, E., 2009. Plastics recycling: Challenges and opportunities. *Philosophical Transactions of the Royal Society B: Biological Sciences*, 364(1526), pp. 2115–2126.
37. Bui, T.D., Tseng, J.W., Tseng, M.L. and Lim, M.K., 2022. Opportunities and challenges for solid waste reuse and recycling in emerging economies: A hybrid analysis. *Resources, Conservation and Recycling*, 177, p. 105968.

14 Microplastics in the Environment
Current Consensus and Future Trends

Komal Saini

14.1 INTRODUCTION

The activities of humans continuously and increasingly put strain on coastal and marine regions. Numerous pollutants, such as pesticides, hydrocarbons, heavy metals, plastics, and microplastics (MPs), influence marine ecosystems. Because coastal areas are so dynamic, the physicochemical properties of freshwater ecosystems, estuaries, and lagoons blend with the oceanographic features of nearby seas. Therefore, one of the trickiest and most pressing problems in ecotoxicology and environmental management today is assessing pollution and remediating coastal and marine settings.

All regions of our oceans are now impacted by the worldwide environmental issue of marine litter [1]. Land-based activities are the primary cause, and poor waste management practices and a lack of standard infrastructure lead to problems with the environment, economy, and public health. According to Alomar et al., the weight of marine litter in the oceans is increasing, with increasing consequences on biology and ecology [2]. It has been discovered that MPs, tiny plastic particles smaller than 5 mm, significantly harm both the environment and human health.

The term "microplastics" was initially used 19 years ago by Thompson et al., researchers studying plastic pollution in the UK's oceans. Since then, governments, nongovernmental organizations, the scientific community, and others have become interested in MPs [3]. Plastics are relatively recent materials, having been developed in the second half of the 20th century [4]. However, the environment is seriously threatened by their excessive production and usage in a wide range of products and sectors [5, 6].

Primary MPs are defined as those that are purposefully manufactured small, such as the cosmetic microbeads found in face washes [7]. Because of their higher abundance and reactivity than MPs, nanoplastics (NPs) are especially dangerous to living things. Their potential harm is increased by their small size, which makes it easy for them to enter live cells and travel to far-off places [8]. In this chapter, we also discuss the causes of MPs, their impacts on the environment and human health, international efforts and solutions to lessen their release, understanding of MPs, and many

DOI: 10.1201/9781003468141-14

strategies that can be used to change this. We contrast several methods of treatment and preventive measures to lessen MP pollution and promote plastic recycling and reuse.

14.2 MP SOURCES

In 2018, a study concerning MP contamination in agricultural soils was published. To find out how common MPs are in agricultural contexts, Piehl et al. examined soil from a farm in Middle Franconia in southeast Germany [9]. Because they were ignorant of the usage of any plastic or organic additives that could introduce MPs into the soil, the farmers continued to practice traditional agriculture. In topsoil samples, the researchers discovered 0.34 ± 0.36 MP particles per kilogram, even in the absence of evident plastic use on farms.

The rising frequency of MPs in agricultural soils despite no apparent usage of plastic items or treatments of organic amendments raised new scientific concerns: What happens when agricultural management approaches raise the demand for pollutants? How can different farming techniques add MPs to the soil? The various routes that MPs can take to enter agricultural soils are covered in this section. First, we enumerate the direct sources of MPs as well as the ways that using plastic products usually results in higher MP contamination. Next, we cover the indirect sources of agricultural techniques that may mask MPs. Finally, we address indirect sources linked to environmental variables for MPs to passively infiltrate soil.

14.2.1 Direct Sources: Agricultural Supplies Made of Plastic

Plastics are necessary for agriculture to manufacture raw materials. Crop and animal production systems both use plastic goods [10]. Plastics are used in crop production in a variety of designs, including irrigation pipes, mulching films, protective netting, tiny tunnels, and greenhouses. Comparably, plastics are also introduced into animal production through feed. Plastic is typically used to make bale nets, stretch films, and twine that are meant to preserve straw.

What does the use of these polymers mean in the grand scheme of things? The agricultural industry accounted for 3.4% of the 50.7 million plastic tons used in Europe in 2019, according to Plastic Europe [11]. Although their percentage is not very large, these polymers are employed in direct contact with soil and water. The most often utilized polymers in this segment were low-density polyethylene (LDPE) and polypropylene (PP), which are primarily found in plastic films meant for small tunnels, stretch films, mulching films, and greenhouses. The primary component of drippers and irrigation tubes, polyvinyl chloride (PVC), came next. Polystyrene (PS), which is used in twine and netting, completes the list [11].

14.2.1.1 Mulching Films

Farmers employ plastic mulches to control weed growth, enhance soil warmth, and optimize water utilization [12]. One downside is that farmers who are unable to thoroughly remove the mulch after the growing season leave tiny fragments of plastic behind, posing an environmental concern. According to Qi et al., farmers who use plastic mulches on a regular basis increase the amount of plastic debris in the soil and the likelihood that it will invade surrounding ecosystems [13]. In recent years, scientists have issued warnings against the use of plastic mulches in agriculture, claiming that doing so increases the number of MPs in soils [7].

Most of the data comes from China, one of the countries with the highest use of plastic mulches. In 2015, more than 1.5 million tons of plastic mulch were spread across over 20 million hectares of arable land in China. Liu et al. were among the first to document the accumulation of MPs caused by plastic mulching [14]. A correlation was detected between the quantity of plastics in the soil and the use of plastic mulch (R2: 0.6, $p < 0.05$), and additional researchers corroborated their findings with data from China [15–17] and other nations, like Spain [18] and the USA [19], where plastic mulching is a widely used practice.

Beriot et al. [18] and Meng et al. [20] discovered the greatest concentrations, with values reaching 2,200 MP particles per kilogram of soil. Other investigators reported lower values of up to 80 MP particles per kilogram, which is significantly higher than the entirety of baseline measurements for plastic. Farmers can use a range of plastic polymers as mulch in their fields. The most popular of these polymers is LDPE [21]. Since the 1990s, LDPE plastic mulches have become more and more popular due to their low cost and ability to boost agricultural yields [22]. LDPE is a completely saturated hydrocarbon polymer with great weather resistance [23], but LDPE mulches must be removed after harvest to prevent them from accumulating in the fields. Liu et al. called the usage of plastic mulch to contaminate soil with plastics a "white revolution to white pollution" [24].

14.2.1.2 Plastics used to Protect Plants

MPs are more common in soil samples when plastic films and nonwoven fabrics made of plastic are employed in shade nets, wind barriers, polytunnels, and greenhouses. The contributions of plastics that protect plants have made to the amounts of MPs in the environment have been the subject of numerous studies. Studies of a different kind sought to measure the quantities of MPs anticipated in soils close to producing systems that employ these kinds of plastics, but few studies explicitly address the emission of MPs from polymers used to protect plants.

Li et al. revealed that for the same productive area, soil samples from farms without greenhouses contained fewer MPs than soil samples from greenhouse farms [25]. The authors reported that on farms where greenhouses were first implemented in the 1980s, the range of MP particles per kilogram was 1,000 to 3,786. In a similar vein, Kim et al. found that the median soil sample collected inside greenhouses had 2,110 MPs kg^{-1}, while the median soil sample collected outside had 310 MPs/kg^{-1} [26]. The significant variation noted by Kim et al. could have had several causes [26].

For example, in greenhouses, producers employ twine and plastic mulches to secure developing shoots in addition to plant-protecting plastics; research on the number of MPs in agricultural soils mentions or includes greenhouses [17, 27, 28]. They all emphasize that the more plastic items are utilized, the more MPs are found in soil samples. Dahl et al. developed a chrono-sequence of plastics in sediments from a Spanish bay to link the growth in the inland surface covered by greenhouses to the rise in MPs observed in sediments [29]. Seldom is research on MP pollution associated with other plant-protecting polymers. Martínez Silva and Nanny discovered a connection between the usage of nonwoven plastic fabrics and elevated quantities of MPs in the Magdalena River in Colombia [30]. However, they did not collect any samples of soil. This is the only instance of a scientific study that considers shade nets, polytunnels, plastic wind barriers, and other plant-protecting polymers to be highly likely to contaminate soils.

14.2.1.3 Silage and Bale Nets

The first study to address the issue of MPs in agricultural soils was published by Piehl et al., who observed that covering grass bales and silage with plastics could contaminate the feed with MPs, which animals could consume and excrete into the soil [9]. Despite the lack of research to date that quantifies this route of entry for MP into arable lands, one research team did discover that sheep excrete MPs following ingestion of pieces of plastic mulch [18].

14.2.1.4 Twine

In 2019, 11% of the EU's demand for agricultural plastic was met by plastic twine [31]. The largest amount of plastic rope used in agriculture is made of PP [32]. Twine is used to tie plants to stakes in major crops like tomatoes, sweet peppers, and hops. It has applications in banana agriculture to tie plants to each other and keep them from dropping over [33]. Hay bales can also be wrapped with twine. After being cut during harvest, the twine is frequently carelessly thrown out in the fields, where it may become compost or MPs in the soil. Initiatives are in place to encourage the use of compostable twine that can be collected and composted alongside plant residues [32, 34]. Plastic twine's role in the buildup of MPs in soil has not been examined in any research until now.

14.2.1.5 Coated Fertilizers and Pesticides

Plastic polymers are used in the fertilizer industry as coating agents for insecticides and fertilizers with gradual releases. Following the successful delivery of these components, the fertilizer pellet's plastic coating stays in the soil. This could be a source of MPs in agricultural areas, but only one research team has collected data regarding this route of MP entry into soils. The group discovered that farms treated with coated fertilizers contained 6–369 mg of plastics per kg^{-1} of soil [35]. Additionally, they noticed that throughout the irrigation season, the amount of MPs in estuary waters increased, with 90% of the MPs seen matching those in fertilizer coatings [36].

14.2.1.6 Plastic Pipes

Plastic pipes are commonly utilized in drip and sprinkler irrigation systems to distribute water across fields [37], and they must be replaced periodically. The dilemma is, how do farmers handle the old pipes? Do they carefully remove them to avoid fragmentation and plastic trash, or do they bury them and run a new line parallel to the old one? The possible penetration of MPs into soils remains largely investigated. It is possible that this is not the primary channel for the penetration of plastics into arable land. More research is needed to identify the level of deterioration or fragmentation in plastic pipes used in agriculture.

14.2.1.7 Improper Storage

Farms contain large amounts of plastic, as detailed in previous sections. Farmers who do not always have the time or technical capacity to clean and store discarded plastics must conduct extra work to guarantee that they are properly collected and kept. To keep plastic rubbish clean and from being blown away, keep it dry and out of the wind in a covered place [38]. Because of the improper disposal of plastic waste, vertebrates such as sheep, cows, dogs, and vultures feed and swallow plastic particles.

14.2.2 Indirect Sources: MPs Hidden in Agricultural Supplies

14.2.2.1 Sewage Sludge and Biosolids

Surface water pollution from urban wastewater is efficiently prevented by plants that treat wastewater [39]. The plants use a thick organic matrix known as sludge to filter MPs from the incoming water. Urban sources of MPs are collected by the sludge, including transportation [40], grey water from homes [41], and industrial processes [42]. Sludge carries MPs to agricultural regions and is utilized as a crop fertilizer in several nations. In Chilean agricultural areas, Corradini et al. discovered 1,100–3,500 MP particles per kilogram [43]. In 2020, Van der Berg examined farms in Spain and discovered 5,190 ± 2,630 MP particles per kilogram. A lower mean range of 87.6–545.9 MP particles per kg was discovered in a third study conducted in China using the same parameters [16]. In a similar line, Crossman et al. determined that the amount of biosolids used as crop fertilizer increased with the number of MPs gathered in the soil profile [44].

14.2.2.2 Compost

Scientists have long believed that organic fertilizers are a source of MPs in soil, even though little research has investigated the plastic composition of compost [45]. Weithmann et al. identified 24 MP particles per kilogram in German compost made up of municipal organic waste and green clippings, ranging in size from 1 mm to 5 mm [46]. In a recent study, Gui et al. found that compost generated in Zhejiang Province, China, contained an average of 2,400 ± 358 MP particles per kg, ranging in size from 50 μm to 5 mm [47]. Van Schothorst et al. observed 1,750 ± 930 MP particles per kg in Dutch compost, ranging from 30 μm to 2 mm [48]. According to Gui et al., the most common polymers are polyester, PP, polyethylene (PE), and polyethylene terephthalate (PET), which account for 70%–80% of the MPs in compost [47]. These records correspond to Europe's requirement for plastic [49].

14.2.2.3 Manure

One unknown pathway of MPs into soils is through manure. Ultimately, how exactly do MPs end up in manure? As of March 2021, there has only been one study released, and the authors did not mention possible routes [50]. However, they revealed that the number of MPs in soil samples from farms fertilized with pig dung for 22 years was nearly twice as high. The authors found an annual average of 1,250 ± 640 MP particles per kilogram of dry manure, often known as MPs. Does this occur frequently? Additional research is needed to determine the various origins of MPs in manure.

14.2.2.4 Irrigation

In a detailed analysis of agricultural soils in Hangzhou Bay, China, Zhou et al. found that irrigation water introduced MPs into farm soils [17]. It makes sense that agricultural soil will eventually be contaminated with MPs from surface water bodies if farmers use the contaminated water bodies for irrigation. This hypothesis aligns well with the global plastic cycle [51], a recently created theory that allows for a better understanding of the difficulties caused by plastic pollution. Unfortunately, there is no information available regarding the number of MPs that irrigation could bring into a field.

14.2.2.5 Agriculture as a Phase in the Plastic Cycle: Unregulated Disposal and Oversight

Historically, most research has focused on the effects of plastic pollution caused by general littering, the dumping of plastic debris, and poor landfill site management [52]. Unauthorized plastic disposal sites are believed to be one of the primary ways MPs enter the environment [53]. Since illegal dumping typically occurs in wilderness areas, it is not anticipated that this pathway will have a direct influence on agricultural soils. However, using one of the methods outlined in the ensuing sections, transport agencies may move the plastics to agricultural land. Baensch-Baltruschat et al. reported about 1,400–2,800 tons of MPs per acre, originating from tire and road wear [54].

14.3 DISTRIBUTION AND TRANSPORT

MPs may migrate across atmospheric or aquatic currents in the environment, depending on their weight and density, from the pedosphere to the hydrosphere via rainfall, surface runoff, and ocean circulation; they travel from land to water as well as from water to land due to ocean circulation [55]. Smaller and lighter MPs can also travel long distances, including to glacier zones and high mountains, by becoming airborne and carried by the wind. More compact MPs can accumulate or be buried in the pedosphere (soil), whereas lighter MPs can be carried by wind across the pedosphere [56].

Surface runoff from towns and agriculture, along with heavy rains, can bring MPs into surface rivers (the hydrosphere). According to studies [57, 58], MPs can be introduced into the soil by the application of plastic mulches to boost crop development or the addition of residential sewage sludge as a soil supplement in agricultural activities. In addition, the MPs generated by tires degrading naturally are carried to adjacent surface waters by rainwater runoff [56]. Moreover, atmospheric fallout allows airborne MPs composed of light fibers from landfills, incinerator waste, and garments to travel considerable distances and deposit themselves in far-off places [59, 60]. Figure 14.1 shows a schematic representation of the worldwide distribution of MPs in the environment. The hydrosphere, pedosphere, and atmospheric portions of the environment detail what happens to MPs in each compartment.

14.3.1 Fate of MPs in the Hydrosphere

Human activities such as tourism and wastewater treatment mostly deposit MPs in the hydrosphere [61]. For example, wastewater treatment plant effluents produce

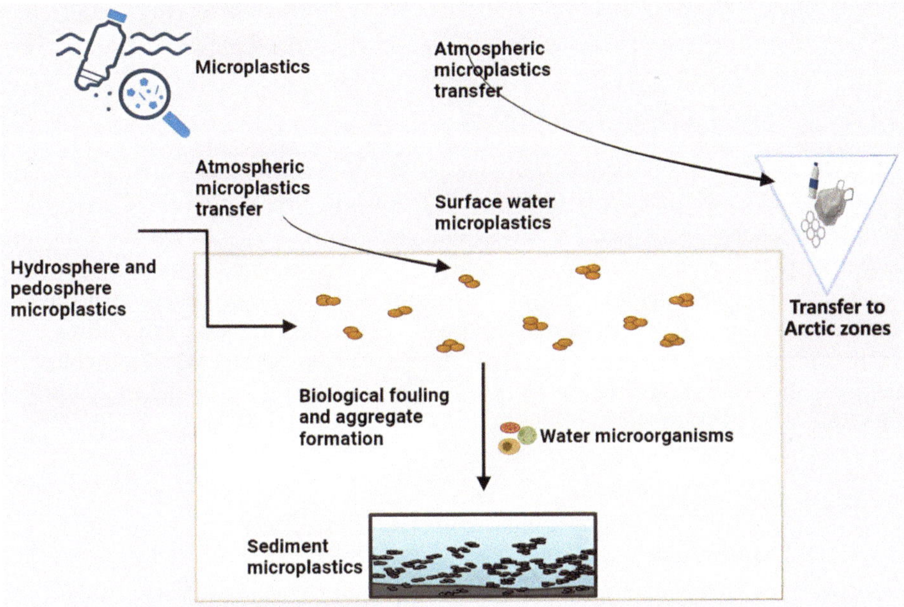

FIGURE 14.1 Microplastic distribution in the environment. The schematic represents the horizontal and vertical distribution of the microplastics in the hydrosphere. Water currents and wind currents cause microplastic to be transferred to the hydrosphere and atmosphere, respectively, and to far-off places like the Arctic zones.

seven million MPs items daily into the aquatic habitat [19, 61]. MPs can either accumulate in the hydrosphere, which is their immediate source of disposal, or they can use different transport mechanisms to travel to other remote environments, like glacial zones, depending on their physical properties (such as shape and size), density, and chemical composition [32, 47, 62].

About 15% of the MPs in the hydrosphere are still suspended, and 70% of the MPs in the hydrosphere are retained in the aquatic body's sediments [63]. The degree of variability in the many pathways of plastic degradation and water dynamics, such as wind and ocean currents, whose velocities and intensities depend on climatic circumstances, makes it difficult to establish the mechanisms for the fate and dispersion of MPs in the hydrosphere. However, vertical and horizontal distributions have been suggested as the two most likely routes for MPs dispersal in the hydrosphere.

The horizontal spread of plastic trash from the pedosphere to the hydrosphere is controlled by wind current, precipitation, and water circulation velocity [64–66]. Plastic litter is subject to biotic degradation, abiotic disintegration, and ocean current abrasion when it enters the aquatic environment. MPs are available in a wide range of shapes, sizes, densities, and chemical makeups. Lighter MPs float on the surface of the water, but denser MPs sink to the bottom [29]. Vertical dispersion is an alternative method of dispersal wherein bigger MPs sink to the ocean floor, riverbed, or sea floor. The presence of organisms such as bacteria, plankton, algae, and microorganisms in the hydrosphere controls biological fouling and aggregate/cluster formation.

Water velocity and direction of flow regulate vertical turbulent mixing. Finally, biota transmission is dependent on the mobility of aquatic creatures [29, 67, 68].

14.4 IMPACT ON ECOSYSTEMS

MPs, which are tiny plastic particles smaller than five millimeters, are a global environmental concern that are affecting ecosystems all over the world. These microscopic substances, which arise from a number of sources such as the decomposition of larger plastic items, microbeads in personal care products, and emissions from factories, are particularly harmful to wildlife and the fragile ecological balance. Examining how MPs affect ecosystems with a specific emphasis on wildlife, especially marine and terrestrial species, is the goal of this study. We also explore the ecological ramifications, including bioaccumulation and the breaking down of food chains.

14.4.1 Wildlife

14.4.1.1 Marine Life

Because MPs have gotten into marine environments, marine life is seriously threatened. Marine species frequently consume these particles for food, which causes ingestion at the base of the food chain; MPs are then accidentally consumed by phytoplankton, zooplankton, and small fish, which has a negative domino effect. When marine life consumes MPs, it can result in physical injury, obstructions in their digestive tracts, and the release of harmful compounds found in plastics. This affects the well-being of individual creatures and may have long-term effects on marine populations [69]. Moreover, MPs transport other contaminants, such as persistent organic pollutants. These toxins can build up in the tissues of marine life, which can worsen the impact they have on their health and possibly reach levels high enough to endanger higher trophic-level predators [69].

14.4.1.2 Terrestrial Fauna

Terrestrial ecosystems are not immune to MP contamination, even though marine environments suffer most of it. There are several ways that MPs might move from aquatic habitats to terrestrial ones, such as wind, water currents, and even air deposition. Insects, birds, and mammals are among the terrestrial animals that may unintentionally ingest MPs through contaminated food sources or directly [70]. Although research on the effects of MP intake on terrestrial creatures is still in its infancy, preliminary findings point to possible effects on nutritional absorption, digestive systems, and general health.

14.4.2 Ecological Consequences

14.4.2.1 Bioaccumulation

MPs are part of the ecological phenomenon known as bioaccumulation, in which particles gradually build up in the tissues of living things. The MP concentrations rise as they pass from primary producers to consumers and ultimately to predators at higher trophic levels in the food chain [71]. Apex predators and marine animals are among the species most at risk from the bioaccumulation of MPs. Because they are at the top of the food chain and frequently have longer lifespans than other creatures, they might be more severely impacted by the bioaccumulation of MPs and related pollutants [72].

14.4.2.2 Disruption of Food Chains

By changing the quantity and quality of food sources available to different species, MPs have the potential to upset food chains. Primary producers like phytoplankton and algae may experience growth and nutrient uptake disruptions if they consume MPs. Consequently, herbivores who consume these primary producers are impacted. MPs can affect the behavior, reproduction, and general fitness of species at various trophic levels as they pass down the food chain. Furthermore, the movement of MPs from prey to predators might have a domino impact on ecosystems [73].

14.5 HUMAN HEALTH CONCERNS

Because MPs are often found in the environment and have been shown to have harmful effects, they could be dangerous to human health. It's critical to comprehend the mechanisms by which humans are exposed to MPs. The most common ways are skin contact, inhalation, and oral ingestion (Figure 14.1). Oral consumption is the most common mode of exposure [74]. In fact, there are numerous ways in which humans are simultaneously exposed to MPs. The term "plastic cycle" coined by Rillig and Lehmann in science, refers to the ability of MPs to move between various environmental mediums [75]. Human exposure risk is increased by the mobility of MPs.

14.5.1 Oral Intake

MPs are present in many commonplace items, including milk, tea bags, seafood, bottled water, salt, and sugar. According to Lee et al., eating shellfish exposes Europeans to about 11,000 MP particles annually per person, and each person's body absorbs between 39,000 and 52,000 plastic particles annually based on dietary consumption [76]. It is probable that MPs are widely spread in soil, particularly in agriculture. If they carry a negative charge, they can enter the plant's water transport system and go to the roots, stems, leaves, and fruits [77, 78].

When MPs infiltrate agricultural systems via composting, plastic mulching, sewage sludge, and other means, they contaminate food and increase the risk of human consumption [76]. Because these common polymer materials (PP, PS, PE, and PET) are used so frequently, take-out food containers contain MPs; people who order takeaway four to seven times per week in containers consume 12–203 MP particles. Additionally, the surface of silicone rubber baby teats deteriorates, and microscopic plastic particles are released into the atmosphere during the steam sterilization procedure; researchers estimate that during a baby's first year of consistent bottle feeding, 0.66 million MP particles will enter their body [76].

14.5.2 Inhalation

The majority of PE, PS, and PET particles and fibers in the air have sizes between 10 and 8,000 μm. The road is the main source of MPs in the atmosphere, accounting for 84% of the total. Scientists found that the average concentration of MP fibers in Parisian air is 5.4 fibers/m^3 outside and 0.9 fibers/m^3 inside, whereas in Shanghai's outdoor air, the average MP concentration is 1.42 particles/m3, ranging in size from 23 to 5,000 μm [79]. Between 74,000 and 121,000 MP particles are thought to be consumed annually considering oral intake and inhalation. Amato-Lourenço et al. revealed MP fibers, mostly made of PE and PP, with a size range of 8.12–16.8 μm and particles smaller than 5.5 μm in human lungs [80]. Lung tissue has MPs that are smaller in size than those found in the atmosphere. This highlights the possible harm to the human body and provides additional evidence that humans might be exposed to MPs by inhalation.

14.5.3 Skin Contact

Although MPs have no ability to penetrate the skin barrier, their collection on the skin can increase the risk of exposure. For example, using commercial goods that contain MPs, such as face cream and cleanser, increases the likelihood of PE exposure [67, 81, 82]. Protective mobile phone covers can also produce MPs that end up in human hands, and children can encounter MPs on the ground while they crawl or play. Certain common plastic additives, such as triclosan, bisphenols, brominated flame retardants, and phthalates, can also be absorbed through the skin [79, 83].

14.6 MONITORING AND DETECTION

MPs are typically thought to be eliminated from the human body through the biliary and digestive tracts, but scientists have found evidence of MPs in human blood [84]. People are starting to reevaluate the threat that MPs pose to human health. More studies are being conducted on the ingestion, distribution, storage, and metabolism of MPs in the human body. Prior to examining MPs' potential detrimental effects, it is critical to understand the amount of these particles in the body. In a recent assessment, MPs were found in human biological components such as the spleen, liver, colon, lung, feces, placenta, and breastfeeding. They are also transported around the body via the bloodstream [85]. The colon and liver had the highest concentrations (28.1 and 4.6 particles/gm, respectively). MPs found include PE, PET, PP, PS, PVC, and PC.

Newborn feces (5,700–82,000 ng/g, median: 36,000 ng/g) had ten times more PET than adult feces (2,200–16,000 ng/g, median: 2,600 ng/g), reflecting that infants were exposed to much more MPs than adults [86]. First discovered in the human placenta, there were 12 MP pieces identified ranging in size from 5 to 10 μm [87], and PVC and PP MPs ranging in size from 2 to 12 μm were found in human breastmilk [88]. MPs have been found in breastfeeding newborns, meconium, and the placenta in subsequent research.

If MPs from landfill leachate are not properly controlled, they can contaminate nearby regions. Often, leachate remediation uses biological, physical, or chemical methods. Common treatment procedures include reverse osmosis (RO), oxidation, coagulation and flocculation, activated carbon, stripping, evaporation, membrane bioreactors (MBR), sequencing batch reactors, and aeration. Leachate cycling in landfills and leachate transfer to wastewater treatment plants are also used. Depending on the treatment strategy used, leachate treatment processes can reduce microparticle content by 3% to 100%.

Microbial breakdown of polymers in landfill leachate is another biological activity that removes MPs from the environment. Several microbes, such as *Bacillus cereus, B. gottheilii, Alcaligenes faecalis, B. amyloliquefacien, B. brevis, Cyanobacterium*, and *Anabaena spiroides*, can break down PE, PS, and PP polymers for food. Microbial degradation of MPs and nanoparticles (NPs) in leachate is a less expensive and generally safe method of application. However, the effectiveness of this approach is dependent on the contact duration between possible microorganisms and the targeted polymer.

Certain physical processes in a leachate treatment system, such as filtering and sedimentation, might be able to get rid of MPs. The three most common leachate treatment procedures based on physical separation are microfiltration, ultrafiltration, and nanofiltration; all of these filtration systems can separate MPs and leachate based on their size. According to Renou et al., RO (pore size 1–0.1 nm) is another promising physical separation technology that can be utilized to extract MPs from landfill leachate [89–91].

Chemical oxidation and coagulation–flocculation are two possible methods for separating and eliminating suspended particles from water samples. Even though coagulation–flocculation is not designed to separate MPs, it can remove them from the leachate. Numerous studies have also documented the removal of significant MPs from wastewater samples [92, 93]. Electro-coagulation is a different technique that successfully eliminates MPs from the influent. In addition to coagulation–flocculation, which is commonly employed in landfill leachate treatment systems, several other chemical treatment methods can extract MPs from leachate effluent. For instance, disinfection treatments such as ozonation, UV radiation, and chlorination can induce MPs to dissolve into smaller sizes, resulting in NPs [91, 92, 94]. However, MPs in the leachate may limit disinfection efficiency by consuming disinfectants or sheltering bacteria from attack [95].

14.7 FUTURE REMARKS AND CONCLUSION

The threats of plastic pollution, known for its impacts on the environment, economy, society, public safety, and health, have led to diverse regulatory tools globally [96]. This top-down governance, whether mandatory or optional, has made strides in reducing plastic pollution worldwide, often through levies or bans at national and regional levels. However, current tools are seen as inadequate, with producers and businesses sometimes opposing legislative proposals. Gathering more research on the prevalence and impacts is crucial for developing effective laws.

Multitiered strategies blending hard and soft laws at various governmental levels are necessary along with infrastructure improvements and a shift toward a circular economy approach in product design [97, 98]. This holistic approach is vital to addressing the urgent need for sustainable alternatives to combat the pervasive issue of MPs. These minute particles, less than 5 mm, are heavily present in our environment, with over 8 million tons entering oceans yearly, endangering marine life and potentially human health. The urgency for finding solutions is clear, with promising alternatives like biodegradable plastics from plant-based sources, mycelium-based materials, and algae-based plastics offering hope for a future where plastic pollution is minimized. Innovation and widespread adoption of these eco-friendly materials are crucial to mitigate the harmful effects of microplastics on our planet.

REFERENCES

1. Shim, W.J. and Thomposon, R.C., 2015. Microplastics in the ocean. *Archives of Environmental Contamination and Toxicology*, *69*, pp. 265–268.
2. Alomar, C., Estarellas, F. and Deudero, S., 2016. Microplastics in the Mediterranean sea: Deposition in coastal shallow sediments, spatial variation and preferential grain size. *Marine Environmental Research*, *115*, pp. 1–10.
3. Thompson, R.C., Olsen, Y., Mitchell, R.P., Davis, A., Rowland, S.J., John, A.W., McGonigle, D. and Russell, A.E., 2004. Lost at sea: Where is all the plastic? *Science*, *304*(5672), pp. 838–838.
4. Gündoğdu, S. and Çevik, C., 2017. Micro-and mesoplastics in Northeast Levantine coast of Turkey: The preliminary results from surface samples. *Marine Pollution Bulletin*, *118*(1–2), pp. 341–347.
5. Osman, A.I., Farrell, C., Al-Muhtaseb, A.H., Al-Fatesh, A.S., Harrison, J. and Rooney, D.W., 2020. Pyrolysis kinetic modelling of abundant plastic waste (PET) and in-situ emission monitoring. *Environmental Sciences Europe*, *32*, pp. 1–12.
6. Qasim, U., Osman, A.I., Al-Muhtaseb, A.A.H., Farrell, C., Al-Abri, M., Ali, M., Vo, D.V.N., Jamil, F. and Rooney, D.W., 2021. Renewable cellulosic nanocomposites for food packaging to avoid fossil fuel plastic pollution: A review. *Environmental Chemistry Letters*, *19*, pp. 613–641.
7. Wang, J., Liu, X., Li, Y., Powell, T., Wang, X., Wang, G. and Zhang, P., 2019. Microplastics as contaminants in the soil environment: A mini-review. *Science of the Total Environment*, *691*, pp. 848–857.
8. Sharma, V.K., Ma, X., Lichtfouse, E. and Robert, D., 2023. Nanoplastics are potentially more dangerous than microplastics. *Environmental Chemistry Letters*, *21*(4), pp. 1933–1936.
9. Piehl, S., Leibner, A., Löder, M.G., Dris, R., Bogner, C. and Laforsch, C., 2018. Identification and quantification of macro-and microplastics on an agricultural farmland. *Scientific Reports*, *8*(1), p. 17950.
10. Miles, C., DeVetter, L., Ghimire, S. and Hayes, D.G., 2017. Suitability of biodegradable plastic mulches for organic and sustainable agricultural production systems. *HortScience*, *52*(1), pp. 10–15.
11. Briassoulis, D., Hiskakis, M. and Babou, E., 2013. Technical specifications for mechanical recycling of agricultural plastic waste. *Waste Management*, *33*(6), pp. 1516–1530.
12. Espi, E., Salmerón, A., Fontecha, A., García, Y. and Real, A.I., 2006. Plastic films for agricultural applications. *Journal of Plastic Film & Sheeting*, *22*(2), pp. 85–102.
13. Qi, R., Jones, D.L., Li, Z., Liu, Q. and Yan, C., 2020. Behavior of microplastics and plastic film residues in the soil environment: A critical review. *Science of the Total Environment*, *703*, p. 134722.
14. Liu, M., Lu, S., Song, Y., Lei, L., Hu, J., Lv, W., Zhou, W., Cao, C., Shi, H., Yang, X. and He, D., 2018. Microplastic and mesoplastic pollution in farmland soils in suburbs of Shanghai, China. *Environmental Pollution*, *242*, pp. 855–862.

15. Huang, Y., Liu, Q., Jia, W., Yan, C. and Wang, J., 2020. Agricultural plastic mulching as a source of microplastics in the terrestrial environment. *Environmental Pollution, 260*, p. 114096.
16. Zhang, L., Xie, Y., Liu, J., Zhong, S., Qian, Y. and Gao, P., 2020. An overlooked entry pathway of microplastics into agricultural soils from application of sludge-based fertilizers. *Environmental Science & Technology, 54*(7), pp. 4248–4255.
17. Zhou, B., Wang, J., Zhang, H., Shi, H., Fei, Y., Huang, S., Tong, Y., Wen, D., Luo, Y. and Barceló, D., 2020. Microplastics in agricultural soils on the coastal plain of Hangzhou Bay, east China: Multiple sources other than plastic mulching film. *Journal of Hazardous Materials, 388*, p. 121814.
18. Beriot, N., Peek, J., Zornoza, R., Geissen, V. and Lwanga, E.H., 2021. Low density-microplastics detected in sheep faeces and soil: A case study from the intensive vegetable farming in Southeast Spain. *Science of the Total Environment, 755*, p. 142653.
19. Ghimire, S., Flury, M., Scheenstra, E.J. and Miles, C.A., 2020. Sampling and degradation of biodegradable plastic and paper mulches in field after tillage incorporation. *Science of the Total Environment, 703*, p. 135577.
20. Meng, F., Fan, T., Yang, X., Riksen, M., Xu, M. and Geissen, V., 2020. Effects of plastic mulching on the accumulation and distribution of macro and micro plastics in soils of two farming systems in Northwest China. *PeerJ, 8*, p. e10375.
21. Kasirajan, S. and Ngouajio, M., 2012. Polyethylene and biodegradable mulches for agricultural applications: A review. *Agronomy for Sustainable Development, 32*, pp. 501–529.
22. Chopra, M. and Koul, B., 2020. Comparative assessment of different types of mulching in various crops: A review. *Plant Archives, 20*, pp. 1620–1626.
23. Crawford, C.B. and Quinn, B., 2017. Microplastic separation techniques. *Microplastic Pollutants*, pp. 203–218.
24. Liu, E.K., He, W.Q. and Yan, C.R., 2014. 'White revolution' to 'white pollution'—agricultural plastic film mulch in China. *Environmental Research Letters, 9*(9), p. 091001.
25. Li, Q., Zeng, A., Jiang, X. and Gu, X., 2021. Are microplastics correlated to phthalates in facility agriculture soil? *Journal of Hazardous Materials, 412*, p. 125164.
26. Kim, S.K., Kim, J.S., Lee, H. and Lee, H.J., 2021. Abundance and characteristics of microplastics in soils with different agricultural practices: Importance of sources with internal origin and environmental fate. *Journal of Hazardous Materials, 403*, p. 123997.
27. Wang, C., Zhao, J. and Xing, B., 2021. Environmental source, fate, and toxicity of microplastics. *Journal of Hazardous Materials, 407*, p. 124357.
28. Ding, L., Zhang, S., Wang, X., Yang, X., Zhang, C., Qi, Y. and Guo, X., 2020. The occurrence and distribution characteristics of microplastics in the agricultural soils of Shaanxi Province, in north-western China. *Science of the Total Environment, 720*, p. 137525.
29. Dahl, M., Bergman, S., Björk, M., Diaz-Almela, E., Granberg, M., Gullström, M., Leiva-Dueñas, C., Magnusson, K., Marco-Méndez, C., Piñeiro-Juncal, N. and Mateo, M.Á., 2021. A temporal record of microplastic pollution in Mediterranean seagrass soils. *Environmental Pollution, 273*, p. 116451.
30. Martínez Silva, P. and Nanny, M.A., 2020. Impact of microplastic fibers from the degradation of nonwoven synthetic textiles to the Magdalena River water column and river sediments by the city of Neiva, Huila (Colombia). *Water, 12*(4), p. 1210.
31. APE-Europe. *Plasticulture in Europe—statistics*. https://apeeurope.eu/statistics/. Accessed March 2021.
32. Guerrini, S., Borreani, G. and Voojis, H., 2017. Biodegradable materials in agriculture: Case histories and perspectives. *Soil Degradable Bioplastics for a Sustainable Modern Agriculture*, pp. 35–65.

33. Hernandez, C.E. and Witter, S.G., 1996. Evaluating and managing the environmental impact of banana production in Costa Rica: A systems approach. *Ambio*, pp. 171–178.
34. Biothop. *Biotwine hop waste transformation into novel product assortments for packaging and horticulture sector.* www.life-biothop.eu/wp-content/uploads/2019/12/BioTHOP-FLYER_ENG_Czech.pdf. Accessed May 2021.
35. Katsumi, N., Kusube, T., Nagao, S. and Okochi, H., 2020. The role of coated fertilizer used in paddy fields as a source of microplastics in the marine environment. *Marine Pollution Bulletin*, *161*, p. 111727.
36. Katsumi, N., Kusube, T., Nagao, S. and Okochi, H., 2021. Accumulation of microcapsules derived from coated fertilizer in paddy fields. *Chemosphere*, *267*, p. 129185.
37. Beieler, R.W., 2012. *Pipelines for water conveyance and drainage.* American Society of Civil Engineers (ASCE).
38. Svensk Ensilageplast Retur AB. http://svepretur.se/en/our-services/proceed-as-follows/. Accessed June 2021.
39. Ziajahromi, S., Neale, P.A., Rintoul, L. and Leusch, F.D., 2017. Wastewater treatment plants as a pathway for microplastics: Development of a new approach to sample wastewater-based microplastics. *Water Research*, *112*, pp. 93–99.
40. Liu, F., Olesen, K.B., Borregaard, A.R. and Vollertsen, J., 2019. Microplastics in urban and highway stormwater retention ponds. *Science of the Total Environment*, *671*, pp. 992–1000.
41. Kärkkäinen, N. and Sillanpää, M., 2021. Quantification of different microplastic fibres discharged from textiles in machine wash and tumble drying. *Environmental Science and Pollution Research*, *28*, pp. 16253–16263.
42. Du, F., Cai, H., Zhang, Q., Chen, Q. and Shi, H., 2020. Microplastics in take-out food containers. *Journal of Hazardous Materials*, *399*, p. 122969.
43. Corradini, F., Casado, F., Leiva, V., Huerta-Lwanga, E. and Geissen, V., 2021. Microplastics occurrence and frequency in soils under different land uses on a regional scale. *Science of the Total Environment*, *752*, p. 141917.
44. Crossman, J., Hurley, R.R., Futter, M. and Nizzetto, L., 2020. Transfer and transport of microplastics from biosolids to agricultural soils and the wider environment. *Science of The Total Environment*, *724*, p. 138334.
45. Rillig, M.C., 2012. Microplastic in terrestrial ecosystems and the soil? *Environmental Science & Technology*, *46*(12), pp. 6453–6454.
46. Weithmann, N., Möller, J.N., Löder, M.G., Piehl, S., Laforsch, C. and Freitag, R., 2018. Organic fertilizer as a vehicle for the entry of microplastic into the environment. *Science Advances*, *4*(4), p. eaap8060.
47. Gui, J., Sun, Y., Wang, J., Chen, X., Zhang, S. and Wu, D., 2021. Microplastics in composting of rural domestic waste: Abundance, characteristics, and release from the surface of macroplastics. *Environmental Pollution*, *274*, p. 116553.
48. Van Schothorst, B., Beriot, N., Huerta Lwanga, E. and Geissen, V., 2021. Sources of light density microplastic related to two agricultural practices: The use of compost and plastic mulch. *Environments*, *8*(4), p. 36.
49. Plastic Europe. Plastics—the Facts, 2020. An analysis of European plastics production, demand and waste data. *Plastics Europe.org*. www.plasticseurope.org/es/resources/publications/4312-plastics-facts-2020. Accessed 24 May 2021.
50. Yang, J., Li, R., Zhou, Q., Li, L., Li, Y., Tu, C., Zhao, X., Xiong, K., Christie, P. and Luo, Y., 2021. Abundance and morphology of microplastics in an agricultural soil following long-term repeated application of pig manure. *Environmental Pollution*, *272*, p. 116028.
51. Rochman, C.M. and Hoellein, T., 2020. The global odyssey of plastic pollution. *Science*, *368*(6496), pp. 1184–1185.

52. Duis, K. and Coors, A., 2016. Microplastics in the aquatic and terrestrial environment: Sources (with a specific focus on personal care products), fate and effects. *Environmental Sciences Europe*, *28*(1), pp. 1–25.
53. Jambeck, J.R., Geyer, R., Wilcox, C., Siegler, T.R., Perryman, M., Andrady, A., Narayan, R. and Law, K.L., 2015. Plastic waste inputs from land into the ocean. *Science*, *347*(6223), pp. 768–771.
54. Baensch-Baltruschat, B., Kocher, B., Kochleus, C., Stock, F. and Reifferscheid, G., 2021. Tyre and road wear particles-a calculation of generation, transport and release to water and soil with special regard to German roads. *Science of the Total Environment*, *752*, p. 141939.
55. Fred-Ahmadu, O.H., Bhagwat, G., Oluyoye, I., Benson, N.U., Ayejuyo, O.O. and Palanisami, T., 2020. Interaction of chemical contaminants with microplastics: Principles and perspectives. *Science of the Total Environment*, *706*, p. 135978.
56. Bergmann, M., Peeken, I., Beyer, B., Krumpen, T., Primpke, S., Tekman, M.B. and Gerdts, G., 2016. Vast quantities of microplastics in Arctic Sea Ice—a prime temporary sink for plastic litter and a medium of transport. *Fate Impact Microplastics Marine Ecosystems*, pp. 75–76.
57. Fucic, A., Galea, K.S., Duca, R.C., El Yamani, M., Frery, N., Godderis, L., Halldorsson, T.I., Iavicoli, I., Ndaw, S., Ribeiro, E. and Viegas, S., 2018. Potential health risk of endocrine disruptors in construction sector and plastics industry: A new paradigm in occupational health. *International Journal of Environmental Research and Public Health*, *15*(6), p. 1229.
58. Talang, R.P.N., Polruang, S., Sirivithayapakorn, S., 2024. Influencing factors of microplastic generation and microplastic contamination in urban freshwater. *Heliyon*, *10*(9), p. e30021.
59. Gaonkar, O.D., Nambi, I.M. and Govindarajan, S.K., 2019. Soil organic amendments: Impacts on sorption of organophosphate pesticides on an alluvial soil. *Journal of Soils and Sediments*, *19*, pp. 566–578.
60. García-Delgado, C., Marín-Benito, J.M., Sánchez-Martín, M.J. and Rodríguez-Cruz, M.S., 2020. Organic carbon nature determines the capacity of organic amendments to adsorb pesticides in soil. *Journal of Hazardous Materials*, *390*, p. 122162.
61. Ge, J., Li, H., Liu, P., Zhang, Z., Ouyang, Z. and Guo, X., 2021. Review of the toxic effect of microplastics on terrestrial and aquatic plants. *Science of the Total Environment*, *791*, p. 148333.
62. Gigault, J., Ter Halle, A., Baudrimont, M., Pascal, P.Y., Gauffre, F., Phi, T.L., El Hadri, H., Grassl, B. and Reynaud, S., 2018. Current opinion: What is a nanoplastic? *Environmental Pollution*, *235*, pp. 1030–1034.
63. Corradini, F., Meza, P., Eguiluz, R., Casado, F., Huerta-Lwanga, E. and Geissen, V., 2019. Evidence of microplastic accumulation in agricultural soils from sewage sludge disposal. *Science of the Total Environment*, *671*, pp. 411–420.
64. Guo, L., Jury, W.A., Wagenet, R.J. and Flury, M., 2000. Dependence of pesticide degradation on sorption: Nonequilibrium model and application to soil reactors. *Journal of Contaminant Hydrology*, *43*(1), pp. 45–62.
65. Haider, T.P., Völker, C., Kramm, J., Landfester, K. and Wurm, F.R., 2019. Plastics of the future? The impact of biodegradable polymers on the environment and on society. *Angewandte Chemie International Edition*, *58*(1), pp. 50–62.
66. Hanachi, P., Kazemi, S., Zivary, S., Karbalaei, S. and Ghadami, S.A., 2021. The effect of polyethylene terephthalate and abamectin on oxidative damages and expression of vtg and cyp1a genes in juvenile zebrafish. *Environmental Nanotechnology, Monitoring & Management*, *16*, p. 100565.
67. Hernandez, L.M., Yousefi, N. and Tufenkji, N., 2017. Are there nanoplastics in your personal care products? *Environmental Science & Technology Letters*, *4*(7), pp. 280–285.

68. Hodson, M.E., Duffus-Hodson, C.A., Clark, A., Prendergast-Miller, M.T. and Thorpe, K.L., 2017. Plastic bag derived-microplastics as a vector for metal exposure in terrestrial invertebrates. *Environmental Science & Technology*, *51*(8), pp. 4714–4721.
69. Ghosh, S., Sinha, J.K., Vashisth, K., Han, S. and Bhaskar, R., 2023. Microplastics as an emerging threat to the global environment and human health. *Sustainability*, *15*(14), p. 10821.
70. Boyle, K. and Örmeci, B., 2020. Microplastics and nanoplastics in the freshwater and terrestrial environment: A review. *Water*, *12*(9), p. 2633.
71. Guo, J.J., Huang, X.P., Xiang, L., Wang, Y.Z., Li, Y.W., Li, H., Cai, Q.Y., Mo, C.H. and Wong, M.H., 2020. Source, migration and toxicology of microplastics in soil. *Environment International*, *137*, p. 105263.
72. Miller, M.E., Hamann, M. and Kroon, F.J., 2020. Bioaccumulation and biomagnification of microplastics in marine organisms: A review and meta-analysis of current data. *PLoS One*, *15*(10), p. e0240792.
73. Carbery, M., O'Connor, W. and Palanisami, T., 2018. Trophic transfer of microplastics and mixed contaminants in the marine food web and implications for human health. *Environment International*, *115*, pp. 400–409.
74. Prata, J.C., da Costa, J.P., Lopes, I., Duarte, A.C. and Rocha-Santos, T., 2020. Environmental exposure to microplastics: An overview on possible human health effects. *Science of the Total Environment*, *702*, p. 134455.
75. Rillig, M.C. and Lehmann, A., 2020. Microplastic in terrestrial ecosystems. *Science*, *368*(6498), pp. 1430–1431.
76. Lee, Y., Cho, J., Sohn, J. and Kim, C., 2023. Health effects of microplastic exposures: Current issues and perspectives in South Korea. *Yonsei Medical Journal*, *64*(5), p. 301.
77. Schwab, F., Rothen-Rutishauser, B. and Petri-Fink, A., 2020. When plants and plastic interact. *Nature Nanotechnology*, *15*(9), pp. 729–730.
78. Rillig, M.C., 2020. Plastic and plants. *Nature Sustainability*, *3*, 887–888.
79. Li, Y., Tao, L., Wang, Q., Wang, F., Li, G. and Song, M., 2023. Potential health impact of microplastics: A review of environmental distribution, human exposure, and toxic effects. *Environment & Health*, *1*(4), pp. 249–257.
80. Amato-Lourenço, L.F., Carvalho-Oliveira, R., Júnior, G.R., dos Santos Galvão, L., Ando, R.A. and Mauad, T., 2021. Presence of airborne microplastics in human lung tissue. *Journal of Hazardous Materials*, *416*, p. 126124.
81. Schneider, M., Stracke, F., Hansen, S. and Schaefer, U.F., 2009. Nanoparticles and their interactions with the dermal barrier. *Dermato-Endocrinology*, *1*(4), pp. 197–206.
82. Prata, J.C., 2018. Airborne microplastics: Consequences to human health? *Environmental Pollution*, *234*, pp. 115–126.
83. Wu, P., Lin, S., Cao, G., Wu, J., Jin, H., Wang, C., Wong, M.H., Yang, Z. and Cai, Z., 2022. Absorption, distribution, metabolism, excretion, and toxicity of microplastics in the human body and health implications. *Journal of Hazardous Materials*, *437*, p. 129361.
84. Leslie, H.A., Van Velzen, M.J., Brandsma, S.H., Vethaak, A.D., Garcia-Vallejo, J.J. and Lamoree, M.H., 2022. Discovery and quantification of plastic particle pollution in human blood. *Environment International*, *163*, p. 107199.
85. Kutralam-Muniasamy, G., Shruti, V.C., Pérez-Guevara, F. and Roy, P.D., 2023. Microplastic diagnostics in humans: "The 3Ps" progress, problems, and prospects. *Science of The Total Environment*, *856*, p. 159164.
86. Zhang, J., Wang, L., Trasande, L. and Kannan, K., 2021. Occurrence of polyethylene terephthalate and polycarbonate microplastics in infant and adult feces. *Environmental Science & Technology Letters*, *8*(11), pp. 989–994.

87. Ragusa, A., Svelato, A., Santacroce, C., Catalano, P., Notarstefano, V., Carnevali, O., Papa, F., Rongioletti, M.C.A., Baiocco, F., Draghi, S. and D'Amore, E., 2021. Plasticenta: First evidence of microplastics in human placenta. *Environment International*, *146*, p. 106274.
88. Ragusa, A., Notarstefano, V., Svelato, A., Belloni, A., Gioacchini, G., Blondeel, C., Zucchelli, E., De Luca, C., D'Avino, S., Gulotta, A. and Carnevali, O., 2022. Raman microspectroscopy detection and characterisation of microplastics in human breastmilk. *Polymers*, *14*(13), p. 2700.
89. Sun, J., Zhu, Z.R., Li, W.H., Yan, X., Wang, L.K., Zhang, L., Jin, J., Dai, X. and Ni, B.J., 2021. Revisiting microplastics in landfill leachate: Unnoticed tiny microplastics and their fate in treatment works. *Water Research*, *190*, p. 116784.
90. Renou, S., Givaudan, J.G., Poulain, S., Dirassouyan, F. and Moulin, P.J.J.O.H.M., 2008. Landfill leachate treatment: Review and opportunity. *Journal of Hazardous Materials*, *150*(3), pp. 468–493.
91. Narevski, A.C., Novaković, M.I., Petrović, M.Z., Mihajlović, I.J., Maoduš, N.B. and Vujić, G.V., 2021. Occurrence of bisphenol A and microplastics in landfill leachate: Lessons from South East Europe. *Environmental Science and Pollution Research*, *28*, pp. 42196–42203.
92. Hidayaturrahman, H. and Lee, T.G., 2019. A study on characteristics of microplastic in wastewater of South Korea: Identification, quantification, and fate of microplastics during treatment process. *Marine Pollution Bulletin*, *146*, pp. 696–702.
93. Lawrence, J., Tosine, H.M., Zimmermann, H.W. and Pang, T.W.S., 1975. Removal of asbestos fibres from potable water by coagulation and filtration. *Water Research*, *9*(4), pp. 397–400.
94. Lv, X., Dong, Q., Zuo, Z., Liu, Y., Huang, X. and Wu, W.M., 2019. Microplastics in a municipal wastewater treatment plant: Fate, dynamic distribution, removal efficiencies, and control strategies. *Journal of Cleaner Production*, *225*, pp. 579–586.
95. Enfrin, M., Dumée, L.F. and Lee, J., 2019. Nano/microplastics in water and wastewater treatment processes–origin, impact, and potential solutions. *Water Research*, *161*, pp. 621–638.
96. Ziani, K., Ioniță-Mîndrican, C.B., Mititelu, M., Neacşu, S.M., Negrei, C., Moroşan, E., Drăgănescu, D. and Preda, O.T., 2023. Microplastics: A real global threat for environment and food safety: A state of the art review. *Nutrients*, *15*(3), p. 617.
97. Mitrano, D.M. and Wohlleben, W., 2020. Microplastic regulation should be more precise to incentivize both innovation and environmental safety. *Nature Communications*, *11*(1), p. 5324.
98. Osman, A.I., Hosny, M., Eltaweil, A.S., Omar, S., Elgarahy, A.M., Farghali, M., Yap, P.S., Wu, Y.S., Nagandran, S., Batumalaie, K. and Gopinath, S.C., 2023. Microplastic sources, formation, toxicity and remediation: A review. *Environmental Chemistry Letters*, *21*(4), pp. 2129–2169.

15 Closing the Loop on Microplastics
Sources, Risks, Analysis, and the Path Forward

Bunty Sharma, Pradeep Kaushik, Moondeep Chauhan, Jaspreet Dhau, and Ganga Ram Chaudhary

15.1 INTRODUCTION

In *Microplastics in Environment: Source, Potential Risks, Analytical Challenges, and Remediation Strategies*, we have delved deeply into the intricate world of MPs and their far-reaching consequences on the environment. We and our contributors have explored the origins and sources of MPs; highlighted the potential risks they pose to ecosystems, organisms, and human health; and shed light on the analytical challenges in detecting and quantifying these minuscule particles. Moreover, we have illuminated various remediation strategies aimed at mitigating the pervasive presence of MPs in our environment [1]. The key takeaway from this comprehensive exploration is that MPs are ubiquitous: They have infiltrated every corner of our planet, from the deepest ocean trenches to the most remote wilderness. Their ubiquity underscores the urgency of addressing this issue.

15.1.1 Ecological and health risks: MPs have demonstrated the potential to disrupt ecosystems and affect organisms, including humans. The risks they pose to wildlife and human health demand our attention and action [2].

15.1.2 Analytical challenges: Accurately identifying and quantifying MPs is a formidable analytical challenge. Overcoming these obstacles is vital for research, policymaking, and environmental management [3].

15.1.3 Remediation strategies: Several remediation strategies have been explored, ranging from prevention at the source to innovative cleanup methods. Successful mitigation requires collaboration across scientific disciplines, government policies, and public awareness [4].

DOI: 10.1201/9781003468141-15

15.2 FUTURE PERSPECTIVES

As we conclude this book, it is evident that the issue of MPs in the environment will remain a pressing concern for the foreseeable future. To pave the way for a cleaner and healthier environment, it is essential to consider the following future perspectives:

15.2.1 Interdisciplinary collaboration: The battle against MPs necessitates collaboration among scientists, policymakers, industry leaders, and the public. Interdisciplinary efforts are crucial for the development of effective mitigation strategies [5].

15.2.2 Technological advancements: Continued research and development in analytical techniques are required to improve our ability to detect and quantify MPs. This includes developing more sensitive and cost-effective methodologies [6].

15.2.3 Regulatory frameworks: Governments and international bodies should work together to establish robust regulations and standards for MP content in products, as well as guidelines for disposal and recycling [7].

15.2.4 Education and awareness: Public awareness is key in reducing the sources of MPs. Education campaigns can empower individuals to make informed choices about their consumption and waste disposal practices [8].

15.2.5 Innovative cleanup methods: Research into innovative and sustainable cleanup methods, such as bioenzymatic degradation, photodegradable materials, and natural processes, should continue to address the existing MP pollution.

15.2.6 Long-term monitoring: Establishing long-term monitoring programs to track changes in MP distribution and concentration in various environments at local and regional levels is crucial for assessing the effectiveness of remediation efforts [9].

In the face of the challenges posed by MPs, there is hope that with concerted efforts, research, and international cooperation, we can reduce the impacts of MPs on the environment and safeguard the health of future generations. *Microplastics in Environment* is but a stepping stone in this journey: The real transformation will come from our commitment to preserving our planet and its fragile ecosystems.

15.3 CONCLUSION

In conclusion, the intricate and pervasive issue of microplastics demands urgent attention and strenuous global efforts. The contributors to this book have navigated through the multifaceted dimensions of microplastic pollution, shedding light on its environmental, ecological, and human health implications. The evidence presented underscores the ubiquity of microplastics across diverse ecosystems, emphasizing the pressing need for comprehensive mitigation strategies.

As we grapple with the enormity of this problem, it is imperative to recognize the importance of interdisciplinary collaboration and groundbreaking solutions. MP pollution frameworks discussed in this book offer a foundation for policymakers to enact measures that curb the production and release of microplastics into the environment. Additionally, heightened public awareness, education, and corporate responsibility are crucial elements in fostering a culture of sustainability.

Looking ahead, continued research into the long-term effects of microplastics on ecosystems and human health is paramount. The book underscores the potential of emerging technologies and sustainable practices as beacons of hope in our collective journey toward a healthier, plastic-conscious future. Through shared responsibility and global cooperation, we have the opportunity to alleviate the impacts of microplastics, safeguard our ecosystems, and safeguard a more sustainable legacy for generations to come.

REFERENCES

1. Koelmans, A.A., Redondo-Hasselerharm, P.E., Nor, N.H.M., de Ruijter, V.N., Mintenig, S.M. and Kooi, M., 2022. Risk assessment of microplastic particles. *Nature Reviews Materials*, 7(2), pp. 138–152.
2. Gwenzi, W., Simbanegavi, T.T., Mahdi, H.I., Azelee, N.I.W., Muisa-Zikali, N. and Rangabhashiyam, S., 2022. Occurrence and ecological health risks of microplastics. In *Emerging contaminants in the terrestrial-aquatic-atmosphere continuum* (pp. 243–270). Elsevier.
3. Silva, A.B., Bastos, A.S., Justino, C.I., da Costa, J.P., Duarte, A.C. and Rocha-Santos, T.A., 2018. Microplastics in the environment: Challenges in analytical chemistry-A review. *Analytica Chimica Acta*, 1017, pp. 1–19.
4. Hu, K., Tian, W., Yang, Y., Nie, G., Zhou, P., Wang, Y., Duan, X. and Wang, S., 2021. Microplastics remediation in aqueous systems: Strategies and technologies. *Water Research*, 198, p. 117144.
5. Mitrano, D.M., Bigalke, M., Booth, A.M., Carteny, C.C., Coffin, S., Egger, M., Gondikas, A., Hüffer, T., Koelmans, A.A., Lahive, E. and Mattsson, K., 2023. Training the next generation of plastics pollution researchers: Tools, skills and career perspectives in an interdisciplinary and transdisciplinary field. *Microplastics and Nanoplastics*, 3(1), p. 24.
6. Fang, C., Luo, Y. and Naidu, R., 2023. Microplastics and nanoplastics analysis: Options, imaging, advancements and challenges. *TrAC Trends in Analytical Chemistry*, p. 117158.
7. Mitrano, D.M. and Wohlleben, W., 2020. Microplastic regulation should be more precise to incentivize both innovation and environmental safety. *Nature Communications*, 11(1), p. 5324.
8. Garcia-Vazquez, E. and Garcia-Ael, C., 2021. The invisible enemy. Public knowledge of microplastics is needed to face the current microplastics crisis. *Sustainable Production and Consumption*, 28, pp. 1076–1089.
9. Weber, C.J., Bastijans, J.E., Lügger, K. and Heller, C., 2023. Microplastic in long-term soil monitoring: First spatial and temporal data on plastics in agricultural top soils. *Journal of Soils and Sediments*, pp. 1–15.

Index

L

M

N

O

P

Q

R

S

T

V

W